SCIENCE & RELIGION

SCIENCE & RELIGION

A Critical Survey

HOLMES ROLSTON III

TEMPLETON FOUNDATION PRESS
Philadelphia and London

Templeton Foundation Press
300 Conshohocken State Road, Suite 670
West Conshohocken, PA 19428
www.templetonpress.org

2006 Templeton Foundation Press Edition

Templeton Foundation Press helps intellectual leaders and others learn about science research on aspects of realities, invisible and intangible. Spiritual realities include unlimited love, accelerating creativity, worship, and the benefits of purpose in persons and in the cosmos.

Previous editions:
Random House, Inc., 1987
Temple University Press, 1987
McGraw-Hill, 1989
Harcourt Brace, 1997

LIBRARY OF CONGRESS CATALOGING-IN-PUBLICATION DATA
Rolston, Holmes, 1932–
Science and religion : a critical survey / Holmes Rolston III.
p. cm.
Originally published: 1987. With new introduction by the author.
Includes bibliographical references and index.
ISBN-13: 978-1-59947-099-3 (pbk. : alk. paper)
ISBN-10: 1-59947-099-3 (pbk. : alk. paper)
1. Science–Philosophy. 2. Religion and science. I. Title.
Q175.R545 2006
501–dc22
2006003888

Printed in the United States of America
06 07 08 09 10 11 10 9 8 7 6 5 4 3 2 1

PERMISSIONS ACKNOWLEDGMENTS
Chapter 1, "Methods in Scientific and Religious Inquiry." Reprinted, with revisions, from *Zygon* 16 (1981): 29–63. Reprinted by permission.
Figure 1.4 on p. 15. Adapted from Joseph B. Birdsell, *Human Evolution*, 2d ed. (Chicago: Rand McNally & Co., 1975), p. 337. Copyright © 1975, 1981 by Harper & Row, Publishers, Inc. Reprinted by permission of Harper & Row, Publishers, Inc.
Figures 2.5 and 2.6 on p. 43. Reprinted from James E. Brady and Gerard E. Humiston, *General Chemistry: Principles and Structure* (New York: Wiley, 1975), pp. 61, 75. Copyright © 1975 by John Wiley & Sons, Inc. Reprinted by permission of John Wiley & Sons, Inc.
Figure 2.11 on p. 56. Illustration by Irving Geis. Reprinted from J. Bronowski, "The Clock Paradox," *Scientific American* (February 1963): 140. Copyright © 1963. Reprinted with permission from The Gies Archives.

CONTENTS

PREFACE

This work is a broadly conceived critical survey of the dialogue between science and religion. We survey the sciences to inquire what room they leave for religion. Although the legitimacy of the problem of God in the legacy of the West is, in the end, a crucial issue, we do not intend to establish any particular religion. Our plan is not to do natural theology, much less supernatural theology, but to show that the natural and social sciences are open to theological inquiry beyond. More than this, the sciences themselves, in any form presently known, address incompletely the ultimate questions and must be complemented by religious interpretations.

We take representative issues from the leading sciences and integrate them in a survey that (after an introductory chapter) begins with matter and moves (in succeeding chapters) through life, mind, culture, history, and spirit. In a way this is a logical progression; in a way it recounts the story of development in the world that science and religion both seek to assess. Over historical time, matter-energy has evolved life, which, in its most striking forms, develops mind, producing at length humans and their cultures. Simple things have spun more complicated things; that is the narrative of creation.

We take, in particular, two natural sciences—physics (Chapter 2) and biology (Chapter 3)—and two human sciences—psychology (Chapter 4) and sociology (Chapter 5). These sciences leave us short of adequate historical explanation, to which we then turn (Chapter 6), finding, in conclusion, that the storied universe and the world history, together with our own roles in the drama, raise religious questions. We are forced to ask whether the whole story stands under the narration of the Divine Spirit (Chapter 7). Selected topics thus serve as keys to crucial issues in a logical and historical analysis of the passage of nature through history and culture toward God. Our goal is to be comprehensive in scope by a strategy selective in detail.

We are not interested, then, in all the ranges of science as much as in science where it seems to proscribe, or to prescribe, religion. In many respects the least troublesome chapter is that on physics. But is this because of the abstract simplicity of physics, or because physics is a relatively mature science? Had this book been written a century ago, such a chapter would have been the most difficult one. The most troublesome (and longest!) chapter is on biology, which is the science that most threatens advocates of religion, despite their frequent, all-too-easy reconciling of biology and God in evolutionary theism. Psychology and social science are sometimes congenial and sometimes challenging to religion, though we find there a frequent need to deflate overblown theories that make dogmatic claims exceeding or overlooking available evidence and experience.

The interface between science and religion is, in a certain sense, a no-man's land. No specialized science is competent here, nor does classical theology or academic philosophy really own this territory. This is an interdisciplinary zone where inquirers come from many fields. But this is a land where we increasingly must live. Science is the first fact of modern life, and religion is the perennial carrier of meaning. Seen in depth and in terms of their long-range personal and cultural impacts, science and religion are the two most important forces in today's world. But their relations have been tumultuous, along a spectrum from conflict to complementarity. Few issues are of such moment in their combination of intellectual and existential dimensions, of head and heart, of the person and his or her culture, of the present and the past in shaping the course of the future.

The risks involved in such a survey must be taken either by conscious choice or by default. The dangers of cutting across so many disciplines are exceeded only by the dangers of adhering to religious beliefs in ignorance of these sciences, or (what is equally common, and more a danger) abandoning religious beliefs in unexamined presumptions about what these sciences permit and prohibit in a viable faith. We have to make judgments for which we are never well enough equipped. Every college student, every professor, every literate religious layperson, every pastor, every scientist does in fact, for better or worse, make the sorts of judgments we undertake here.

Even so, after admitting that such judgments are unavoidable, readers may wonder whether this survey by a single author is not overbold. Can I expect to be more than semiskilled? Would it not be better for you to consult a variety of experts? Although the latter approach is not to be disparaged, it does not guarantee a unified outlook or competence at judgments over the whole. One can be expert at some particular junction of science and religion, but there are no experts who bridge all the critical areas. Local judgments are likely to be fragmented, each made without benefit of an overall gestalt. Skills in a specialized field have a way of shrinking to technical skills in proportion as they become expert. Such narrow skills can be deployed over larger areas with no special authority. In making the broad-scale evaluations that fit the puzzle pieces into the big picture, we are all apprentices at a craft in which there are no masters.

Just this fact can give us some boldness. The specialized concepts and vocabularies of the multiple sciences contain great knowledge. But when stretched from those domains where they are scientifically operational across to more philosophical domains, they can be used to feign knowledge where there is actually only speculative opinion that is hardly better grounded than are the beliefs of critical nonscientists and nonexperts. Specialization in some area of science (electron microscopy or meteorology, let us say), which increasingly characterizes scientists today, does nothing to help a scientist make broadly synthetic judgments about the nature of science, its epistemology and metaphysics, its social settings and revolutionary moments, or to synthesize science and religion. We need to ask questions *about* science, not simply *within* science. On the other hand, we cannot make these synthetic and critical judgments without some feeling for what the technical scientists are doing and what concepts of nature and human nature their discoveries support.

On the scale that interests us here, it is possible to distill the key issues from technical science and set them forth for critical review by intelligent readers. After learning to plow through the intricacies of physics or psychology to some extent, we can articulate the impact of these sciences on religion for any literate person with a serious interest in religious and scientific worldviews. We hope duly to appreciate the sciences but also to maintain toward them a level of skepticism that is healthy alike in theology and in science.

The religion that is married to science today will be a widow tomorrow. The sciences in their multiple theories and forms come and go. Biology in the year 2050 may be as different from the biology of today as the religion of today is from the religion of 1850. But the religion that is divorced from science today will leave no offspring tomorrow. From here onward, no religion can reproduce itself in succeeding generations unless it has faced the operations of nature and the claims about human nature with which science confronts us. The problem is somewhat like the one that confronts a living biological species fitting itself into its niche in the changing environment: There must be a good fit for survival, and yet overspecialization is an almost certain route to extinction. Religion that has too thoroughly accommodated to any science will soon be obsolete. It needs to keep its autonomous integrity and resilience. Yet religion cannot live without fitting into the intellectual world that is its environment. Here too the fittest survive.

A felicitous skill for getting good conclusions from premises that are partly faulty is a sign of genius both in science and in religion, and I covet that skill. Some of my data, observations, and conceptions will prove incorrect and distorted. Given the flux in both science and theology, this work will soon prove dated. It was not the scientists first but certain reforming theologians who resolved to put every account under a formula of "revisability" *(semper reformanda!)*. Yet scientists and theologians alike try to give as systematic an account as is possible, given the state of the art and their own capacities. I believe that my conclusions will stand in overview, even though the supporting details may shift. There is ample logical and experiential room for religious belief after science. Indeed, science makes religious judgments more urgent than ever. Never in the histories of science and religion have the opportunities been greater for fertile interaction between these fields, with mutual benefits to both.

My thanks go to all who have preceded me in this effort, a host of forerunners whose cumulative efforts have made this era a stimulating time in which to be alive and this study the most exciting (if also exacting) thing I have done in my life. Particular thanks go to readers of portions of the manuscript: to Sanford Kern in physics; Richard T. Ward in biology; A. Wayne Viney and Richard F. Kitchener in psychology; Ronny Turner and David M. Freeman in sociology; Robert Theodoratus in anthropology; James W. Boyd in Oriental religions; and, above all, Donald A. Crosby for his painstaking attention to the entire manuscript. I thank Karl E. Peters and *Zygon* for permission to reprint here as Chapter 1 material that first appeared in that journal.

Introduction to the 2006 Edition

Human Uniqueness and Human Responsibility

Science and Religion in a New Millennium

On the scale of decades and centuries, ongoing science is reconfigured into human history that must be interpreted. So I concluded two decades back: "Progressively reforming and developing theories are erected over observations. . . . This leads at a larger scale to progressively reforming and developing narrative models. . . . The story is ever reforming" (pp. 338–39). I faced the future with hopes and fears about the escalating powers of science for good and evil, finding it simultaneously powerless for the meaningful guidance of life, the classical province of religion. So now, having turned the millennium, what's new in the story?

If I had to give a sound-bite answer (as we now must do with the media), my reply would be: increasing concern about *human uniqueness and human responsibility*. We are reforming our accounts of who on Earth we are, where on Earth we are, and what on Earth we must and ought do. Paradoxically, the more we naturalize ourselves (as many scientists wish to do), finding ourselves products of an evolutionary process, descended from the apes, incarnate in flesh, we find ourselves the only species that knows its origins. We demonstrate our continuity with natural history, and in so doing find that the capacity to demonstrate this, requiring paleontology, genomics, cladistics, disrupts the continuity demonstrated. And, if this epistemic crisis were not enough, the scientific knowledge so gained simultaneously generates a moral crisis.

Homo sapiens, the wise species—so we have named ourselves—is the one species in the history of the planet that, now in this new millennium, has more power than ever for good or evil, for justice and injustice; indeed, the one species that puts both its own well-being and that of life on Earth in jeopardy. I was right, closing that book, that today we find ourselves seeking and "doing the truth on the cutting edge of nature and history" (p. 344). The media metaphor for this is "playing God"; the archaic metaphor is whether this creature rising from dust can act in the "image of God."

I began two decades ago: "Science is the first fact of modern life, and religion is the perennial carrier of meaning. . . . Science and religion are the two most important forces in today's world" (p. vii). That has not changed. Not so, my critics may reply: science is the new millennium; religion has been left in the last millennium. The last millennium does already seem history, almost antique. Who would open a physics or a biology text from the 1980s if one from the 2000s were at hand? Neither should one bother with a dated science and religion book. Great novels, such as *War and Peace*, can be classics, as can philosophical treatises, such as Immanuel Kant's *Critiques*. But in the science and religion debate one needs to be on the front lines.

In the original preface I wrote: "The religion that is married to science today will be a widow tomorrow. The sciences in their multiple theories and forms come and go. . . . But the religion that is divorced from science today will leave no offspring tomorrow. . . . Religion cannot live without fitting into the intellectual world that is its environment. Here too the fittest survive" (p. ix). After two decades, nearly a generation later, what are the winds of change in the disciplines of science? Is religion still a good adapted fit? What's new in the new millennium?

In the biological sciences, for organisms to survive and reproduce in the next generation is mostly a matter of conserving genetic information from the past, embodied in present organisms. This is likely to be useful today and tomorrow. But surviving also requires genetic variation, innovations that yield still better adapted fit, selected when these novel forms and behaviors critically track changes in the surrounding world. I would not be reprinting this book if I did not think that there is much of value here to be conserved in the next generation(s). Mostly, my arguments still stand. All of them are part of our survival in past and present. But neither can one stand still. One must generate and test new ideas. What are the critical innovations that, in a new millennium, we must track?

1. AN OPEN FUTURE

A major innovation is a sense of accelerating change and open future. Yes, we conserve the past; but today, as never before in the past, there is new possibility space, opened up by advances in science and technology, by industrial and agricultural capitalism, by global communication and exchange, by an explosion of ideas and activities. With the turn of the millennium, there is a sense that what is past is more past than when turning previous centuries, more behind us than ever. It's a new day, a new epoch, a new millennium, an exponential future. This is good, bad, and uncertain news. The possibilities of good and evil are escalating. Across multiple levels, from the quantum world through chaos theory, from genetic mutations and engineering, from medical advances to military prowess, to stock markets and terrorist attacks, we face uncertainties and unknowns. The discoveries and powers of science, once supposed to give us rationality, law, and predictability, now feed as much into this openness. We face at once promise and peril.

Looking into the future is hazardous. The sun will rise tomorrow because it rose yesterday and the day before. Induction is reasonably reliable if one is predicting simple systems. But induction is notoriously problematic, both for logical and empirical reasons, the more so if one is dealing with complex systems, exponential not linear, as our world is increasingly becoming. If one predicts on the basis of past and present, one will be right much, even most of the time—at least that has previously been the case. But one will be wrong at the times of critical innovation, the most important times of all, when the future is unlike the past. That has been so before; but, such a time, as never before, is now.

The future develops from what is already seeded into the present. That is true in biology, in science, in theology. But, when we survey such a past in natural and human history, in science and in religion, the one thing certain is that there will be surprises, the more so in more complicated systems. It is easier to predict eclipses a century hence than to predict tomorrow's headlines. We have been living through a century of change in our ideas about how determinacy and contingency, design and chance, order and chaos fit together to make up the world. These changes shape religion in its account of both science and nature. It now looks as though that reshaping will continue. There are no laws, plus initial conditions, by which we can predict the new millennium; but there are stories that we will enact and tell. Science deals with causes, and religion deals with meanings (pp. 22–31); we can be sure that both causes and meanings will be ingredients perennially interwoven in the fabric of history. And, dramatically, we write the next chapters of the story. Today, the future is open, as never before in the history of the planet.

Anticipating the future relations between science and theology, we can only extrapolate and wonder. In this century, we humans have come to know who we are and where we are in ways unprecedented in all past millennia. We know the size, age, and extent of our universe; we know the deep evolutionary history of our planet, and ourselves as part of this story. We know our molecular biology; we have sequenced our genome. These facts of science have required integration into our classical religious worldviews; and this blending of theory and principle in science and religion will continue. In this century, we humans have gained, through science and technology, more power than ever before to affect, for better or worse, our own well being, that of the human and natural worlds, and even planetary history. That is power, promise, hope, as never before. But caution! Danger! The fate of the Earth, the fate of all who dwell thereon, depends, in the next century, on the responsible use of that power. Everything depends on how we join science, ethics, and religion, with policy, with economics in practice.

Right on the edge of the new millennium (1999), I concluded:

> These three features of our human life on Earth—knowledge, power, and duty—are especially puzzling. How does reason, the mind with its knowledge, fit into the biological picture? Does it produce only more survival power? Does not this human mind gain some new power, manifest in cumulative transmissible cultures, that changes the evolutionary story? (Rolston 1999, p. 215)

Such power, if cumulating over thousands of years, seems, in the last century turning into the new millennium, to have crossed a new threshold.

Editing a *Scientific American* issue on managing planet Earth, William Clark writes that humans are moving toward consciously managing the Earth. Here "two central questions must be addressed? What kind of planet do we want? What kind of planet can we get?" (Clark 1989, p. 48). This, they say, will be the principal novelty of the new millennium: Earth will be a managed planet. Nature will be increasingly humanized. Humans have, now and increasingly, the power to impose their will on nature, re-making it to their preferences. The era when nature was the principal determinant of events on Earth is coming to an end; we are entering a post-natural world.

We will manage the planet, so the enthusiasts say, at the global level. Such resourceful management continues at all scales, down to microlevels, even to nanotechnology. The breakthrough is epitomized in genetic decoding and modification. Edward Yoxen writes:

> This is not just a change of technique, it is a new way of seeing. . . . The limitations of species can be transcended by splicing organisms, combining functions, dovetailing abilities and linking together chains of properties. The living world can now be viewed as a vast organic Lego kit inviting combination, hybridisation, and continual rebuilding. Life is manipulability. . . . Thus our image of nature is coming more and more to emphasise human intervention through a process of design. (Yoxen 1983, pp. 2, 15)

But who will manage the managers? Can the managers manage themselves? When they intervene, what ought they to design? Is the only human relationship to nature one of engineering it for the better? Of manipulating a vast Lego kit? Ought these engineers be impressed with the spontaneous self-assembling of this Lego kit that has generated biodiversity and biocomplexity over millennia? Do we have a duty of caring for a good, a godly creation? Ought humans make nature an end in itself, complementary to their own human ends? Yes, we humans are dominant, but what responsibility comes with this unique role? Invoking those contemporary and archaic metaphors again, are we playing God? Ought we to do this without concern whether we are imaging God?

These questions remain unanswered. In the last two decades, the increase of human knowledge and power has increased the paradox of our evolving out of nature with novel capacities to change natural history. This is nowhere better illustrated than in *Homo sapiens* decoding its own genome, surprised to find how few genes we had, confirming our kinship with other primates, unable as yet to discover what makes us so different, and wondering whether in our genetics laboratories we ought to remake ourselves (treating ourselves as a Lego kit, and asking whether to use stem cells to discover how to do so). This future, with its increasing knowledge opening up new possibility space, confirms human uniqueness and human responsibility.

2. MATTER, ENERGY, INFORMATION

At cosmological scales there is deep space and time: this a matter-energy question. At evolutionary scales Earth is a marvelous planet, a living wonderland in this deep space-time: this is an information question. The wonder is coded in the vital know-how in DNA. There is an originating matter/energy big bang, when the universe is formed. There is a second big bang, when biological information explodes on Earth. Both the first and the second big bangs result in us: the *Homo* that is so *sapiens*. A third takes place within us, the mind's big bang in the explosion of cultures with radical capacities for the generation and cumulative transmission of ideas—knowledge, wisdom. And we humans today, passing into a new millennium, seem poised on the edge of yet another combinatorial information explosion, with escalating possibilities in science and technology, evidenced in that recent decoding (and possible transforming) of our own genome or in unprecedented information storing and sharing on the Internet.

I wrote two decades ago:

> We humans do not live at the range of the infinitely small, nor at that of the infinitely large, but we may well live at the range of the infinitely complex. . . . The human being is the most sophisticated of known natural products. In our hundred and fifty pounds of protoplasm, in our three pounds of brain, there may be more operational organization than in the whole of the Andromeda galaxy. . . . The most significant thing in the known universe is still immediately behind the eyes of the astronomer. On a gross cosmic scale, humans are minuscule atoms. Yet the brain is so curiously a microcosm of this macrocosm, since the mind can contain so much of nature within thought and thus mirror the world. (p. 66)

If that has changed, it has become truer than ever. Our minds contain more of nature within (from quarks to antimatter, fossil history to the sequence of our own genome), and we continue to search out our world (its biodiversity, natural history) and our place within it (unique minds on Earth, planetary manager, sustainable developer, Earth's trustee, self-transforming species). Here is my current assessment:

> Alone among the other species on Earth, *Homo sapiens* is cognitively remarkable for being a spirited self and for self-transcendence. . . . *Homo sapiens* is the only part of the world free to orient itself with a view of the whole. That makes us, if you like, free spirits; it also makes us self-transcending spirits. Consider this transcendence first in the sciences. . . . With our instrumented intelligences and constructed theories, we now know of phenomena at structural levels from quarks to quasars. We measure distances from picometers to the extent of the visible universe in light years, across 40 orders of magnitude. We measure the strengths of the four major binding forces in nature (gravity, electromagnetism, the strong and weak nuclear forces), again across 40 orders of magnitude. We measure time at ranges across 34 orders of magnitude, from attoseconds to the billions-of-years age of the universe.
>
> Nature gave us our mind-sponsoring brains; nature gave us our hands. Nature did not give us radiotelescopes with which to "see" pulsars, or relativity theory with which to compute time dilation. These come from human genius cumulated in our transmissible cultures (though we do not forget that nature supplies these marvelous processes analyzed by radiotelemetry and relativity theory). (Rolston 2005, pp. 30–31)

We continue to find these cosmic processes marvelous, amazed more than ever by how they are "fine tuned" or "anthropic" in such a way that the natural history we know takes place (for a view open to theism, see Barr 2003; for a multiverse view, see Rees 2001). The subsequent decades have buttressed what I said two decades back:

> There seem to be in fact all kinds of connections between cosmology on the grandest scale and atomic theory on the minutest scale, and we may well suppose that we humans, who lie in between, stand on the spectrum of these connections. The way the universe is built and the way micronature is built are of a piece with the way humans are built. The shape of the rest of the universe, of all the levels above and below, is crucial to what is now taking place close at hand. In its own haunting way, the physical structure of the astronomical and microphysical world is as prolife as anything we later find in the prolife biological urges. Prelife events can have, and have had, prolife consequences. The universe is a biocosmos. (p. 70)

It is difficult to envision any cosmology that does not require creation of the complex out of the simple, more out of less, something somehow out of nothing. It is difficult to imagine that all of the remarkable phenomena that have worked together to make our universe possible will disappear. It is difficult to imagine a universe more staggering, dramatic, and mysterious, for all its rationality. It is difficult to imagine a universe that starts simpler (perhaps as quantum fluctuation in a vacuum) and becomes more complex (*Homo sapiens* sequencing its own genome; moral debates about the war in Iraq). The universe story, the Earth story, is a phenomenal tale of more and more later on out of less and less earlier on. As events move from quarks to protons, from amino acids to protozoans, to trilobites, to dinosaurs, to persons, from spinning electrons to sentient animals, from suffering beasts to sinful persons, the tale gets taller and taller. No doubt there will be surprises in cosmology in the next century, a taller tale still, but it would be even more surprising if these were wholly uncongenial to theology.

In astronomical nature and micronature, at both ends of the spectrum of size, nature lacks the complexity that it demonstrates at the mesolevels, found at our native ranges on Earth. On Earth, the surprises compound, and this is vitally keyed to its genetics, as we see next. The Earth-system is a kind of cooking pot sufficient to make life possible. Spontaneously, natural history organizes itself. The system proves to be prolife; the story goes from zero to five million species (more or less) in five billion years, passing through five billion species (more or less) that have come and gone en route, impressively adding diversity and complexity to simpler forms of life. That life support and promise evolve richness in biodiversity and biocomplexity that, after eons of millennia, reaches humans with their uniqueness and responsibility. Earth is a kind of providing ground for life, a planet with promise, a promise offering still more fulfillment in this oncoming millennium.

There are other planets. In the two decades since the first edition, the presence of over a hundred possible planets has been detected, though none suitable for life is yet known. If there proves to be a second (or prior) genesis of life elsewhere, that will be welcome. But Earth will not on that account cease to be remarkable, nor will its particular natural history—trilobites, dinosaurs, primates—and social history—Israel, Europe, China, global capitalism, the Internet—cease to be unique in the universe.

Two decades ago what needed to be explained was the generation of *complexity*. In recent decades scientists have come more to focus on the *information* required for specifying and generating such complexity. In the beginning, and, continuing in the astronomical and geophysical sciences, there have been two metaphysical fundamentals: matter and energy. At deeper levels, physicists reduced these two to one: matter-energy. Later, cyberneticists began to insist that there were still two fundamentals: matter-energy and information. Norbert Wiener, a founder of cybernetics, insisted, "Information is information, not matter or energy" (Wiener 1948, p. 155). Hans Christian von Baeyer, a physicist, anticipates, "If we can understand the nature of information, and incorporate it into our model of the physical world . . . then physics will truly enter the information age" (von Baeyer 2003, p. 17).

The term "information" is complex and has been used variously in differing sciences. There is information on the surface of the moon, in the sense that a geologist can read some of the history of the moon from the overlay of meteoric impacts there. Mathematical information (or communication) theory, Shannon information, deals with reliable signal transmission, without regard to the significance, the semantic content, of the signal transmitted. Relevant information in addition has both signal reliability and signal significance. Any science, physics included, is a question of information gained.

The discovery that information is a critical determinant of history has thrown the causal/contingency debate into a new light. The world is composed of matter and energy, with the two united in relativity theory—so physics and chemistry have insisted. But the earthen world, biologists now insist, is composed by information that superintends the uses of matter and energy. The biological sense of information is more proactive, agentive than the physical sense. Such vital information is carried in the genes. What makes the critical difference is not the matter, not the energy, necessary though these are; what makes the critical difference is the information breakthrough with resulting capacity for agency, for *doing* something. Something can be discovered, learned, conserved, reproduced on Earth, but not on the moon.

Afterward, as before, there are no causal gaps from the viewpoint of a physicist or chemist, but there is something more: novel information that makes possible the achievement of increasing order, maintained out of the disorder. The same energy budget can be put to very different historical uses, depending on the information in the system. The miraculous is not the punctuation of natural order with supernatural intent—God sneaking into the causal gaps. The miraculous is the more out of less that the coupling of natural order with disorder generates, with nature wonderfully, surprisingly, regularly breaking through to new discoveries because there is new information emergent in the life codings. These achievements are, if you like, fully natural: they are not unnatural; they do not violate nature. But they also are novel achievements of "know-how," of agentive power. Something higher is reached, and in that sense there is something "super" to the precedents, something superimposed, superintending, supervening on what went before; there is more where once there was less, something "super" to the previously natural. The "super" for scientists is "cybernetic." For the theologians, what is added to matter-energy is "logos."

This was already on my horizon two decades back:

> An evolutionary life stream . . . conserves basic chemistries and maintains an information flow for millions and even billions of years. . . . Life is a kind of fire that outlasts the sticks that feed it, except that with ordinary fire the flame is merely a chemical product of the fuel burned, devoid of heredity, while the characteristics of the life "fire" are coded in an information flow from the parental fires that light it. The food that fuels the fire is taken over and "informed" into this life form. The information persists and increases over time, is more or less as long-enduring as the particles it employs, and is no less real or significant. In this world in which the atoms present early on have organized themselves into life and mind, the total tale of the pattern states of these atoms hardly seems told until these later levels have been given their place. . . . If one is to understand what is going on, one has to rest explanations at the appropriate level of informational control. (pp. 73–75)

One finds as fact of the matter that human affairs and astronomical and microphysical affairs are not irrelevant to each other, with the superposing of increasing information on matter-energy, and the apical manifestation of this in ourselves. One wonders at this hint that there must be some great Cause adequate to this great effect, something that infuses meaning across the whole and suffuses meaning in ourselves. The macrophysics and the microphysics are affecting our metaphysics.

3. GENETICS: PAST, PRESENT, FUTURE

The last two decades have been spectacular for genetics, and if I were re-writing Chapter 3, on life, I would now feature genetics as much as evolutionary theory, the intertwining of the genetic molecular scale with the scale of evolutionary natural history (as I have done in my *Genes, Genesis, and God*). Biologists have not only agreed but even more insisted on these two metaphysical fundamentals: matter-energy and information, with a more complex sense of information distinctive to life. Genes do not contain simply descriptive information "about" but prescriptive information "for" the vital processes of life. There is natural selection "for" what a gene does contributing to adaptive fit. Stored in their coding, genes have a "telos," an "end." Magmas crystallizing into rocks and rivers flowing downhill have results but no such "end." Genes are *teleosemantic*.

That differentiates physics from biology; and, biologists argue, biologists need to be alert to this. George C. Williams is explicit: "Evolutionary biologists have failed to realize that they work with two more or less incommensurable domains: that of information and that of matter. . . . The gene is a package of information" (in Brockman 1995, p. 43). John Maynard Smith says: "Heredity is about the transmission, not of matter or energy, but of information" (Maynard Smith 1995,

p. 28). Massive amounts of information are coded in DNA, a sort of linguistic or cognitive molecule. Now the semantic content of such information is critical, as it was not in the minimal, mathematical, physical sense. "Life is guided by information and inorganic processes are not. . . . The sequence hypothesis in the genome and in the proteome is a new axiom in molecular biology . . . unique to biology for there is no trace of a sequence determining the structure of a chemical or of a code between such sequences in the physical and chemical world" (Yockey 2005, pp. 8, 183).

Some leading theoretical biologists are now calling this genetic information "intentional," using that word in a nonconscious sense. John Maynard Smith claims: "In biology, the use of informational terms implies intentionality" (Maynard Smith 2000, p. 177). That word has too much of a "deliberative" component for most users, but what is intended by "intentional" is the directed life process, going back to the Latin: *intendo*, with the sense of "stretch toward," or "aim at." Genes have both descriptive and prescriptive "aboutness"; they stretch toward, "attend to" what they are about. This is tropistic, tensed behavior. Kim Sterelny and Paul E. Griffiths speak of "intentional information" in contrast to "causal information." "Intentional information seems like a better candidate for the sense in which genes carry developmental information and nothing else does" (Sterelny and Griffiths 1999, p. 104). Genes specify an ordered trajectory that produces highly complex organized, functional organisms.

Intentional or semantic information is for the purpose of ("about") producing a functional unit that does not yet exist. Here there arises the possibility of mistakes, of error, and genes have some machinery for "error correction." Sometimes that is by using gene copies that are "redundant." None of these ideas make any sense in chemistry or physics, geology or meteorology. Atoms, crystals, rocks, and weather fronts do not "intend" anything and therefore cannot "err." A mere "cause" is pushy but not forward looking. A developing crystal has the form, shape, and location it has because of, caused by, preceding factors. A genetic code is a "code for" something. The code is set for "control" of the upcoming molecules that it will participate in forming. There is proactive "intention" about the future. This line of analysis confirms the actively cybernetic nature of biology.

Although dominant throughout biology, evolutionary theory has continued to prove quite problematic itself (independently of any theological agenda). Many of these issues surround the questions of what to make of this natural history of increase of biodiversity and biocomplexity resulting in humans, and what to make of human uniqueness and human responsibility. There are disagreements involving the relative degrees of order and contingency, repeatability, predictability, the role of sexuality, competition and symbiosis, the extent of social construction in evolutionary theory, the evolutionary origins of mind, especially the human mind, and differences between nature and culture.

Does the Earth set-up make life probable, even inevitable? Already in my discussion of evolutionary history two decades back, I was apprehensive about accounts that find natural history to be essentially "a random walk" (pp. 104–09) and I was searching for "an alternative paradigm" (pp. 109-119). In the subsequent two decades, biologists have continued to spread themselves across a spectrum thinking that natural history is random, contingent, caused, unlikely, likely, determined, inevitable, open. Many continue to hold, with John Maynard Smith, that we need "to put an arrow on evolutionary time" but get no help from evolutionary theory (Maynard Smith 1972, p. 98; cited p. 120). Others think that any trends toward biodiversity or biocomplexity are read into, not out of, the evolutionary record. A somewhat surprising trend in the last two decades has been to interpret any such reading as nothing but social construction.

Michael Ruse, a philosopher of biology, reports: "A major conclusion of this study is that some of the most significant of today's evolutionists are Progressionists, and . . . we find (absolute) progressivism alive and well in their work" (Ruse 1996, p. 536). Nevertheless, Ruse thinks, they are all wrong, because, biased, they are reading progress into the evolutionary record. They have slipped into "pseudo-science." "For nigh two centuries, evolution functioned as an ideology, as a secular religion, that of Progress" (p. 526). In fact, he argues, today more "mature" scientists, unbiased, "expelled progress" from evolutionary history (p. 534). "Evolution is going nowhere–and rather slowly at that" (Ruse 1986, p. 203). One of the reasons Ruse's book is so long is that he has to argue away what most biologists have believed and continue to believe. Similarly, Daniel McShea finds a clear consensus among evolutionary biologists that there has been increasing complexity over evolutionary time, but he suspects this arises from cultural bias interpreting the fossil evidence (McShea 1991).

Other biologists argue the contrary with equal conviction: Simon Conway Morris is recently the most vigorous paleontologist arguing that human life has appeared only on Earth but did so here as a law of the universe: We are "inevitable humans in a lonely universe." "The science of evolution does not belittle us. . . . Something like ourselves is an evolutionary inevitability" (Conway Morris 2003, p. xv). Christian de Duve, Nobel laureate, argues, "Life was bound to arise under the prevailing conditions. . . . I view this universe [as] . . . made in such a way as to generate life and mind, bound to give birth to thinking beings" (de Duve 1995 pp. xv, xviii).

Many evolutionary theorists doubt that the Darwinian theory predicts the long-term historical innovations that have in fact occurred. John Maynard Smith and Eörs Szathmáry analyze "the major transitions in evolution" with the resulting complexity, asking "how and why this complexity has increased in the course of evolution." "Our thesis is that the increase has depended on a small number of major transitions in the way in which genetic information is transmitted between generations." Critical innovations have included the origin of the genetic code itself, the origin of eukaryotes from prokaryotes, meiotic sex, multicellular life, animal societies, and language, especially human language. But, contrary to Conway Morris and de Duve, they find "no reason to regard the unique transitions as the inevitable result of some general law"; to the contrary, these events might not have happened at all (Maynard Smith and Szathmáry 1995, p. 3).

So what makes the critical difference in evolutionary history is increase in the information possibility space, which is not something inherent in the precursor materials, nor in the evolutionary system. This is not something for which biology has an evident explanation, although these events, when they happen, are retrospectively interpretable in biological categories. The biological explanation is modestly incomplete, recognizing the importance of the genesis of new information channels.

A major feature of genetic natural history is co-option generating novel possibilities. This introduces new possibilities of order, layer by layer. The biological constructions are historical, but they are not simply linear combinatorial processes. In the DNA molecules the coding is linear, and the changes are incremental in the linear sequences. But these changes also involve reassorting blocks that reshuffle to produce surprises. A few changes in the linear sequence produce quite different folding patterns at tertiary and quaternary levels in the finished protein. Novel possibilities open up whole new regions of search space; old molecules recombine to learn new tricks in unprecedented circumstances.

Such composition is not linear because, at critical turnings, it involves co-option: An existing gene and its product are recruited to a new function. Such co-options open up new possibility space, and the new genetic information achieved proves valuable in an evolutionary search for better environmental information. For example, lens crystallins used in eyes first evolved in an altogether different role, as heat stress proteins. They happened to be transparent, irrelevant to their original use. Surprisingly, they get used to make eye lenses (Wistow 1993). Hearing evolves from body pressure cells sensitive to touch, greatly elaborating and modifying such cells, developing complex ears able to receive information at a distance. With continuing co-option, vertebrate ears open up the possibility of animal communication (Bear, Conners, Paradiso 2001, Chapter 11).

Spoken language requires simultaneously the evolution of genes for speech, and such genes, differentiating humans from other primates, arose at a highly critical period in our evolution. The FOXP2 gene, called a speech gene, arose less than 200,000 years ago and became the subject of strong selection, making language and culture possible. But if one is to have something to say, ideas to communicate, one needs a complex brain/mind. Acetylcholine, an ancient molecule, was around for millennia doing other things in plants and bacteria; when nerves appear it gets co-opted for use in synaptic transmission, which makes mental life possible (along with other neurotransmitters). Ideas pass from mind to mind, and for this, hearing is more important than sight—at least until the invention of writing. Millennia later, written language (needing those eyes and their crystallins) transforms cultures by making possible the transmission of thoughts non-orally, across centuries and peoples. Printing makes possible massive public communication, followed by radio, television, electronic communication, the Internet. Escalating co-option drives the information explosion.

There are remarkable forks off pre-existing pathways. Previously disconnected parts working along unrelated pathways are co-opted off and put together to start serving a novel function, perhaps only slightly well at the first. Radically different selection pressures begin to work in new directions that are completely unanticipated when they occur. Once launched, the novel functions may improve steadily and completely transform the course of natural and human history. Perhaps it all takes place by slight modifications of a precursor system. These incremental changes keep "bootstrapping" on themselves and hence the self-organization.

But these slight modifications are sometimes made in new, unprecedented directions. The co-opting modification is not improving the initial function but angles off in a new direction.

The change is not iterative; it is metamorphic. Co-option breaks up channelized and entrenched developmental lines (more and better pressure cells) and opens up new directions (hearing at a distance, meaningful sounds). Restriction enzymes, one of the most important features of genetic innovation and a principal tool in genetic engineering, were first invented by bacteria to cut their parasites into pieces. They turned out to be useful for organisms to cut their own genomes into pieces and reshuffle them in the search for co-options.

Evolutionists can make *ex post facto* explanations. After the events have taken place, the paleontologist can say, well, this is what happened, and this is what resulted. But prior to the co-opting events, if asked what would be the result, if such and such happened, one could seldom, from the knowledge of the constituent parts, say in advance what the results of co-option would be. Much less could one predict that such results had to happen. Perhaps one will say, since it has so often happened in evolutionary history, that there must be some tendency in biological nature to co-opt, a disposition to improvise, to be opportunistic. But where is such tendency located? Hardly from "bottom up" in the precursor materials. Hardly either from "top down" in the planetary system.

One can say that evolution is disposed to exciting serendipity. In such cases of co-opted emergence, repeatedly compounding, something that is genuinely new pops out, pops up. The novelty is, of course, based on the precedents, but there is genuine novelty not present in any of the precedents. What emerged required the precedents, but the presence of the prior organisms did not determine or make inevitable these results. There are critical turning points in the history of life that hinge on events more idiographic (unique, one-off events) than nomothetic (law-like, inevitable, repeatable trends). Things get recruited for new roles. Novel possibilities open up whole new regions of search space; old molecules recombine to learn new tricks.

Sometimes the explanatory account is by laws applied to initial conditions, and the same laws again reapplied to the resulting outcomes, now treated as further initial conditions. But sometimes, with co-options, endosymbioses, lateral genetic transfers, and mutations, the outcomes are not just further sets of initial conditions. The novel outcomes revise the previous laws; the rules of the game change, and the future is like no previous past. One can say that all this surprising serendipity is somehow "inherent" from the start; but the explanatory power of such a claim is rather vague. The main idea in co-option is the unpredictable and unexpected; co-option is as revolutionary as it is evolutionary.

Stuart Kauffman ponders this ongoing co-option of what he calls pre-adaptations, adaptations previously used for some other function:

> Consider the concept of Darwinian pre-adaptation, the idea that a feature that was selected for one purpose turns out to be useful for a second purpose. . . . Do you think you could ever say ahead of time what all possible Darwinian pre-adaptations are? . . . We can never say ahead of time what the relevant variables are in the evolu-

> tion of the biosphere. This means the biosphere keeps inventing new functionalities and we can't say ahead of time what they are. That's a radical new kind of failure to predict. It's not quantum uncertainty, and it's not chaos theory. Still, it's the kind of uncertainty that seems central. Life keeps inventing things. (Kauffman 2002)

Kauffman calls this "the mystery of the emergence of novel functionalities in evolution where none existed before: hearing, sight, flight, language. Whence this novelty? I was led to doubt that we could prestate the 'configuration space' of a biosphere. . . . Life is doing something far richer than we may have dreamed, literally, something incalculable" (Kauffman 2000, pp. 5, 7). With the opening up of new possibility space, the future is unanticipated. In my own account of a "supercharged nature" two decades ago, I had termed this "an inexhausible open-endedness . . . greater than we now know, or can foreseeably know" (p. 301).

The system rings the changes (rings the chances!) until there is high probability, even near certainty, that something creative will happen. Those discoveries are coded in the genetics, and the adventure continues next generation. The randomizing element begins to look different. It does not need to be taken away, at least not all of it, but it can remain as openness and possibility. This puzzling mixture of both the openness and the cybernetics in biology is what's new in biology with the turn of the millennium. And nothing here precludes going on to ask religious questions about the meaning of life on such an Earth.

It is not just the atomic or astronomical physics, found universally, but the middle-range earthen system, found rarely, that is so remarkable in its zest for complexity. My prediction is that, in the century to come, science will reveal this order achieved on Earth to be even more remarkable still, and that biological science will continue both to support and to underdetermine it. That will keep an active dialogue between biology and theology about the ultimate source of this creative ordering of our world.

The astounding drive that really needs explanation is what transforms chance into order, as the creatures emerge and exploit the opportunities in their environment, and are themselves transcended by later-coming, more highly ordered, more dazzling forms and dynamic processes, more intelligence, passion, consciousness, self-consciousness. Biologists, philosophers, and theologians will continue to need metaphysics adequate to occurrent reality. "Almost anything can happen in a world in which what we see around us has actually managed to happen. The story is already incredible, progressively more so at every emergent level" (pp. 301–02).

Astronaut Edgar Mitchell, a rocket scientist, reports being earthstruck:

> Suddenly from behind the rim of the room, in long, slow-motion moments of immense majesty, there emerges a sparkling blue and white jewel, a light, delicate sky-blue sphere laced with slowly swirling veils of white, rising gradually like a small pearl in a thick sea of black mystery. It takes more than a moment to fully realize this is Earth—home. (Quoted in Kelley 1988, at photograph 42)

The shining pearl is proving to be as much mystery as the surrounding black space. Humans have deep roots in and entwined destinies with this wonderland

Earth, and they and they alone can view their planet. So humans too are entwined in the mystery, remarkable on this remarkable planet, a wonder on wonderland Earth. To their nature, role, and place we next turn.

4. HUMAN UNIQUENESS: BRAIN, MIND, CULTURE

One of the ways my *Science and Religion* differs from similar books is in its attention to the human sciences, particularly psychology and sociology, also (if this be a science) to history. Twenty years ago, I began that inquiry worried about the possibility and limits of a human science (pp. 151–59): "Physics established a science of matter, from the seventeenth and eighteenth centuries onward. Biology established a science of life, from the nineteenth century onward. But how much science of persons can we have? This twentieth-century question is as yet unanswered" (p. 152). It still remains unanswered in the twenty-first century; indeed it has become more of an open question than ever—and that in spite of decoding our own genetics, despite sociobiology and evolutionary psychology, despite breakthroughs in neuroscience, cognitive science, and linguistics.

The challenge is what to make of the human mind, of the sharing, conserving, elaborating, criticizing, and acting upon these ideas, made possible by the symbolic and abstractive powers of language, and resulting in our cumulative transmissible cultures. We know time and space across 40 orders of magnitude, and we build cultures that, if we may phrase it this way, exceed anything known in animal cultures by 40 orders of magnitude. Compare a chimpanzee imitating its parent using an ant-fishing stick, with Einstein and his theory of relativity, used to build a nuclear bomb. The chimpanzee copied what he saw immediately before him. Einstein constructed his theory with mental genius, achieved as he stood on the shoulders of thinkers standing on the shoulders of other thinkers for five thousand years—with ideas passing from mind to mind, critically evaluated in each new generation.

Here's the way I posed that paradox and dialectic two decades back.

> The most baffling symbolic logic that we confront is not that of the mathematicians who write equations and metricize things, not that of the logicians who abstract into symbols portions of our thought processes. The ultimate symbolic logic is language itself, ordinary language, the languages of science and of religion, where words and texts become such powerful symbols of the world, of the world-logos, and of our place in the world. Humans have a double-level orienting system: one in the genes, shared with animals in considerable part; another in the mental world of ideas, as this flowers forth from mind, for which there is really no illuminating biological analogue. (p. 155)

The power of ideas in human life is as baffling as ever. The nature and origins of language is proving, according to some experts in the field, "the hardest problem in science" (Christiansen and Kirby 2003, p. 1). Paleoanthropologists can only speculate about when and how human language arrived; estimates vary over a million and a half years. Perhaps even recent hominids such as Neanderthals had quite limited capacities for speech. Genes went molecular (discovering DNA molecules) to discover that what is really of interest is not the chemical matter-energy transformations but the information stored in such transformations about how to cope, how to survive in a niche in an ecology. Neuroscience likewise went molecular (acetylcholine in synaptic junctions, voltage-gated potassium channels triggering synapsizing) to discover that what is really of interest is how these synaptic connections are configured by the information stored there, enabling function in the inhabited world.

Neuroscience has imaged much of the brain, to realize that it was imaging brains, or, more accurately, blood flow in brains, and not thoughts in minds. There is little or no success in correlating the flow of mental representations (as in a novel) with the details of neural architecture. What will neuroscientists think when, imaging their own thinking brains, they discuss with one another how it is that one species has gained the capacity to do this, to discuss the significance of such neuroscience, and watch the brain images of their discussion?

Animal brains are already impressive. But this cognitive development has reached a striking expression point in the hominid lines leading to *Homo sapiens*, going from about 300 to 1,400 cubic centimeters of cranial capacity in a few million years. There is only one line that leads to persons, but in that line at least the steady growth of cranial capacity makes it difficult to think that intelligence is not being selected for. "No organ in the history of life has grown faster" (E. O. Wilson 1978, p. 87). One can first think that in humans enlarging brains are to be expected, since intelligence conveys obvious survival advantage. But then again, that is not so obvious, since all the other five million or so presently existing species survive well enough without advanced intelligence, as did all the other billions of species that have come and gone over the millennia. In only one of these myriads of species does a transmissible culture develop; and in this one it develops explosively, with radical innovations that eventually have little to do with survival (such as symbolic logic or the equations of relativity theory).

The human brain has a cortex 3,000 times larger than that of the mouse. Our protein molecules are 97 percent identical (more or less) to those in chimpanzees, only 3 percent different. But we have three times (300 percent) their cranial cortex. The connecting fibers in a human brain, extended, would wrap around the Earth forty times. The human brain is of such complexity that descriptive numbers are astronomical and difficult to fathom. A typical estimate is 10^{13} neurons, each with several thousand synapses (possibly tens of thousands). Each neuron can "talk" to many others. This network, formed and re-formed, makes possible virtually endless mental activity. The result of such combinatorial explosion is that the human brain is capable of forming more possible thoughts than there are atoms in the universe.

Even more of interest is how thoughts in the conscious mind form, re-form, or, most accurately, in-form events in this brain space. We neuroimage blood brain flow to find that such thoughts can re-shape the brains in which they arise. Our ideas and our practices configure and re-configure our own sponsoring brain structures. In the vocabulary of neuroscience, we map brains to discover we have "mutable maps." For example, with the decision to play a violin well, and resolute practice, string musicians alter the structural configuration of their brains to facilitate fingering the strings with one arm and drawing the bow with the other (Elbert et al. 1995).

With the decision to become a taxi driver in London, and several years of experience driving about the city, drivers likewise alter their brain structures, devoting more space to navigation-related skills than have non-taxi drivers (Maguire et al. 2000). Similarly, researchers have found that "the structure of the human brain is altered by the experience of acquiring a second language" (Mechelli et al. 2004). Or by learning to juggle (Draganski et al. 2004). The human brain is as open as it is wired up. No doubt our brains shape our minds, but also our minds shape our brains. The process is as top down as it is bottom up.

Some trans-genetic threshold seems to have been crossed. Humans have made an exodus from determination by genetics and natural selection and passed into a mental and social realm with new freedoms. Humans may differ in their protein molecules from chimpanzees by only a fraction of a percent. But the startling successes of humans doing biological sciences can as readily prove human distinctiveness. Chimpanzees have no capacities for cumulative transmissible cultures leading to a science by which they can decode their own genes; much less can they debate the ethics of cloning or have their religious convictions challenged by reading Darwin's *The Origin of Species*.

Richard Lewontin, Harvard biologist, puts it this way:

> Our DNA is a powerful influence on our anatomies and physiologies. In particular, it makes possible the complex brain that characterizes human beings. But having made that brain possible, the genes have made possible human nature, a social nature whose limitations and possible shapes we do not know except insofar as we know what human consciousness has already made possible. . . . History far transcends any narrow limitations that are claimed for either the power of the genes or the power of the environment to circumscribe us. . . . The genes, in making possible the development of human consciousness, have surrendered their power both to determine the individual and its environment. They have been replaced by an entirely new level of causation, that of social interaction with its own laws and its own nature. (Lewontin 1991, p. 123)

The genes outdo themselves in culture (Tomasello 1999). Human societies are a spectacular anomaly in the animal world (Richardson and Boyd 1995, p. 195). Mind of the human kind seems to require incredible opening up of new possibility space. An information explosion gets pinpointed in humans. Humans alone have "a theory of mind"; they know that there are ideas in other minds. Linguistic cultures make possible these escalating achievements as ideas pass from mind to mind.

The surprise is that human intelligence becomes reflectively self-conscious as it builds these cumulative transmissible cultures. What is really exciting is that human intelligence is now "spirited," an ego with felt, self-reflective psychological inwardness. In the most organized structure in the universe, so far as is known, molecules, trillions of them, spin round in this astronomically complex webwork and generate the unified, centrally focused experience of mind. For this process neuroscience can as yet scarcely imagine a theory. A multiple net of billions of neurons objectively supports one unified mental subject, a singular center of experience. Synapses, neurotransmitters, axon growth—all these can and must be viewed as objects from the "outside" when neuroscience studies them. But what we also know, immediately, is that these events have "insides" to them, subjective experience. There is "somebody there," already in the higher animals, but this becomes especially "spirited" in human persons (Russell et al. 1999).

The self-actualizing and self-organizing characteristic of all living organisms in humans now doubles back on itself in this reflexive animal with the qualitative emergence of what the Germans call "Geist," what existentialists call "Existenz," what philosophers and theologians often call "spirit." This sense of existential self, the Cartesian "I think, therefore I am," is present in all persons and remains at once our central certainty and the great unknown. An object, the brained body, becomes a spirited subject. A team of neuroscientists concludes: "It is difficult to study the brain without developing a sense of awe about how well it works." They also concede: "Exactly how the parallel streams of sensory data are melded into perception, images, and ideas remains the Holy Grail of neuroscience" (Bear, Connors, and Paradiso 2001, pp. 740, 434).

If language is not the hardest problem in science, then what to make of the human mind in nature is. In the deeper philosophical senses—how mind evolved, whether the evolution of mind was inevitable, probable, contingent, the uniqueness of human mind, with its cumulative transmissible cultures, why there is a universe that (on Earth at least, elsewhere? but rarely?) evolves mind—remain the deepest mysteries we face. This has now become the main agenda: the place of this spirit awakened in nature. What does human uniqueness imply for human responsibility? Science and religion are equally challenged, and stressed, to answer. And we, with self-reflective consciousness, are right at the center of such mystery.

At this point much of what is in Chapter 4 can now seem quaint. Freudianism has lapsed, behaviorism seems archaic. Personality theory continues to seek self-actualizing persons but now psychologists may prefer Prozac to do so (affecting levels of the neurotransmitter serotonin with effects on personality). Still, these thinkers are part of the history one needs to learn when recounting the dialogue of science with religion. The story has moved on to cognitive science, to neuroscience. But what is of more interest to those interested in relating religion to the human sciences is how the same issues return in new form.

Sigmund Freud thought unconscious determinants were buried in the *id*; B. F. Skinner discounted mind as determinant of behavior and replaced this with empirically verifiable stimulus-response patterns. Neither allowed for personal freedom, for agency. Two decades later the questions of determinants within are now genetic, biological, physiological, or sociobiological, but parallel to those of Freud. The question of determinants without is now environmental, sociological, or economic, but parallel to those of behaviorism. How to relate both determinants within and without to human self-making (as in humanistic psychology) is still a central issue, only the location has shifted. "Mind: Religion and the Psychological Sciences" (Chapter 4) would be differently written, were I to start today. But those who read it will be smarter today at detecting science passing into scientism, then and now.

The sciences often claim that we humans in our behaviors are motivated by causes that are largely invisible to us. Consider, for instance, contemporary claims that we are genetically determined. "Now we know, in large measure, our fate is in our genes" (replacing what Freud would have said: unconscious mind). That comes with great authority from one of the discoverers of the genetic code, Nobel laureate James Watson, first director of the Human Genome Project (quoted in Jaroff 1989, p. 67).

But many geneticists demur: J. Craig Venter and over 200 co-authors, completing the Celera Genomics sequencing of the human genome, caution that genetic "determinism, the idea that all characteristics of the person are 'hard-wired' by the genome" and accompanying "reductionism" "are two fallacies to be avoided." They continue, in their concluding paragraph:

> In organisms with complex nervous systems, neither gene number, neuron number, nor number of cell types correlates in any meaningful manner with even simplistic measures of structural or behavioral complexity. . . . Between humans and chimpanzees, the gene number, gene structural function, chromosomal and genomic organization, and cell types and neuroanatomies are almost indistinguishable, yet the development modifications that predisposed human lineages to cortical expansion and development of the larynx, giving rise to language culminated in a massive singularity that by even the simplest of criteria made humans more complex in a behavioral sense. . . . The real challenge of human biology, beyond the task of finding out how genes orchestrate the construction and maintenance of the miraculous mechanism of our bodies, will lie ahead as we seek to explain how our minds have come to organize thoughts sufficiently well to investigate our own existence. (Venter et al. 2001, pp. 1347–48)

The "massive singularity" is this self-investigating creature so full of ideas about both self and world. Natural selection passed over into something else. Nature transcended itself in culture, with radical new chapters in the ongoing story of the evolution of information, cognition, and history. The world moved into a future quite unlike its past—the "wise" (*sapiens*) species rebuilding, exploiting nature, "playing God" with unprecedented powers for good and evil, and putting the community of life on Earth into jeopardy. Indeed, this species decoding its own genome and pondering remaking itself is again "playing God," choosing good from evil, quite unprecedented on Earth. Such escalating powers of agency we might still call "natural," not "supernatural," but they are evidently "super" to anything previously called natural.

Another effort to deny such "massive singularity" in humans and to bind the mind to natural selection is found in evolutionary psychology. Humans have what John Tooby and Leda Cosmides call an "adapted mind" made up of a set of "complex adaptations" that, over our evolutionary history, have promoted survival. "What is special about the human mind is not that it gave up 'instinct' in order to become flexible, but that it proliferated 'instincts'—that is, content-specific problem-solving specializations" (Tooby and Cosmides 1992, pp. 61, 69, 113). "These evolved psychological mechanisms are adaptations, constructed by natural selection over evolutionary time" (Cosmides, Tooby, and Barkow 1992, p. 5). These channelled reaction patterns form a set of behavioral subroutines more like a "Swiss army knife," tools for survival, rather than a general purpose learning device (Cosmides and Tooby 1994, p. 60). These "Darwinian algorithms" are each dedicated to task-specific functions such as picking mates, desiring many children, eating fats and sweets, or helping family, or obeying parents, defending one's tribe, fight or flight, being suspicious of strangers, dealing with non-cooperators by ostracizing them, or preferring savannah-type landscapes. In picking mates, for example, men are disposed to select younger women, likely to be fertile. Women select men of social status, likely to be good providers.

The human mind is indeed complex, and various "automatic" subroutines to which we are genetically programmed may indeed be convenient shortcuts to survival, reliable modes of operating whether or not persons have made much rational reflection over these behaviors. Nevertheless, these need to be figured back into a more generalized intelligence. Genetically programmed algorithms seem unlikely for the detail of such decisions under changing cultural conditions. Capacities to select a good mate across decades of marriage and over diverse cultures are perhaps somewhat "instinctive," but they are unlikely to be an adaptive mechanism isolated from general intelligence and moral sensitivity.

Even by accounts of evolutionary psychologists, the mind is not so compartmentalized that humans—modern ones who read this literature at least—cannot make a critical appraisal of what behavioral subroutines they suppose they inherit by genetic disposition, and choose, if they wish, to offset these. Cosmides and Tooby call for "conceptual integration" of the diverse academic disciplines studying humans, their behavior, and their minds. These include "evolutionary biology, cognitive science, behavioral ecology, psychology, hunter-gatherer studies, social anthropology, biological anthropology, primatology, and neurobiology," among others (Cosmides, Tooby, and Barkow 1992, pp. 4, 23–24). These are not disciplines in which one becomes expert by behavioral mechanisms in a Swiss-army-knife mind.

They and their readers must have quite broadly analytical and synoptic minds. The mind is fully capable of evaluating any such behavioral modules, and of recommending appropriate education so as to reshape these dispositions in result. When evolutionary psychologists wonder whether to re-adapt by critical thought their own adapted minds, we as their colleagues wonder whether they are alone in this capacity. We see the Freudian problem returning, the behaviorist problem returning, the selfish gene problem returning—the perennial problem when any science that proposes to explain human behavior is made self-referential. No one—scientist, philosopher, ethicist, theologian—can be free to evaluate such theories unless they are free to think and to act on the basis of such evaluations.

Cognitive science, switching from genetics and neuroscience to computer science, often models mind as some kind of computer. I worried about that in 1987, and that worry has become prophetic. The mind is capable of information processing; there is no doubt about that. The mind is less capable than computers in doing some kinds of information processing: scanning vast data sets or making complex calculations (on the basis of algorithms programmed into the computer by smart computer engineers). But the mind, embodied in flesh, incarnate, works at different levels of information processing: experiential, self-consciously reflective, narrative story lines, biography, facing death, disease, giving birth to and rearing a next generation, educated into and re-evaluating a cumulative transmissible culture. People wonder who they are, where they are, what they ought to do. All this seems vastly more than the capacity of computers of whatever kind.

Indeed, computer analogies can be quite misleading. That danger comes first from the genetics. No computers reproduce themselves by passing a single set of minute coding sequences from one generation of computers to the next, like sperm and egg, with the next generation of computers self-organizing from this single transferred information set. Such danger comes, secondly, from the neuroscience. Software and hardware, which can be easily separated in a computer, are completely interwoven in brains. Brains generate minds that re-form these brains, both during development and across adult life, modifying, rebuilding synaptic connections, and even generating new neurons. Although some computer programs have open search programs, none of those mutable cognitive maps reformed by resolute decision (those violin players) have any significant parallels in existing computers. Computers do not have minds with which to reconfigure themselves.

The computer model is misleading because of those mutable maps, of minds forming brains, but even more because minds inhabit incarnate flesh. The problem is not hardware, not software, but "wetware." Now we do need both the biology and the spirit, both the blood and the *Geist*. We are self-actualizing persons who can both think and suffer.

> Have we reached a model competent to the whole human person? At this point limitations in the cybernetic model begin to appear. Storing, retrieving, and using information are certainly important. But are these the only, or the central, features of personality? Even in terms of a biological model, cognitive processors as such do not suffer, and here we could wish for more awareness within this psychology of that dimension in experience of which evolutionary biology has left the organic world almost too full, of the cruciform nature of life. In terms of a human model, cognitive processors do not feel ashamed or proud; they do not have angst, self-respect, fear, or hope. They do not get excited about a job well done, pass the buck for failures, have identity crises, or deceive themselves to avoid self-censure. They do not resolve to dissent before an immoral social practice and pay the price of civil disobedience in the hope of reforming their society. They do not weep or say grace at meals. . . . They do not have heroes or saviors. They do not die for the sins of the world, launch the Kingdom of God, or fall into other passionate ideologies about the meanings of life and history. The model of the cognitive processor, while necessary, is yet insufficient for the human personality. (pp. 182–83)

Humans anticipate death; they sense their finitude. They face limit questions, sense the sacred, worry about communion with the ultimate or atonement of their sins. They know guilt, forgiveness, shame, remorse, glory, and pride. They suffer angst and alienation. They build symbols with which they interpret their place and role in their world. They create ideologies, affirm creeds, and debate their rights and responsibilities. They are capable of religious faith. They worship God. All of this can be summed up in the one word: *spirit*. In this life of the spirit, humans, arriving late on the planet, remain remarkably distinctive from the other millions of species, indeed the billions that have come and gone over evolutionary time. They also remain remarkably distinct from any machines they have yet built. The most complex thing in the known universe, the "massive singularity" (Venter), is still right behind our eyes. We humans live at the center of the most genesis yet known.

5. MIND KNOWING NATURE: REALISM AND SOCIAL CONSTRUCTION

Two decades ago, I recognized the "pervasive and persuasive characteristics of paradigms," alike in science and religion, in keeping with the insights of Thomas Kuhn (pp. 9–15). I worried about "observer involvement in science and religion (pp. 19–22). I sought "universal intent" (pp. 16–17), and I searched for good paradigms and feared lest they become "bliks." "The theory that begins as a synthetic judgment about the world can get subtly transformed into an analytic prejudgment brought to the world, so that variant experience can no longer transform the theory but rather the theory transforms the experience. A blik is a theory grown arrogant, too hard to be softened by experience" (p. 11). I resisted the claim that religions are nothing but a "social projection" (pp. 219–225). I closed the book with the claim that science must be fitted into a culture's ongoing sense of historical narrative.

Philosophers of science have continued to soften the realism in science in favor of more historical and culture-bound accounts. Science is an interactive activity between humans and a nature out there that we know only through the lenses, theories, and equipment that we humans have constructed. Much of this is welcome, I had already said. But I did not anticipate the "culture wars"—the way in which postmodernists, deconstructionists, pragmatists, feminists, and others would come to claim that all our knowing, science included, is little more than a "social construction."

Critics of science have pressed these claims about the social construction of science and theology further than most scientists wish (Hacking 1999; and with waffling, Ruse 1999). Theologians are of mixed opinions whether to welcome these developments. What Thomas Kuhn taught us a generation ago about science, Alasdair MacIntyre (1981) afterward applied to ethics, then George Lindbeck (1984) said much the same thing about theology. Each discipline is embedded in a conceptual framework so comprehensive that it shapes its own criteria of adequacy. These paradigmatic communities govern what all thinkers within them look for and how they interpret what they find. The search for any autonomous, universal truth, defended by a neutral rationality, has failed. All our claims are in a "web of belief" (Quine and Ullian 1978). They attach to dynamic, culturally conditioned, historical worldviews.

Earlier the problem was an adapted mind, so full of genetically programmed stereotyped behavioral modules that humans were not free to think for themselves. Now the problem is different, but similar: a socially constructed mind, so full of culturally imposed filters that humans struggle to be free to think independently. It can almost seem as though Freud and Skinner have been reincarnated. Both adapted minds and socially constructed minds challenge human freedom, human rationality, human uniqueness, and human responsibility. And both are challenged in turn by efforts to think through and think beyond our evolutionary and our cultural legacies.

Keep pushing these claims, for instance, and natural science loses all its objectivity, its powers to describe the natural world. Alexander Wilson claims, "We should by no means exempt science from social discussions of nature. . . . In fact, the whole idea of nature as something separate from human existence is a lie. Humans and nature construct one another" (A. Wilson 1992, p. 13). Don Cupitt puts this quite bluntly: "Science is at no point privileged. It is itself just another cultural activity. Interpretation reaches all the way down, and we have no 'pure' and extra-historical access to Nature. We have no basis for distinguishing between Nature itself and our own changing historically-produced representations of nature. . . . Nature is a cultural product" (Cupitt 1993, p. 35).

David Pepper, urging a postmodern science, insists "that there is no one, objective, monolithic truth about society-nature/environment relationships, as some [scientists] might have us believe. There are different truths for different groups of people and with different ideologies. . . . Each myth functions as a cultural filter, so that adherents are predisposed to learn different things about the environment and to construct different knowledges about it" (Pepper 1996, pp. 3–4). Science is one more myth, a Western cultural filter, no better (maybe at times worse) than the other, classically religious filters.

From a pragmatist perspective, Richard Rorty asks whether "science describes a world already there?" No, says Rorty, "we must resist the temptation to think that the redescriptions of reality offered by contemporary physical or biological science are somehow closer to 'the things themselves.'" The big mistake is "to think that the point of language is to represent a hidden reality which lies outside us" (Rorty 1989, pp. 16, 19). We must not think that "Reason" offers "a transcultural human ability to correspond to reality"; the best that reason can do is ask "about what self-image society should have of itself" (Rorty 1991, p. 28).

In the passing scientific fashions, Rorty concludes, "We may have no more than conformity to the norms of the day. . . . This century's 'superstition' was last century's triumph of reason . . . the latest vocabulary, borrowed from the latest scientific achievement, may not express privileged representations of essences, but be just another of the potential infinity of vocabularies in which the world can be described" (Rorty 1979, p. 367). Science only provides makeshift sketches that we will replace, after more explorations, with a new round of cartoons. At the turn of the millennium, philosophers found themselves back in their ancient epistemological prison, in Plato's cave, or like one of the Indian blind men groping at the elephant. If it isn't the genes that keep us in the cave, then it is our modular mind; if not those, then our cultures do.

This seems to make impossible any successful discovery of what is out there in astronomy, geology, in evolutionary natural history. It renders suspect all accounts of "nature for real" distinguished from "nature for us," "nature that I can functionally cope with." At this point, the deconstructionists may back off and explain that the items scientists see (stars, planets, meter readings, rocks, fossils, lions) may be there, but the interpretive framework is a mythical construction. Hence, as we saw earlier, Michael Ruse dismissed the conclusions of most biologists that biodiversity and biocomplexity have increased over evolutionary time, because, biased, they let evolution function as an ideology. They read British ideas of progress into the evolutionary record and thereby slipped into "pseudoscience." Ruse claims that his new evolutionary paradigm (the "going nowhere" one) is pushing out the old European cultural biases, but what Ruse cannot see is that his allegedly "more mature" scientists today, a minority but the best of them (so he claims), are just using the latest secular gestalt to read the record differently. If we take his book seriously as being self-referential, it too undermines itself. His new view is just the next passing scientific fashion. We will never get past different spokes for different folks, unless somebody can regain convictions about truth in science (Rolston 1997), for which Ruse himself is groping.

So there is an epistemic crisis, which, on some readings, can seem to have reached consummate sophistication, and, the next moment, can reveal debilitating failure of nerve. Taken to an extreme, postmodernist, pragmatist, deconstructionist views of science, those of the "strong program," seem to have become bliks, theories grown arrogant. Even Thomas Kuhn, though in some sense launching the postmodernist movement, came radically to reject where it was going.

> Nature itself, whatever that may be, has seemed to have no part in the development of beliefs about it. Talk of evidence, or the rationality of claims drawn from it, and of the truth or probability of those claims has been seen as simply the rhetoric behind which the victorious party cloaks its power. What passes for scientific knowledge becomes, then, simply the belief of the winners. I am among those who have found the claims of the strong program absurd: an example of deconstruction gone mad. (Kuhn 1992, pp. 8–9)

The extreme claim from the academic left that there are only differing ideologies, and that nobody is objectively right about anything, and that this finally gets the truth question right, if made self-referential, destroys itself. That claim is itself as spongy as all those they deconstruct. Illusion evaluating illusion is nonsense on stilts and soon collapses. These skeptics go the way of Freud, Skinner, genetic determinists, and evolutionary psychologists with their modular minds, since they leave no one, including themselves, free to think with any plausible rational powers.

Biologists all have their cultures, their personal backgrounds, their preferences, their biases, their worldviews. They can frame up the results of their observations differently, as we have already noticed in the spectrum across which they line up wondering about order, disorder, contingency, probability, predictability in evolutionary natural history; these frame-ups can reflect their educations. But meanwhile, biologists make many observations, construct many concepts, using many theories and instruments. Scientists do discover some things about surrounding phenomena, transcending their cultures, claims about events past and present that are true because they successfully (if also approximately) describe the phenomena as these exist in themselves.

There is no unmediated nature; therefore, we know nothing of nature as it is in itself. But this assumes that media cannot, reliably, descriptively, transmit truths about what is there. Biologists do abstract, and this can result in failing to see what is left out of the abstractions. They invent the theories with which they see, and these may blind them to other things. But inventions can also help us see. Science can regularly check its constructs against causal sequences in nature. Scientists can regularly cross-check each other. Some scientific claims will be revised; scientists work at that constantly.

But the general cluster of advancing scientific discoveries is not going to fail as passing cultural myth. As science progresses, scientists get clearer about what they are studying. Concepts are dynamic because scientists find out what was previously unknown. Older concepts will be used in new ways that align with the advances in the field; atoms, composed of electrons, protons, neutrons, can be broken apart. They can be split and relativity theory illuminates the distribution of matter and energy in their splitting. Darwin transformed the concept of fixed species into evolving species. Older concepts may also be entirely abandoned: phlogiston and entelechy.

In result, in Plato's famous phrase, scientists learn to "carve nature at the joints" (*Phaedrus* 265e). The sporophyte generation of mosses is haploid. Malaria is carried by *Plasmodium* in mosquitoes. Neither of those facts is likely to change with a new cultural filter. Golgi apparatus and mitochondria are here to stay. There is no feasible theory by which life on Earth is not carbon based and energized by photosynthesis, nor by which water is not composed of hydrogen and oxygen, whose properties depend on its being a polar molecule. Glycolysis and the Krebs cycle, ATP and ADP, will be taught in biology textbooks centuries hence, as well as lipid bilayers and immunoglobulin molecules. Oxygen will be carried by hemoglobin. Biologists are right that CO_2 is released in oxidative phosphorylation and that this cycles through photosynthesis II and photosynthesis I, so that in the world there is a symbiotic relationship between plants and animals and that this a vital ecosystemic interdependence (cf. Rolston 1999, pp. 187–88).

We have made progress in knowing who we are and where we are. Humans now know a round planet, orbiting the sun; we know something of its circulations, evolutionary origins, ecosystemic connections, fauna and flora. There is no more flat Earth, no turtle island cosmology, no more Earth created in 4004 BC with a garden planted in Eden in the Middle East, no Izanagi and Izanami stirring up the Japanese islands, or Amaterasu bringing order to them. There is no more enchanted world, populated with fairies and demons, though perhaps there remains, as much as ever, a sacred or numinous world. Any truth in these prescientific views, other cultural filters, will have to be de-mythologized. If one insists that this is re-mythologizing, then know that the right worldviews, the "true myths," must be trans-scientific, trans-humanist, trans-cultural. Science, humans, and culture must take reference points outside themselves in these planetary events if ever we are to describe them, much less make sense of their significance. Or know what ethic to construct.

6. SCIENCE AND CONSCIENCE

Hard science has a soft underbelly: conscience. We could almost say, provocatively, the harder the science, the softer the underbelly. The unavoidable question is what do scientists care about? What do those to whom their science becomes available care about? This probes the logic of science and worries about its zest for mastery, fearful lest this become a lust for mastery. If this seems unkind, then turn to Lord Acton: "Power tends to corrupt and absolute power corrupts absolutely" (Acton 1887, 1949, p. 364). He was absolutely right.

I worried about this two decades ago. I worried about it more a decade ago, in my *Genes, Genesis and God*:

> With this knowledge comes power. More than any people before, as a result of our technological prowess through science and industry, we humans today have the capacity to do good and evil, to make war or to feed others, to act in justice and in love. Nor is it only the human fate that lies in our hands. We are altering the natural history of the planet, threatening alike the future of life, the fauna and the flora, and human life. With such increasing knowledge and power comes increasing duty. Science demands conscience. . . .
>
> One can hardly claim that modern science has figured out ethics, either its historical origins or a current evaluation. The more usual account is that ethics is not science, nor science ethics; the one is a descriptive discipline, what *is* (was, or will be) the case; the other an evaluative discipline, what *ought* to be. "Good and evil" . . . are not categories that appear in science textbooks. . . . So, although there is a profound sense in which we humans now know who and where we are, there is an equally deep puzzlement about what we ought to do, and the grounds of its justification. Science has made us increasingly competent in knowledge and power, but it has also left us decreasingly confident about right and wrong. . . .
>
> The same science that demands a conscience has difficulty explaining and authorizing conscience, for we struggle to understand how amoral nature evolved the moral animal, how even now *Homo sapiens* has duties, humans to fellow humans, and humans to the community of life on Earth. The value questions in the twentieth century remain as sharp and as painful as ever in our history. (Rolston 1999, pp. 213–15)

If one needs proof of that, read the newspapers: The Iraqi war, 9/11, Enron, protests at G-8 summits, health care for the poor, corruption in government, deforestation, global warming, and so on.

Scientists may reply that these are not issues in science, though they may deal with its application in economics, technology, and public policy. But there is a rising and revealing critique of science, one that is likely to prove still more forceful in the decades ahead. Science presents itself as detached and objective, capable of describing the world as it is in itself. That first seems plausible. I was just defending such an account of science against the extremes of social construction. Yes, the claims of physics about the big bang and the expanding universe, or those of biology about evolutionary history, are claims about what once took place on Earth, long before humans arrived. The genetic coding in the DNA and the protein synthesis by which organisms are produced and maintained, the food chains in ecosystems, the adapted fitness of organisms, their capacities for coping as they make a way through the world—all these seem to be descriptive claims. Science seems to have its independent authority warranting these claims.

But look more deeply. Science is the quest for knowledge and knowledge is power. Even pure science is driven by a desire to understand, and that, ipso facto, is a desire to conquer, seldom pure. The fundamental posture of science is one of analysis, the discovery of laws and generalizations, theory with implications, prediction, testability, repeatability. One wants better probes, better techniques, higher resolution detectors, more computing power. This always invites control; but more than that, this very approach to nature is driven by the desire to control. The underlying premise of all scientific logic is mastery, and with that insight the claims to detachment, objectivity, and independence take on a different color. Allegedly objective science is inevitably bent, sooner or later, into the service of technology, and such scientific knowledge coupled with technological power is neither detached nor objective. Willy-nilly, such information will be put to use for some better or worse ends. Thus relativity theory is used to make nuclear weapons; the human genome, mapped, invites first medical therapy and later genetic engineering. Such utility is not simply an outcome of science: it runs in, with, and under its worldview.

Such an account sees not only the outcome but the presumption of science in the escalating consumerism of the First World and in the disproportionately distributed wealth between First and Third Worlds, or, as we increasingly say, between North and South. These are symptoms of a fundamentally misplaced caring. Science is the product of the powerful urge to dominate nature, and those who have it are ready enough to colonize elsewhere and harvest whatever resources they can wherever they can, to build machines of industry and of war, to dominate other peoples and races.

The scientist, to be sure, when moving from pure to applied science, pretends to care; the benefits of science in the service of humans are preached incessantly. No doubt such caring and benefits are often true; but it is equally certain that science lacking critical caring for others on behalf of the scientists, or those who exploit their science, is what has produced the present distributional crisis. And caring for others—loving one's neighbor—is the central claim in religious ethics. Science is not religion. Religion cannot suggest the content of any science, but religion can notice the forms into which such content is being poured; it can also defend a content of its own. One can do science without adverting to theology, but one cannot live by science alone.

Indeed, science cannot teach us what we most need to know—that about which we most should care. In that sense, science is not independent. There is an information gap, this time not in the causal chains of science, but in the very logic of science itself. More computing power is not likely to give us the information we need here. There are no algorithms for good and evil. Nor is more analysis of our brain-behavioral modules going to give us an answer, nor genomic analysis of our protein similarities with the chimpanzees. All this suggests that the dialogue between science and religion is likely to continue. There will be a humane future only if we can integrate the two.

Science could be part of the problem, not part of the solution. Science can, and often does, serve noble interests. Science can, and often does, become self-serving, a means of perpetuating injustice, of violating human rights, of making war, of degrading the environment. Science is used for Western dominion over nature. Science is equally used for Western domination of other nations. The values surrounding the pursuit of science, as well as those that govern the uses to which science is put, are not generated out of the sciences, not even the human sciences, much less the biological or physical sciences (cf. pp. 339–43).

Where science seeks to control, dominate, manipulate either persons or nature or both, it blinds quite as much as it guides. Nothing in science ensures against philosophical confusions, against rationalizing, against mistaking evil for good, against loving the wrong gods. "The whole scientific enterprise of the last four centuries could yet prove demonic. We may be caught in a Faustian bargain, in a scientific sink" (p. 342). As good an indication as any of that is our ecological crisis.

Not only has a science-based technology failed to solve the deeper problems of developed nations, but a larger problem looms globally. There are about five billion persons in the world. Approximately one fifth, those in the developed nations, produce and consume about four fifths of the material goods that a science-based industry provides; about four fifths of the world divide the remaining one-fifth of the wealth, and about half of these live in poverty (*World Development Report 2004*). There are more poor persons today than ever before; there will be more yet in decades ahead. For every person added to the population of the developed nations, twenty individuals are added in the less developed ones. For every dollar of economic growth per person in the one, twenty dollars accrue to each individual in the other. Of the 90 million new people on Earth this year, 85 million will appear in the Third World, the countries least able to support them. Meanwhile, the 5 million new people in the industrial countries will put as much strain on the natural resources and cause as much environmental degradation as the 85 million new poor.

There are three problems: overpopulation, overconsumption, and underdistribution. The reasons for these outcomes are complex, but whatever explanations one finds for this mal-distribution of wealth, the outcome hardly seems either just or loving. We in the West may say, with some justification, that we have earned or merited our wealth. There is a first tendency to say the problem is that too many of the Earth's peoples are unblessed by the fruits of science and technology; we need to teach everyone how to produce up to Western standards. The distribution patterns reflect achievement; what the other nations need to do is to imitate this.

For solving this problem, science is necessary, since providing for human needs in the next century without science and technology is unthinkable. But science is not sufficient without conscience that shapes the uses to which science is put, informing policy. Science and religion together must face the impending disaster of today's trends projected cumulatively into tomorrow: population explosion, dwindling food supply, climate change, soil erosion and drought, deforestation, desertification, declining reserves of fossil fuels and other natural resources, toxic wastes, the growing gap between concentrated wealth and increasing poverty, and the militarism, nationalism, and industrialism that seek to keep the systems of exploitation in place. Few problems or none loom more foreboding on the horizon than these, and I predict that these value problems are, in the coming century, likely to become more acute than ever.

Religion has been the classical informer of conscience, and still remains a powerful force in moral life. Ethics can be autonomous—independent of religion—but such ethical systems have not yet proved themselves capable of shaping cultural reformations over generations. Here the religious ideologies do persist over changing science. It is much safer to predict that the Golden Rule will be an imperative in ethics a century hence than it is to predict that cosmology will still have a big bang with an inflationary period in the first few seconds. It is also, alas, much safer to predict that the seven deadly sins will still be present a century hence, with human life needing to be redeemed from these sins, than that biologists will be emphasizing the contingency in natural history over against a tendency toward increasing complexity over evolutionary time. Whether the Golden Rule or covetousness will have done more to shape the future is not safe to predict; that outcome depends, in significant part, on the extent of the dialogue between science and religion.

The radical differences between nature and culture, if not already evident, will become yet more evident as the speed of cultural innovation increases, owing in large part to the powers of science. In the more recent centuries, and in the most recent decades of this century, information accumulates and travels in culture at logarithmically increasing speeds. The pace of the story steps up; and now, as we turn from the long evolutionary and cultural past to face the future, there is a certain feeling that the pace of the action is accelerating, both with excitement and danger. The computer revolution exemplifies this, with its dramatic capacities for extending the human computational power, for information storage and processing of data, including scientific data, for long-distance communication and networking (not to mention possibilities for exploitation and invasion of personal privacy). Discoveries in physics and chemistry show us how the world was made. Discoveries in the biosciences—mapping, for instance, the human genome, with the further possibilities of genetic engineering—offer us the possibility of remaking the world. We humans too are agents, increasingly powerful agents, but will that bring more blessing or corruption?

We seem to have reached a turning point in the long, accumulating story of cognition actualizing itself. We are now coming around to oversee the world and to face the prospect of our own self-engineering, to the genesis of a higher-level ordering of the world in the midst of its threatening disorder. Increasingly we are like gods. But we need the wisdom of God, and that need programs poorly on computers and is not found in physics, chemistry, or biology textbooks. There is an information gap about good and evil.

Though biologists are typically uncertain whether life has arrived on Earth by divine intention, they are almost unanimous in their respect for life and seek biological conservation on an endangered planet. Earth's impressive and unique biodiversity, evolved and created, warrants wonder and care. In that sense, many of the discoveries in biology sensitize us to the need for caring for life on Earth. Biologists and theologians, though they may continue to argue about the past origins of life, are likely to reach consensus that humans, facing the future, ought to care responsibly for this wonderland Earth. Pressing these questions of caring for Earth against those of using the Earth with justice and charity, choosing the right path and finding resolution to follow it, motivating such behavior, will demand all the resources and insight we can muster in both science and religion.

Crises lie ahead of us, not for the lack of science but for the lack of wisdom, a wisdom that only religion in the broad sense can supply—worldviews that orient us philosophically and that can redeem our human nature from its perennial failings. The need for justice, for love, for caring will remain undiminished, and science will need conscience in the next century more than ever before. What on Earth are we doing? What on Earth ought we to be doing? There is no figuring this out without both science and religion; there is no doing it right without integration of the two.

This is the Earth in which we live and move and have our being, and we owe this Earth system the highest allegiance of which we are capable, under God, in whom also we live and move and have our being. Biologists, again, may not share the monotheism, but they are coming to share the concern for the Earth, and concern for the springs of human motivation. When they do, the mentality of dominance in science, about which we have worried, can itself become regenerated, and science put in the service of responsible care for this only home planet. Scientists as much as anyone else, theologians included, wish a sustainable harmony between humans and this very special planet. The foreboding challenge is that these spectacular humans, the sole moral agents on Earth, now jeopardize both themselves and their planet. Science and religion are equally needed, and strained, to bring salvation (to use a religious term), to keep life on Earth sustainable (to use a secular, scientific term).

> Science, ethics, and religion all have to do with sharing what is valuable; science is itself valuable and enables us to generate more value. But science alone does not teach us all we need to know about sharing values. For all its recent brilliance, science has proved penultimate to ethics and religion. . . . Science is know-how without know-whether. Science describes what is (or was, or will be), not what ought to be. Scientists, qua scientists alone, are not ipso facto wise. After science, we still need help deciding what to value; what is right and wrong, good and evil; how to behave as we cope. The end of life still lies in its meaning, the domain of religion and ethics. (Rolston, 1999, pp. 161–62)

7. HISTORICAL AND CRUCIFORM NATURE: LIFE PERSISTING IN PERISHING

Life on Earth is indisputably historical. "Nature after science is *historical* to the core, more historical after science than before" (p. 246). Where once there were no species on Earth, there are today five to ten million. Prokaryotes dominated the living world more than three billion years ago; there later appeared eukaryotes, with their well-organized nucleus and cytoplasmic organelles. Single-celled eukaryotes evolved into multicelled plants and animals with highly specialized organ systems. First there were cold-blooded animals at the mercy of climate, later warm-blooded animals with more energetic metabolisms. From small brains emerge large central nervous systems. From primates emerge humans, the one primate with cumulatively transmissible culture. Biologists do find a need to put some kind of an "arrow on evolutionary time"; there are cumulative achievements that can reset initial conditions. To use the pejorative term of the deconstructionists, evolutionary natural history producing these humans with their cultural stories is quite a "grand narrative."

The evolutionary psychologists will join the deconstructionists to see "storytelling" as one more subroutine in our modular mind, helping us to survive by concocting stories, myths, and narratives that orient us to cope in the world. And, once more, such an account cannot be made self-reflexive, because if we all have a mind predisposed to self-serving storytelling, this undermines both their capacity and ours to evaluate whether and how far earthen natural history is indeed historical. One is going to need a more inclusive intelligence—capable of astronomy, geology, paleontology, climatology, botany, zoology, genetics, genomics, cladistics, radiometric dating, cognitive science, neuroscience, physical and cultural anthropology, comparative religion—a "conceptual integration" of disciplines to tell this story, much less to discover how much theory confirmation, narrative confirmation, mental bias, or "mythology" is being read into and out of the earthen "facts."

Surely this is among the commanding facts: Life persists in struggle, generated and regenerated, yesterday, today, tomorrow—from the dawn of life until now, with our value questions as sharp and as painful as ever, confronting the promise and peril of our open future. This claim is quite corroborated by evolutionary biology, by the very existence of social construction, and by evolutionary psychology, even if their results are increasing doubt about our human competence. The last two decades well underscore my closing claims about how science is configured and re-configured into an ongoing historical cultural narrative, an ongoing struggle to make sense of nature and culture on this wonderland Earth.

In retrospect across two decades, if asked to judge what has proved most insightful in my *Science and Religion*, this may well prove to be my analysis of a "cruciform nature." Life has its logic, its history; in the course of that history, life has its *pathos*. The story we have from Darwinian natural history echoes classical religious themes of death and regeneration. In the midst of its struggles, life has been ever "conserved," as biologists find; life has been perpetually "redeemed," as theologians find. Both in the divine Logos once incarnate in Palestine and in the life incarnate on Earth for millennia before that: "Light shines in the darkness and the darkness has not overcome it" (John 1:5).

I celebrate "green pastures in the shadow of death, a table prepared in the midst of mine enemies" (Psalm 23). Now the science reinforces native range experience. All living things are caught up in the struggle for life; we humans, too. Seen from space, Earth is a shining pearl in a sea of black mystery; seen on the ground, life shines across the eons of natural history, perennially renewed in its perpetual perishing. There is "abundant life" in the midst of death. This "secret of life" continues to challenge both science and religion.

I closed the chapter on biology twenty years back:

> To translate from evolutionary science to theology, just as the suffering at Calvary was human, creaturely suffering, out of which new life on Earth was redemptively to come, and yet, seen more deeply by Christian conviction, was the very suffering of God for the creation, so in the natural course there is creaturely suffering, autonomously owned, necessitated by the natural drives, though unselected by those caught in the drama. Yet this drive too may be construed, in the panentheistic whole, as the suffering God with and for the creation, diffused divine omnipresence, since each creature both subsists in the divine ground and is lured on by it. . . .
>
> In some way that we mixedly believe and dimly understand, the biology of the world, not less than the physics of the universe, is a necessary and sufficient habitat for the production of caring sentience and, at length, of suffering love in its freedom. Life is a paradox of suffering and glory, and this "secret of life" remains hidden in God, unresolved by biochemistry or evolutionary theory. The way of nature is the way of the cross; *via naturae est via crucis*. (p. 146)

Drawing the book to a close, I returned to this theme in my chapter on nature and history:

> Every life is chastened and christened, straitened and baptized in struggle. Everywhere there is vicarious suffering. The global Earth is a land of promise, and yet one that has to be died for. All world progress and directional history is ultimately brought under the shadow of a cross. The story is a passion play long before it reaches the Christ. Since the beginning, the myriad creatures have been giving up their lives as a ransom for many. In that sense, Jesus is not the exception to the natural order, but a chief exemplification of it. (p. 291)

That perspective has deepened. The story is of the evolution of suffering; this too is among the emergents. In chemistry, physics, astronomy, geomorphology, meteorology, nothing suffers; in botany life is stressed, but only in zoology does pain emerge. Genes do not suffer; organisms with genes need not suffer, but those with neurons do. Life is indisputably prolific; it is just as indisputably pathetic (Greek: *pathos*), almost as if its logic were pathos. The fertility is close-coupled with the struggle.

I returned to this theme in closing my *Genes, Genesis and God*:

> Suffering is a troubling fact, but the first fact to notice is that suffering is the shadow side of sentience, felt experience, consciousness, pleasure, intention, all the excitement of subjectivity waking up so inexplicably from mere objectivity. Rocks do not suffer, but the stuff of rocks has organized itself into animals who experience pains and pleasures, into humans whose *Existenz* includes anxiety and affliction. We may wonder why we suffer, but it is also quite a wonder that we are able to suffer. Something stirs in the cold, mathematical beauty of physics, in the heated energies supplied by matter, and there is first an assembling of living objects, and still later of suffering

> subjects. Energy turns into pain. The world begins with causes, mere causes; it rises to generate concern and care. . . .
>
> Struggle is the dark side of creativity, logically and empirically the shadow side of pleasure. One cannot enjoy a world in which one cannot suffer, any more than one can succeed in a world in which one cannot fail. The logic here is not so much formal or universal as it is dialectical and narrative. In natural history, the pathway to psychosomatic consciousness, the only kind of experience we know, is through flesh that can feel its way through that world. An organism can have needs, which is not possible in inert physical nature. If the environment can be a good to it, that brings also the possibility of deprivation as a harm. To be alive is to have problems. Things can go wrong just because they can also go right. Sentience brings the capacity to move about deliberately in the world, and also to get hurt by it. . . . The story is not merely of goings on, but of going concerns, that is, of values that matter.
>
> The system historically uses pain for creative advance. . . . Theologically speaking, this position is not inconsistent with a theistic belief about God's providence; rather, it is in many respects remarkably like it. There is grace sufficient to cope with thorns in the flesh (2 Corinthians 12:7–9). . . . The "birthing" metaphor is at the root of the concept of "nature"; here creativity comes only with "labor" and "travail." . . . In this struggle there is something demanding appropriate respect, something inviting reverence, something divine about the power to suffer through. The cruciform creation is, in the end, deiform, godly, just because of this element of struggle, not in spite of it. Among available theories, there is no coherent alternative model by which, in a painless world, there might have come to pass anything like these dramas of nature and history that have happened, events that in their central thrusts we greatly value. . . . The view here . . . is a tragic view of life, but one in which tragedy is the shadow of prolific creativity. (Rolston 1999, pp. 303–07)

I think I can argue my case here, but my argument is experiential, existential. When, each spring in the Rocky Mountains, I confront the pasqueflower, I am moved by life beset by storms, persisting through winter, and flowering again at Easter. Though plants do not suffer, plants are caught up in the struggle to survive. I confront this "cruciform creation," life dying and regenerated through death—nature as "*via dolorosa*." I find an encouraging beauty in life's perennial regeneration. I know the evolutionary science, I know there is life-death-life-death, but when I encounter the lovely blossoms breaking through the snow, I take the flower, a distributive token, as type for the collective Earth, with its millions of species, continuing after a turnover of billions of species. I put that as my creed in *Natural History*: "The Pasqueflower" (Rolston 1979).

The central fact of the matter biologically is the survival of life over millennia, life-death-life-death-life-death; but such fact of the matter is *ipso facto* valuable, vital. Nature produces matter and energy, then objective life, then subjective life, then mind and culture. The latter movements are increasingly in a minor key—and beautiful for the conflict and resolution. "Experiences of the power of survival, of new life rising out of the old, of the transformative character of suffering, of good resurrected out of evil, are even more forcefully those for which the theory of God has come to provide the most plausible hypothesis" (p. 135).

One must be in this river to sense the flow. "We must live at the eye of the storm" (p. 344). This book, after two decades, will still invite you into a "participatory universe." I guarantee it.

REFERENCES

Acton, Lord (John Emerich Edward Dalberg-Acton). 1949. *Essays on Freedom and Power*. Ed. Gertrude Himmelfarb. Glencoe, IL: Free Press.

Barr, Stephen M. 2003. *Modern Physics and Ancient Faith*. Notre Dame, IN: University of Notre Dame Press.

Bear, Mark F., Barry W. Conners, and Michael A. Paradiso. 2001. *Neuroscience*. 2nd ed. Baltimore: Lippincott Williams & Wilkins.

Brockman, John. 1995. *The Third Culture: Beyond the Scientific Revolution*. New York: Simon & Schuster.

Christiansen, Morten H., and Simon Kirby. 2003. "Language Evolution: The Hardest Problem in Science?" In *Language Evolution*, ed. Christiansen and Kirby, 1–15. New York: Oxford University Press.

Clark, William C. 1989. "Managing Planet Earth." *Scientific American* (September): 46–54.

Conway Morris, Simon. 2003. *Life's Solution: Inevitable Humans in a Lonely Universe*. Cambridge: Cambridge University Press.

Cosmides, Leda, John Tooby, and Jerome H. Barkow. 1992. "Introduction: Evolutionary Psychology and Conceptual Integration." In *The Adapted Mind: Evolutionary Psychology and the Generation of Culture*, ed. Jerome H. Barkow, Leda Cosmides, and John Tooby, 3–15. New York: Oxford University Press.

Cosmides, Leda, and John Tooby. 1994. "Beyond Intuition and Instinct Blindness: Toward an Evolutionarily Rigorous Cognitive Science." *Cognition* 50:41–77.

Cupitt, Don. 1993. "Nature and Culture." In *Humanity, Environment and God*, ed. Neil Spurway, 33–45. Oxford: Blackwell Publishers.

de Duve, Christian. 1995. *Vital Dust: The Origin and Evolution of Life on Earth*. New York: Basic Books.

Draganski, Bogdan, et al. 2004. "Changes in Grey Matter Induced by Training." *Nature* 427 (January 22): 311–12.

Elbert, T., et al. 1995. "Increased Cortical Representation of the Fingers of the Left Hand in String Players." *Science* 270:305–07.

Hacking, Ian. 1999. *The Social Construction of What?* Cambridge, MA: Harvard University Press.

Jaroff, Leon. 1989. "The Gene Hunt." *Time* (March 20): 62–67.

Kauffman, Stuart A. 2000. "Prolegomenon to a General Biology." In *Investigations*, 1–22. Oxford: Oxford University Press.

______. 2002. "Consciousness: It Blows My Mind." Metanexus interview by Jill Neimark. Online at http://www.metanexus.net/metanexus_online/show_article.asp?5605.

Kelley, K. W., ed. 1988. *The Home Planet*. Reading, MA: Addison-Wesley.

Kuhn, Thomas S. 1992. "The Trouble with the Historical Philosophy of Science." Rothschild Lecture, November 19, 1991. Department of the History of Science, Harvard University.

Lewontin, R. C. 1991. *Biology as Ideology: The Doctrine of DNA*. New York: HarperCollins Publishers.

Lindbeck, Gerge. 1984. *The Nature of Doctrine: Religion and Theology in a Postliberal Age*. Philadelphia: Westminster Press.

MacIntyre, Alasdair. 1981. *After Virtue: A Study in Moral Theory*. Notre Dame: University of Notre Dame Press.

Maguire, E. A., et al. 2000. "Navigation-Related Structural Change in the Hippocampi of Taxi Drivers." In *Proceedings of the National Academy of Sciences of the United States of America* 97 (8): 4398–403.

Maynard Smith, John. 1972. *On Evolution*. Edinburgh: University Press.

______. 1995. "Life at the Edge of Chaos?" *New York Review of Books* (March 2): 28–30.

______. 2000. "The Concept of Information in Biology." *Philosophy of Science* 67 (June): 177–94.

Maynard Smith, John, and Eörs Szathmáry. 1995. *The Major Transitions in Evolution*. New York: W. H. Freeman.

McShea, Daniel W. 1991. "Complexity and Evolution: What Everybody Knows." *Biology and Philosophy* 6:303–24.

Mechelli, Andrea, et al. 2004. "Structural Plasticity in the Bilingual Brain." *Nature* 431 (October 14): 757.

Pepper, David. 1996. *Modern Environmentalism: An Introduction*. London: Routledge.

Quine, Willard Van Orman, and J. S. Ullian. 1978. *The Web of Belief*. 2nd ed. New York: McGraw Hill.

Rees, Martin J. 2001. *Our Cosmic Habitat*. Princeton, NJ: Princeton University Press.

Richerson, Peter J., and Robert Boyd. 2005. *Not by Genes Alone: How Culture Transformed Human Evolution*. Chicago: University of Chicago Press.

Rolston, Holmes, III. 1979. "The Pasqueflower." *Natural History* 88 (4): 6–16.

______. 1997. "Nature for Real: Is Nature a Social Construct?" In *The Philosophy of the Environment*, ed. Timothy D. J. Chappell, 38–64. Edinburgh: University of Edinburgh Press.

______. 1999. *Genes, Genesis and God*. Cambridge: Cambridge University Press.

______. 2005. "Genes, Brains, Minds: The Human Complex." In *Soul, Psyche, Brain: New Directions in the Study of Religion and Brain-Mind Science*, ed. Kelly Bulkeley, 10–35. New York: Palgrave Macmillan.

Rorty, Richard. 1979. *Philosophy and the Mirror of Nature*. Princeton: Princeton University Press.

______. 1989. *Contingency, Irony, and Solidarity*. New York: Cambridge University Press.

______. 1991. *Objectivity, Relativism, and Truth: Philosophical Papers*. Vol. 1. New York: Cambridge University Press.

Ruse, Michael, 1986. *Taking Darwin Seriously*. Oxford: Basil Blackwell

______. 1996. *Monad to Man: The Concept of Progress in Evolutionary Biology*. Cambridge, MA: Harvard University Press.

______. 1999. *Mystery of Mysteries: Is Evolution a Social Construction?* Cambridge, MA: Harvard University Press.

Russell, R., N. Murphey, T. Meyering, and M. Arbib, eds. 1999. *Neuroscience and the Person*. Berkeley, CA: Center for Theology and the Natural Sciences.

Sterelny, Kim, and Paul E. Griffiths. 1999. *Sex and Death: An Introduction to Philosophy of Biology*. Chicago: University of Chicago Press.

Tomasello, Michael. 1999. *The Cultural Origins of Human Cognition*. Cambridge, MA: Harvard University Press.

Tooby, John, and Leda Cosmides. 1992. "The Psychological Foundations of Culture." In *The Adapted Mind: Evolutionary Psychology and the Generation of Culture*, ed. Jerome H. Barkow, Leda Cosmides, and John Tooby, 19–136. New York: Oxford University Press.

Venter, J. C., et al. 2001. "The Sequence of the Human Genome." *Science* 291 (February 16): 1304–51.

von Baeyer, Hans Christian. 2003. *Information: The New Language of Science*. London: Weidenfeld & Nicolson.

Wiener, Norbert. 1948. *Cybernetics*. New York: John Wiley.

Wilson, Alexander. 1992. *The Culture of Nature: North American Landscape from Disney to the Exxon Valdez*. Cambridge, MA: Blackwell Publishers.

Wilson, Edward O. 1978. *On Human Nature*. Cambridge, MA: Harvard University Press.

Wistow, Graeme. 1993. "Lens Crystallins: Gene Recruitment and Evolutionary Dynamism." *Trends in Biochemical Sciences* 18:301–06.

World Development Report 2004 [c. 2003]. Oxford: Oxford University Press.

Yockey, Hubert P. 2005. *Information Theory, Evolution, and the Origin of Life*. Cambridge: Cambridge University Press.

Yoxen, Edward. 1983. *The Gene Business: Who Should Control Biotechnology?* New York: Harper & Row.

SCIENCE & RELIGION

Chapter 1

Methods in Scientific and Religious Inquiry

To have a method is to have a disciplined mode of "following after" (μέθοδος) truth, and in science and religion alike one intends an orderly approach to understanding, to be a methodist, but procedures in the two fields may seem very different and even incompatible. In this overview we will broadly assess their operation so as to see whether and how far they are related or opposed. Lest the diversity in religion prove overwhelming, the plan here is to consult mainly Western theistic belief, itself diverse enough, as it has developed in interaction with the sciences, which have a diversity almost equal to that in theism. Despite the pluralism, these two great epistemic lines in the West are cousins, at once kindred and independent. What follows is partly a description characteristic of science and theology, but, so far as I choose good science and good religion for models, it is a prescription of how inquiry there ought to be done, perhaps not always, but at least in the present state of these arts.

The thesis that will emerge is that in generic logical form science and religion, when done well, are more alike than is often supposed, especially at their cores. An implication of this is that positivistic and scientistic views that exalt science and downgrade religion involve serious misunderstanding of the nature of both scientific and religious methods. At the same time, in material content, science and religion typically offer alternative interpretations of experience, the scientific interpretation being based on causality, the religious interpretation based on meaning. There are differing emphases in specific logical form in the rational modes of each. But both disciplines are rational, and both are susceptible to improvement over the centuries; both use governing theoretical paradigms as they confront experience. The conflicts between scientific and religious interpretations arise because the boundary between causality and meaning is semipermeable.

1. THEORIES, CREEDS, AND EXPERIENCE

The Hypothetico-deductive Method and Theory-laden Facts

Whether there exists an overall scientific method is open to question, since the procedures of electronics engineers, plant taxonomists, and social psychologists are so diverse. In a generalized way science mixes observation, theory, and inference, but these ingredients with their blending are more complex than at first appears, and not until something of this complexity is appreciated can one appreciate a scientific method and then profitably ask how far religious inquiry differs from it. Let us begin by saying that a scientist attempts to operate out of theory in an if-then mode "over" the facts. A schematic of this would find a theory (the hypothesis) arising out of the facts, followed by deduction back down to further empirical-level expectations, those then being related back to observations to confirm or disconfirm the theory, more or less, and to generate revised theory, from which new conclusions are drawn, after which the facts are again consulted (Figure 1.1). This is sometimes called the hypothetico-deductive model, but we are using a more expanded version of it than that phrase usually implies, and also noticing already that a theory comes to have a developmental history.[1]

Such facts quickly become theory-laden. When the engineer reports that the current through the meter is ten amperes, or the zoologist discovers that the vertebrates are related to the tunicates, the larval notochord of the latter and the spinal chord of the former having evolved from a long-extinct hypothetical ancestor, their facts come within and are partially products of their theoretical frameworks. Fabricated concepts and laws are used to trace and to classify natural events, and the facts so obtained do not come nakedly but rather filtered through these constructs. In the more theoretical sciences, those likeliest to affect cosmic belief, there is often a tenuous combination of speculative abstraction with sense observation, linked by hundreds of intervening hypotheses, as in the experiments that verify the time dilation of relativity theory by measuring the supposed decay of muons at high velocity, all translated into streaks on photographic plates and meter readings. The geneticist maps a gene by back inference from statistical phenotypic expressions. The biochemist decodes the amino acid sequence in a protein by observing certain colored stains or layers of material in an ultracentrifuge. Molecular biochemistry contains highly theoretical construction of models of unobservable entities and processes—for instance, the lac-operon genetic sequence—to account for observed gross phenomena at great distance from the postulated microentities. Geology has become a unified science only in recent years, with the appearance of plate tectonics, but that supertheory stands at a great inferential distance from the immediate observation of fault lines, subsidence measurements, chart tracings that indicate oceanic ridges, and magnetome-

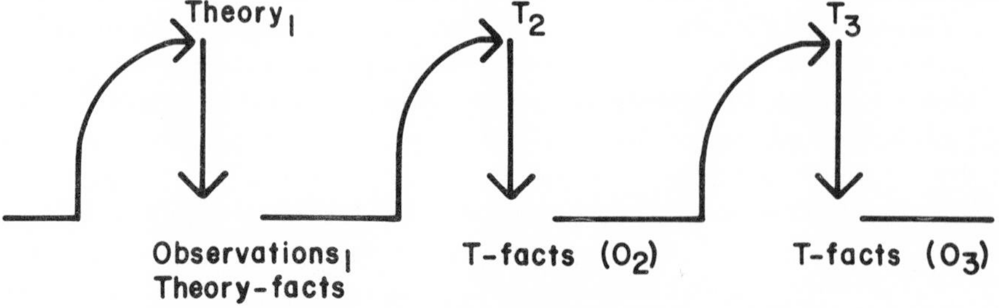

Figure 1.1 The developmental history of a theory

ter readings from which are inferred prehistoric reversals of Earth's magnetic field.

Even in the plainer bare world there are no centimeters, or calories, or lines of latitude and longitude; nor can it be Tuesday, 1:30 P.M. (EST), for these are all conceptual overlays on nature. The center of gravity in a rock is as much assigned as discovered. Still, one may reply, at least there are some evident natural kinds; there are tunicates and genes, there were trilobites in the Cambrian period, and Yosemite's Half Dome is made of quartz monzonite. But even these facts do not come unalloyed with the theories by which they were obtained. There is always some definition or decision about theoretical kinds in what counts as a tunicate, a gene, quartz monzonite, or the Cambrian period, as these are fitted into explanatory theories.

The whole numbers may seem natural enough until we add, divide, and multiply by zero and infinity, and with some artificial innovation must define what these operations will mean. The point in science is to mix theory and fact appropriately, and not to pretend that they can be insulated from each other. The naked fact is mostly a mythical entity; facts are contextual truths. To believe in pure facts is to believe "the dogma of the immaculate perception." The "facts" are always to some extent "artifacts" of the theory. The "facts" are preceded by "acts" that set up the facts. The facts are seldom, if ever, immediately given; they are arranged for, indeed, chased down on long hunts by those armed with powerful theories. Even where theoretical concepts can be cashed in for observations in a fairly straightforward way, the cash-in rules come out of the theory, not the observations, and such rules can change in the course of the development of the theory.

How such theories are originated, as distinct from their subsequent verification, has proved troublesome to analyze, and recently it has seemed that the context of discovery is more important, more interesting, than is the later context of justification. Given a certain set of observations, what theory will fit them? In cataloging natural types or in formulating simple regularities one is tempted to say that science works by induction, a logic that leads in toward a concluded general principle from premised particular occasions. Here the contribution of the scientist can seem minimal, even though the law vastly overprojects what can be verified. But the generating of theories is more complex; the scientist comes up with models and abstractions, such as "lines of force in an electromagnetic field," or "covalent bonding," or "black holes," concepts that no doubt come by mulling over the data, but in which he also contributes creative hypotheses that require the stroke of genius.

These initial ideas may come in the laboratory or at study but are sometimes reported to come in unusual circumstances. While dozing by the fire, August Kekulé dreamed a reverie of gamboling atoms and snakes, one biting its own tail, out of which the great chemist that night developed the chain-linked ring structure of benzene.[2] Fred Hoyle regarded as pivotal in triggering the steady-state theory of the universe a curious personal incident in which he lost a screw or nail and could never find it, as though it had forever vanished. He reversed the experience to conceive of the spontaneous creation of matter.[3] Albert Einstein reported that he initiated his relativity theory, partly at least, "in vision" late one night, and he greatly emphasized the free play of the imagination, first and charismatic, only later to be put sternly to observational test.[4] Hans Adolf Krebs, on the other hand, reported a long and steady step-by-step deciphering of the citric acid cycle.[5] Both elements are present in Charles Darwin and difficult to separate.[6] But if eurekaism is one extreme, dull inductivism is another. There is much inspiration whenever a fertile hypothesis is born. The logic of such inception has proved elusive; it involves something beyond either induction or deduction, and there seems to be no recipe for cooking up theories. This is perhaps necessarily so proportionately as it is creative. Revolutionary science is more chaotic here than is normal science.

Verification and Falsification

Crucial though the question is of how one gains a novel theory, the real test comes with its verification. Given a theory (T), what observations (O) follow? Here deduction is in order, at least in a broad sense; logic leads out from premised general principles to particular conclusions. In the mathematical phases of science, where one has formal laws and initial conditions, this can be exact and necessary deduction, but elsewhere it is less so. Atomic theory is only partially metric, and what could be deduced from the atomic table about the properties of as yet unfound elements was suggestive and imprecise. Often a theory permits the deduction only of a range of possible alternatives, and we must sometimes deduce in a weak, nontight sense. Still, a fertile theory will suggest new observations that can be made to check it. Here we often presume that our logic is paralleling a causal chain, that a law causally produces an observed event, the narrower sense of the hypothetico-deductive or covering-law model. But the principle here is broader than this, including whatever particular events or observational structures follow from general theoretical models:

If T, then O
Given: O
Therefore: T

Alas, however, this procedure commits the logical fallacy of affirming the consequent, since some quite variant theory (T′) might as well or better explain the observations in question, and the history of science is replete with examples of this. On the other hand, if the observations fail (not-O), then the theory is refuted, by *modus tollens,* an elementary principle of valid argument:

If T, then O
Given: not-O
Therefore: not T

Science then first appears to be caught in a rotten asymmetry: no amount of positive observations can prove a theory, while a single negative observation will destroy it. We can be definitely wrong, but only vaguely right! This asymmetry has led some scientists to concentrate on falsification, counting disconfirming instances as more weighty than confirming cases.[7]

What happens in actual science is that positive observations do in some way tend to establish the theory, although it is difficult logically to specify just how. Again, it is tempting to say that positive observations by induction render the theory probable, while conceding that this is never hard proof even in science and recognizing that the rational status of induction is flawed, especially so far as future predictions from the theory involve a kind of backing into the future. Positive observations corroborate or strengthen the theory, although they cannot clinch it. We get no proofs; we get at best plausibility arguments.

On the other hand, on closer inspection, those negative observations that first appear to offer hard disproof also soften. Theories are not tested purely and simply but in conjunction with various presumed or unknown intermediate factors, called auxiliary hypotheses (A), such as those pertaining to instruments, to irrelevant or absent influences, etc., and one can typically adjust for upsetting circumstances so as to salvage the central theory:

If (T + A), then O
Given: not-O
Therefore: not-T and/or not-A

Something has been falsified, but what? Some variant auxiliary hypothesis (A′) will allow deducing the obtained observations while retaining the theory. Thus, the auxiliary belt of surrounding hypotheses becomes a protective cushion. In most practical and theoretical science we are reduced to saying: if T, then probably O. But then not-O no longer refutes the theory, especially where this is an occasional not-O.

But it may be, of course, that the error is rather in the body of the theory itself. Newtonian theory predicted planetary movements reasonably well, except that the orbit of Uranus was irregular, and some astronomers suspected that the theory might be faulty. In a celebrated triumph of mathematical astronomy, John Couch Adams and Urbain Jean Joseph Leverrier introduced the auxiliary hypothesis of an unknown planet that was disturbing Uranus' orbit, and thus Neptune was found and Isaac Newton confirmed. Later, when aberrations in the perihelion of Mercury were found, Leverrier again suggested the auxiliary hypothesis of an innermost planet, Vulcan, whose influence was perturbing Mercury. But no such planet was found. Perhaps it was lost in the solar glare? Eventually the trouble proved to lie in Newtonian theory, and relativity theory came to replace it and to explain these discrepancies in Mercury's behavior. The problem is to know when to "put in some epicycles" to protect a theory and when to suspect the core theory itself.

Every theory is held in the face of certain anomalies, margins of error, and so on. For so simple a law as that for the distance (S) traveled in a specified time (t) under the acceleration due to gravity (a), $S = \frac{1}{2}at^2$, the observations never fit the theory exactly, since the theory specifies a perfect vacuum. We also have to assume that there is no magnetism present countering the gravity, but to check this one needs a theory of magnetism and a measuring device built on the theory. In genetics and biochemistry one is constantly invoking as yet unknown genetic codings, enzymes, or repression or induction effects to explain departures from the norm. The theoretical imbalance in corroboration and falsification abstracted above, by the time it is emplaced in the practice of science, loses much of its asymmetry.

At the same time, really stubborn disconfirmations are more unwelcome than repeated confirmations are welcome. The structural asymmetry probably does mean, contrary to a certain sense of fair play, that in science (and in religion too, we shall soon maintain) you want to try to hit an opposing theory not where it is muscular but rather in the soft underbelly where it is weak; you ought to evaluate a theory (or a creed) more on its weaknesses than on its strengths.

In more complex and partly established theory there are large amounts of confirming and some disconfirming observations, and one has to decide just how good the evidence is. That decision is rational, perhaps progressively corroborated as science settles into a theory, but often it is more discretionary and less tidy than is admitted by those charmed by an ideal of absolute demonstration. Every comprehensive theory has got to argue away some of the evidence it faces. Sometimes we do not believe the theory because it is not confirmed by the facts; but sometimes we do not believe the "facts" because there is no theory that confirms or predicts them and they go against a well-established theory that we have. We could handle this exception, if we had a little more time to deal with it! Meanwhile, an anomaly makes a poor logical fit in what theory we do have. Then again, experiments can be quite repeatable and quite wrong, where the conceptual framework repeatedly gives you the wrong result. You can step on a bathroom scale and get 150 pounds every time, when your weight is really 160 pounds. Hidden faults and errors are repeatable. Theories cast light, but may also put some things in shadow.

Crucial experiments are infrequent, if indeed they exist at all. Hardly anywhere is there a simple verification or falsification, and the more massive the governing theory becomes, the less convenient these procedures are. The evidence for the big theories, which make any metaphysical difference, is never of the here-and-now, before-your-very-eyes sort. What counts for a good theory is its ability to draw together and make sense of the available experiential material, and in this the relationship between theory and observation is often indirect and interactional.

Testing Creeds in Experience

Religion too methodically mixes experience, theory, and inference. There are many disanalogies; often one finds notions of revelation and inspiration, and hence of normative authority, that cannot be easily reconciled with the procedures of science as just sketched. Creeds are not so provisional as scientific theories sometimes are, but more like settled operational assumptions (which scientific theories also can become). And there are many noncognitive elements in religion not present in science. Nevertheless, in a general way religious convictions develop in the face of certain experiences judged to be of ultimate importance, as of suffering or of joy, of sin and salvation, of the holy and the moral. On reflection by theologians there arise cognitive, theoretical notions suggesting certain universal spiritual laws or generalizations, leading to a positing of an underlying ultimate reality in and beyond the world that is sufficient to account for such experiences. God, Brahman, or *śūnyatā* (Emptiness) is then used to interpret ongoing experience, and here, as with science but more so, the subsequent experiences are produced by and come within that framework of convictions that these experiences first spawned.

The later religious experience provides a testing of dogmas, confirming or disconfirming them. The history of religion is strewn with abandoned beliefs, largely overcome by more commanding creeds or made implausible by new ranges of experience. To the contemporary religious mind, primitive fetishes and taboos, superstitions and sacrifices seem quite as quaint as (and perhaps a form of) primitive science. Only a handful of the myriad religious hypotheses of the human race have survived the sifting in experience that makes them classic (that is, verified in experience), and for that handful this durability increases their categorical element. Most earlier religions are extinct; a few are relict. Some will say that it is only modern science that wipes out old creeds, but this is not always the case. Sometimes new creeds wipe out old ones. Witchcraft and astrology were already prohibited in the Scriptures as unbecoming to monotheistic theory (although some belief in them persisted, *per nefas*). The Hebrews disenchanted the universe on the basis of monotheism long before science appeared—a finding that subsequently made science possible.

Even classical theism, though once medieval, has nowhere become modern without dramatic revisions. One central element in the creeds of the Reformation churches is that they are "always reforming" (*semper reformanda*), that is, steadily improving their creeds in the light of contemporary experience that brings a new perspective to the foundations of the tradition, retaining only so much of that classic faith as continues to prove adequate, and that often in a reinterpreted form. The Roman Catholic Church has claimed an irreformable core to its creeds, but in the second half of this century this classical infallibility claim has been found by many who once held it not really to square with experience, and the Roman church is now undergoing hardly less radical revision than those churches that confess to a continuing reformation. Religious belief has to weather a critical thinking out and testing out of the experiences that follow from its creeds, and theologies too are selected for their success over historic time.

Religion does use the if-then mode of deriving consequences from its creeds and testing them in experience. In this, however, religious convictions cannot usually be cast into empirically testable frameworks. Simple events, such as planetary motions or chemical reactions, adapt well to watching with objective instruments, but more complex events, such as guilt and forgiveness, quantify poorly and are difficult to make operational. The instruments for their recording are subjective selves, and the hunting down of those experiences that are found when armed with religious creeds is a matter indeed of experience, of "going through," and not merely of observation, "looking on." In physics and chemistry, material things instantiate laws in a rather tight way, but living things, even in biology, often show only generalizations or statistical trends, hardly rejected by occasional counterexamples. Personal beings, as unique, rational, affective agents, can test religious convictions only experientially, not experimentally; existentially, not operationally.

Low-level generalizations can sometimes be tested empirically, as with "The family that prays together stays together"[8] or "Persons become more religious in adverse times." (Even if verified statistically, the underlying explanatory theory might still be contested.) Intermediate religious generalizations need personal experiences mingled with observation. "Blessed is the man who walks in the law of the Lord" is the judgment that the moral life, as described biblically, yields the good life, and a considerable number of persons have claimed to find this replicable and thus verified. The Buddhist claim that worldly life at its core is eventually unsatisfactory (*duḥkha*), so far as life is driven by uncontrolled desires, is perhaps only part of larger cosmological claims in the first and second noble truths, but this relatively specific claim is at least in some degree subject to experiential verification.[9]

2. MODELS, PATTERNS, PARADIGMS

Scientific and Religious Paradigms

"Seeing" is universally "seeing as." We interpret what we see in order to see it. To tell what is going on, to see what is taking place, our observations are formed within gestalts. We see cows, not red patches, persons rather than bodies, love or hate rather than bare behavior. To notice this is not to deny that philosophically oriented observers can sometimes strip away the coordinating patterns and lay bare rudimentary data. But such naked facts are abstractions artificial to normal experience, which occurs within natural and conventional categories. Routinely in science and in religion alike an event makes sense not merely as our senses register it but as it is found to be intelligible within certain established patterns of expectation.[10] The understanding cannot see and the senses cannot think; cognizing and perceiving are wired up together. This interpretive seeing is sometimes thought to contrast with hypothetico-deductive science, but it is really in keeping with an earlier realization that observations are heavily theory-laden, that we come to see things as instances of types or universals. As these models become increasingly dominant, they become paradigms, and then we are able to give a better account of the revolutionary phases of theory overthrow while retaining our earlier hypothetico-deductive account for the evolutionary development of theories.[11] At times some theory replacements cut clean from previous theory (the heavenly spheres of medieval astronomy were abandoned in Newtonian astronomy); at other times much is conserved, if reinterpreted, in subsequent theory (Newton's laws are a special case within Einstein's relativity theory). Both clean cuts and conservation under radically new theories involve paradigm shifts.

Paradigms are governing models that, in some fairly broad range of experience, set the context of explanation and intelligibility. Their holders wish to conserve these basic referent theories so far as they can by using them to interpret new experience or, in the event of counterexperiences, by introducing subsidiary hypotheses that allow the theory's conservation by peripheral adjustments. Paradigms are abandoned reluctantly, because they have hitherto been highly successful in structuring the data of experience. It has proved difficult in some cases to specify just what qualifies as a paradigm; paradigms have sometimes broader, sometimes narrower scope, and there may be a hierarchical interweaving of major and minor paradigms. But the basic idea here of a controlling patterned seeing does seem to characterize the history of science and religion alike. Prominent examples of dominant or subordinate paradigms in science include the Copernican and Ptolemaic astronomies; the fixity of species and the evolution of species; Newton's absolute space-time and Einstein's relativity; mechanism and teleology; determinism and indeterminism; natural selection and orthogenesis; theories of phlogiston and of the ether; the taxonomic sequence of phylum, class, order, family, genus, and species; geologic uniformitarianism and catastrophism; the Paleozoic, Mesozoic, and Cenozoic periods; the wave and particle theories of light; atomic theory.

Those familiar with the history of science will realize how much of its controversy and upheaval comes at periods of major paradigm shifts. Those engaged in its present practice will notice that many of these examples pervade their work as the assumptions that make it possible, while some of the overthrown paradigms now seem incredible. Notice too that one is not entirely oriented here by cognitive knowing; by following the techniques and methods of his predecessors and peers a scientist gets also a "know how" to do, as well as a "know that" something is so, so that there are tacit as well as explicit elements in our control by a paradigm. As Thomas S. Kuhn argues, a paradigm is a "disciplinary matrix" as well as a theoretical viewpoint.

Religious paradigms are found prominently in creedal affirmations—for example, that Jesus Christ is fully human, fully divine, one person; that God is love; that persons are made in the image of God (the divine character of the person); that an immortal soul resides in the body; that God predestines all; that Israel is a chosen people; that God (Allah) is, and Muhammad is his messenger; that the *ātman* (inmost self) is Brahman, the divine Absolute; that the conventional world (*saṃsāra*) is illusory (*māyā*); that the mundane world (*saṃsāra*) is the transmundane world (*nirvāṇa*) upon enlightenment; and that, short of enlightenment, a law of moral causation (*karma*) operates by which persons are reincarnated from life to life. Here again, a paradigm is not merely cognitive but carries a kind of skill at judgment, some tacit knowledge of how to work with it, from it. Some examples of paradigms in religion that have been entirely abandoned or seriously questioned by modern persons include animism and polytheism, the six-day creation, the fall of an original couple and the subsequent biological transmission of that original sin, the demon-possession theory of disease, the three-story universe (heaven above, earth, hell beneath), medieval accounts of purgation and indulgences, and (much revised if not abandoned) the verbal inerrancy of the Bible.

Pervasive and Persuasive Characteristics of Paradigms

A good paradigm has a maplike character in that reality is selected and represented through it so as to fit into a kind of basic picture: Newtonian mechanism portrays the world as a great machine; Darwinian evolutionary survival of the fittest portrays the world primarily as a jungle; behaviorism sees life-environment interactions as stimuli and responses; physics views protons, electrons, and photons as both waves and particles. God is a Father, Shepherd, and Creator. Jesus is the normative person. The Church is the body of Christ. Persons get "lost" and "saved." Life in the common world is driven by "thirst" (*taṇhā*); essentially this world is a realm of "suffering" (*duḥkha*) that is "empty" (*śūnya*), with one's fortunes in it the result of deeds (*karma*) in present or past lives. Imagery is present alike in science and in religion, and to become aware of the representational or symbolic character here is to realize that these critical affirmations are maps rather than exact pictures of reality. Maps and models organize reality; they are never passive containers for experience, but they actively help us find organization in reality just because they abstract its structures. They tell us what to look for, what to discount, and what to make of what we find; and in this sense they are proposals as well as discoveries.

In this sense, while the outcome of an experiment does not depend on the mental states of those who conduct it, the setup of an experiment, what outcomes are arranged for, does depend on the mental states (the theories) of the experimenters. Arrangements can and should be made for outcomes that falsify or verify our theories, and there are surprises, outcomes that we do not expect or understand. All the same, only those sorts of outcomes can happen that we have advertently or inadvertently arranged for. We never catch black holes, DNA molecules, neurotransmitters, or tectonic plates unless there is, preceding the catch, a mental state that goes looking for them. We catch patterns with a frame of mind.

"If I hadn't believed it, I wouldn't have seen it." Physicists spent decades looking for the neutrino. After repeated failures, they prepared extremely elaborate experiments (sixteen tons of scintillating liquid, 144 photomultiplier tubes, electronic apparatus 120 feet long) finally to catch it—inferring it from rare flashes of certain kinds amidst thousands of other flashes, arranged for with hundreds of thousands of dollars' worth of equipment, all taken two miles underground in a South African gold mine.[12]

On the other hand, when physicists got a theory that suggested that they look for positrons, they looked back to discover that positrons had been appearing for years in cloud chamber photographs and ignored as an anomaly. One can't see what one isn't looking for, even though the evidence is amply present. Often, what we find ourselves looking *at* depends on what we are looking *for* and *with*.

As a paradigm proves to have high deployability it increasingly permeates all that we see, and thus a widely inclusive paradigm has a very low negotiability. We have faith in it. Like a creed, it has a categorical element in practice, although it is in principle a hypothesis. The belief that every event has a necessary and sufficient set of causes is virtually nondebatable, by some of its holders, as the basic assumption of all science. The precise status of this belief—whether it is an a priori claim, an empirical discovery, or a methodological hypothesis—is difficult to uncover. Recent physics has especially had to trouble over it, but scientists find it impossible to work without assuming that it is true sufficiently for the purposes of their research. The paradigm of evolution has rapidly become so entrenched that by its means biologists, geologists, anthropologists, and astronomers explain the origin of species, life, society, landscapes, Earth, matter, and even the universe. In biological phases of evolution, the principle of natural selection has so come to govern accounts of why things happened as they did that adduced counterexamples are likely to be reinterpreted with auxiliary hypotheses protecting the principle that only the fittest survive.

Gestalts, Anomalies, and "Bliks"

This pervasive and persuasive tenacity of a good paradigm raises the fear that they sometimes come to be held "no matter what" and thus degenerate into an ideology or a "blik"—a presupposition with which we view experience, spectacles through which all data will be viewed, with adjustments only in ad hoc hypotheses that are rigged for the sole purpose of saving the theory from refractory facts, and that actually insulate the theory from experience.[13] This is perhaps allied with a law in gestalt theory by which viewers tend to complete a pattern regardless of whether it is completed in the observed reality. Hence, a source of error in theology and in science is a tendency to see causes and meanings, first in ranges of experience where they are readily found, and later to project them onto places where they are missing or incomplete. The theory that begins as a synthetic judgment about the world can get subtly transformed into an analytic prejudgment brought to the world, so that variant experience can no longer transform the theory but rather the theory transforms the experience. A blik is a theory grown arrogant, too hard to be softened by experience.

A humorous illustration is provided by the case of the deluded patient who complains to his physician, "Doctor, I'm dead." The doctor tries to assure him otherwise, with little success, and eventually exclaims in exasperation, "Well, dead men don't bleed, do they?" The patient agrees, "No, they don't." Whereupon the doctor jabs the patient's finger with a needle. As the blood trickles out, the patient sighs, "O.K. I was wrong! Dead men do bleed!" Actual instances of the power of a paradigm are more serious. After rejecting his earlier years in communism, Arthur Koestler reflected over its hold on him: "My Party education had equipped my mind with such elaborate shock-absorbing buffers and elastic defenses that everything seen and heard became automatically transformed to fit the preconceived pattern."[14] Reflecting on an earlier dominance of Freudian ideas in her psychoanalytic theory, Karen Horney recalled how "the system of theories which Freud has gradually developed is so consistent that when one is once entrenched in them it is difficult to make observations unbiased by his way of thinking."[15]

In the judgment of many critics this conversion of a paradigm into an ideological prejudgment happens notoriously in religious belief. Belief in God begins in experience, perhaps that of goodness in creation, or of the numinous, or of sin and salvation; but it thereafter becomes transformed into a blik, which is held by introducing ad hoc revisions so as to allow no evidence to contradict the theory. All good paradigms are self-serving, no doubt, but the trouble arises when they brainwash us. Still, in less fanatical religion criticism is as much encouraged as it is in science and often is as telling. In both fields doubts arise as a result of experience, and these doubts are the first steps toward revised and improved theories and creeds.

We do have an innate thirst to complete an explanation, and our tendency to hold on to available explanations and to press them as far as possible is as often fruitful as it is misleading. What one wants and expects in a fecund paradigmatic theory is massive explanatory power, a capacity to be deployed into ever-widening ranges of experience. A good paradigm can eat up and digest its competitors, and  often absorbs and continues the explanatory power that opposing accounts once had. The paradox of a paradigm, whether in science or in religion, is that the better it is, the longer it survives, the more its resilience, the closer we probably are to the truth, and the more we ought to hang on to it, because it is to be expected that the nearer we are to the truth, the harder a theory will be to overthrow. The ultimate theory will, of course, be unfalsifiable anywhere in practice, because it is entirely true! But just this element of trust that is well justified makes it harder to get a wedge of doubt in, to seek truth in unlikely directions, and to face up to an epistemic crisis. There is a sense in which one needs both to seek disconfirmations and to distrust them.

One does need ever to beware of an ideology, that is, having one's logic (logos) so controlled by a form (idea) as to be oblivious of empirical and experiential input, so that this input is neither supportive nor constitutive of the theory, nor any longer able to reform it. The first part of being reasonable is to hold on to whatever logic you have, conserving a tradition, entering a paradigm, appreciating the best sense that can be made of the phenomena to this point. One keeps an inherited truth so long as it yields clarity without arrogance. The second part, more chaotic and threatening, is to know when to give up the old, to launch into the new. One needs to be able to recognize the kind of exception to a rule that signals the end of the rule. This is the hardest part, because there is no precedent for it, so far as it is a genuinely creative step, although of course unprecedented steps have often been taken in the past.

The operation of paradigms is usefully, but oversimply, illustrated by the young lady–old hag reversible drawing (Figure 1.2). Viewers do not see, except by artificial straining, just black and white lines and patches, hard data, but they see now a young

Figure 1.2 Lady-Hag reversible drawing

lady and again an old hag; and whether a certain line is a necklace or a mouth, or another is a nose or a chin, depends upon the gestalt. The viewer does not just "see" particulars; she "sees as" these are governed by the gestalt. In physics an electron can be seen as a wave or a particle. In biochemistry the conduct of a hunting coyote can be interpreted mechanistically or teleologically. Behavioristic psychology sees the self in a stimulus-response pattern; humanistic psychology sees the inwardness of a centered, creative self. Sometimes these paradigms are complementary, as is suggested by the reversible drawing here, but sometimes they are not. When burning came to be seen as oxidation, the phlogiston theory had to be abandoned. When astronomers came to see the movements of the planets as Copernican theory does, those orbits could no longer be seen as Ptolemaic theory had seen them. On a broader scale, science and religion provide variant grids by which the world can be mapped, but how far they are complementary and how far incommensurable is not easily discovered.

Some of the subjectivity in the example considered here is offset by remembering that the acceptance of a paradigm is collective, not just individual, and that a paradigm is much argued over, sifted through, and tested out, depending not just on what one perceives but on one's capacity to persuade others and to retain community allegiance. A paradigm is intersubjective and must command a community. The example is also too static, in that drawings have no dynamism. Theories and creeds grow and mature, and these epistemic gestalts replace one another in historical succession, so that we have in reality not a frozen picture but a motion picture, a story, a living narrative alike in the development of science and religion. Older theory and creed are sometimes dissolved, judged incoherent, and forgotten, but they are sometimes reallocated and retained under a transformed gestalt. Thus, the past in science and in religion is partly discounted and partly recounted as an earlier, juvenile chapter in an ongoing narrative now become more sophisticated.

The major religious systems offer diverse creeds through which the significance of life in the world may be viewed. But in attempting to appreciate them there is often considerable, if not insurmountable, difficulty in making that gestalt switch which occurs so easily in the lady-hag drawing. This difference arises because the viewing sensitivities of beholders depend on their behavior, experience, and character. Further, these gestalts may overlap in part but be ultimately incommensurable. The Theravada Buddhist and the Christian concur in seeing life as something sacred and to be reverenced, but they disagree over whether to see the world as the creation of a personal God or as a fluid matrix of dependent origination.

The complexity of gaining and defending a religious view can be suggested by another drawing, that of a black-and-white Rorschach picture that contains hidden the figure of a bearded man somewhat like traditional pictures of Jesus (Figure 1.3). The viewer may have to study the picture to detect this pattern, and even then will puzzle over whether it is really there or just an illusion. Still, when one has seen it, the portrait tends to govern what one sees there afterward. Christians are able to see the mystical presence of Jesus hidden in certain ranges of experience, while unbelievers often cannot make sense of the same ranges of experience, or see them under some other paradigm. The gestalt with the hidden

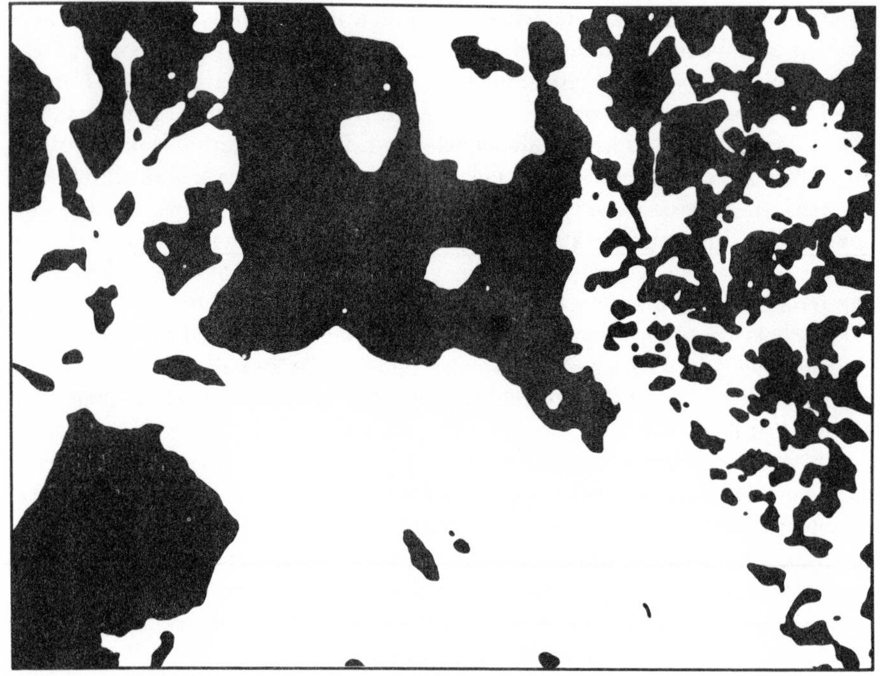

Figure 1.3 The Jesus gestalt

Christ figure here is ambiguous and easily dismissed, but in actual Christian life this hidden presence is often not so easily discounted. It is as though by moral and spiritual experience one's resolving power in the drawing could be sharpened, as though some of the black areas would seem grayer or duller and others more intense and sharper, so that the believer and unbeliever still could see the same gross shapes but attach different weights and intensities to them. Thus, the believer would find more confirmation, while the unbeliever might remain puzzled before equivocal experience. What one is willing to tolerate as static or noise—the meaningless blotches in the background—depends greatly on developed sensitivities. This points up the participatory nature of religion, increased over that in science, which later we have to examine.

In paleoanthropology many anthropoid skulls, which are often partial or in fragments requiring much construction, have been recovered. The brain capacity of these fossil skulls increases with geological time (Figure 1.4). This tends to support the prevailing view that human intelligence evolved out of prehuman forms, through several stages of *Australopithecus* (gracile, robust, and perhaps habiline), with a brain capacity of about 500 cc, reaching *Pithecanthropus,* about 1,000 cc, and later Neanderthal man, 1,500 cc, more or less. But there is one anomalous skull, known as ER-1470, which is removed from the rest, with much greater cranial capacity than the theory allows. What an anthropologist, using his best judgment across all the data, has to decide is what sort of flier, fluke, or hoax ER-1470 is and whether revision in the main theory is called for.

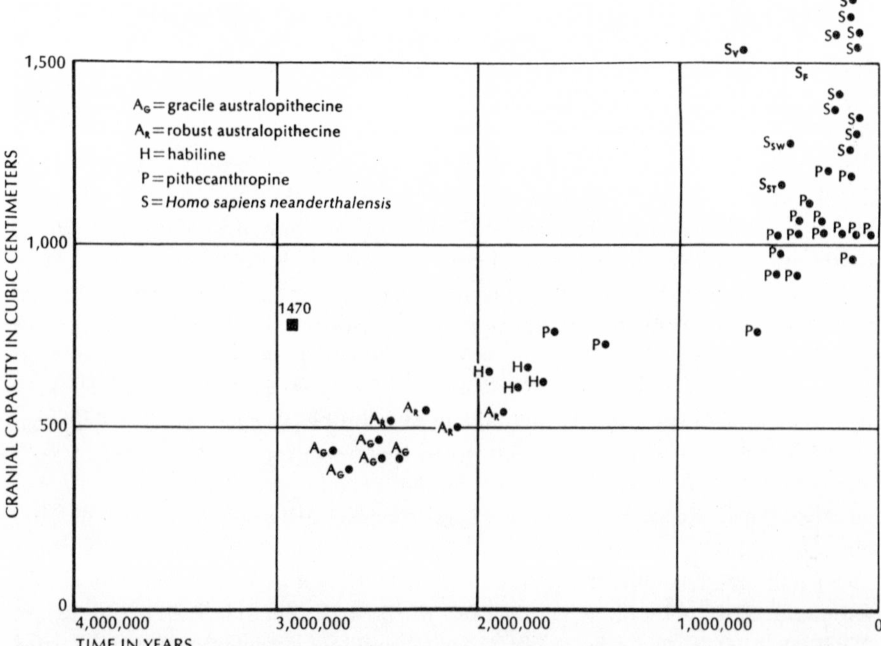

Figure 1.4 Evolution based on cranial capacities (Adapted from J. B. Birdsell, *Human Evolution,* 2d ed. [Chicago: Rand McNally & Co., 1975], p. 337)

In the gospel accounts hundreds of deeds and sayings of Jesus are preserved, written down some years afterward. To some extent these were reshaped by the mind of the early church; to some extent they faithfully portray the historical Jesus. In the Christian mind these portraits warrant viewing him as a perfect human, and Jesus has been a principal, if not the principal, ideal of moral and spiritual character in the Western world. There is also surviving in the gospels an odd story in which Jesus, traveling into Jerusalem while staying at nearby Bethany, sees from a distance a fig tree, precociously leafed out. He reaches it to find it barren, and, although it is not yet the season for figs, curses it for the lack of fruit. As he passes that way later, the blasted fig tree has withered.[16]

On the face of it, and magical elements aside, this conduct seems intemperate; and what the Christian, using her best judgment across all the gospel data, has to decide is whether the normative ideal is a paradigm read into, or out of, the actual historical character of Jesus, whether this account of cursing the fig tree calls merely for an auxiliary hypothesis (such as that Jesus used the tree as an object lesson to condemn the fruitlessness of Israel) or for revision of the main claim that Jesus lived an ideal life, the anomaly surviving as a relic of a churlish side of Jesus' character, glossed over in the prevailing paradigm. In science and in religion alike, one needs to attend to all the appropriate facts, but the sifting of these into the most credible paradigm is never easy. Sometimes it requires the wise neglect of awkward facts, and sometimes those awkward facts that are initially dismissed later prove so serious as to overthrow the theory.

3. OBJECTIVITY AND INVOLVEMENT

Science and religion both exist only as processes in persons. Although nature and God may be out there, science and religion alike are an informing of the subject—personal knowledge. Even if their respective theories and facts are in some degree objective knowledge, representing the real world, they are inescapably also subjective knowledge, information acquired, achieved, and processed by human subjects. The knower is never less present than is the known, since knowing is a relationship. A little reflection here will check the facile assumption that science is or ought to be entirely an objective discipline while religion is altogether a subjective one, an opinion usually ventured by persons flailing religion. It may also warn us to be cautious in divorcing the activity of scientists, who are persons as subject to involvements as anyone else, from the structure of their science, at least until one has seen how far the latter can be factored out of the former. A participatory element is always present in science, although it is true that this element significantly deepens as the nature of the inquiry becomes religious.

No discipline, certainly not science, can proceed without truth telling, and this at once introduces ethical demands. Not only must the investigator tell the truth, but also he depends on the honesty of colleagues and predecessors, since he can personally verify only the thousandth part of what he knows. Researchers occasionally forge or color their reports, and if such dishonesty is not soon detected it often proves seriously disruptive, as with the Piltdown hoax in anthropology. A scientist trusts in the integrity of a community of scholars, and a decision about truth here is rather rarely by replication of experiments but ordinarily comes by judgments about whom to trust. That the world is round, or that a hydrogen atom contains one electron and one proton, or that time dilates with increased velocity, or that the sporophytes of cryptogams are diploid—these are facts in principle verifiable by work with firsthand data, but, since time and talents are limited, they are routinely believed as communicated by the honesty of others.

Dedication and Universal Intent

We speak of a dedicated scientist as we do of a dedicated saint. Being a good scientist is not merely an occupation, it is a calling. Although conscientiousness in the two differs in important respects, which will appear later, there is in both a commitment to an inquiry as genuinely worthwhile and as profitable enough to warrant the sacrifices that are required to pursue it. The pure scientist operates at a level of involvement unreached by those with applied concerns; like the saint, she is so devoted as enormously to invest the self in her discipline. No major accomplishments in either science or religion have been made without commitment.

Einstein remarked that science is driven by "passion" no less than are the humanistic pursuits.[17] This passion ought to enhance the capacity for judgment rather than to prejudice it. It is a passion at a level of involvement advanced from those with applied concerns for human welfare; it is a passion for truth intrinsic in the subject matter. All good scholars so love their disciplines as to hate error in them, especially as contributed by partisan bias. Like a judge who is intensely interested in justice and just so disinterested before disputants, every intellectual needs disinterest fed by concern.

Scientist and theologian alike seek what is called universal intent, a setting aside of private interests so as to promote the single-minded discovery of public truth, what is true at large and for all persons.[18] It is odd to speak of "my science," yet permissible to speak of "my religion," owing to ranges of involvement that we will soon trace. But both scientist and theologian are humans making their way around in the world, and their self-understanding is hooked into their disciplines. It is appropriate to speak of a professor of science, equally with a professor of religion, and mean by that one who values the integrity of his discipline and pledges his life against the truth witnessed to by it. The witness of either professor is personally backed but moves toward public truth. All of the classical faiths would find it deviant for "my religion" to mean a faith intended for myself alone. Their truth is preached for all. Good religion shares with science an interest in truth independent of one's personal stake in it. It is equally bad in either area to start so proving one's private beliefs to others that this defense becomes primary, for then the discovery of a better theory for myself and for others is thwarted. Such willingness to submerge one's own achievements in the advancing tide of knowledge requires a steady humility on the part of theologian and scientist alike. Any who ask which discipline is the most troubled by dogmatism will find the question difficult to answer, and perhaps will conclude only that arrogant self-confidence is becoming in neither. A bare self-interest has to be overcome, the self's concerns aligned with this ultimate truth, which is in this sense objective or universally intersubjective.

Science and theology alike are designed to correct anthropocentric error by inferring how things are in nature or in ultimate reality independently of the peculiarities of our sensory and cognitive faculties. Yet the knower always enters into her descriptions and cannot escape her framework. We cannot think without paradigms, and yet we hope to submit to the facts, do this what it may to our models. Just this willingness to set the compulsion of the truth above a compelling paradigm prevents the latter from becoming an ideology; it enables our paradigms, self-serving as they are, to be self-correcting. Only devotion to truth can accomplish this; and so a willingness not only to give of oneself but to give up one's preconceptions and illusions for the sake of the truth—a determination to hear the whole truth and nothing but the truth, come what may, cost what it may—is as characteristic of good theology as it is of good science. The reforming spirit in theology is just this insistence that a person must not get in the way of the truth, must not bias it, but hear it sensitively and entirely.

Informed Judgment and Decision

The foregoing virtues perhaps can be assigned to the climate in which the scientist works, and thus they are a part of his larger methodology shared with other scholars. But it might be contended that these have little to do with the content of his knowledge. Honesty, truth, commitment, selflessness, and humility in the scientist facilitate the inception and teaching of science, but they do not belong to the finished product that science delivers. They belong to the psychological matrix of science but not to its logical structures. Even this makes them indispensable, but there are, we must notice, some areas where the immediate scientific judgments have an inescapable personal coefficient.

Like religion, science can be communicated only to those who are subjectively prepared, that is, willing and able, to receive its claims. It can be appreciated only by those who value it, and this requires a joining of and an education into a skilled community. Science has its logic, often an impressively rigorous one, but that logic is simply not available without sustained study, critical interaction, and this is both psychologically and logically costly. Science is not so much objectively as it is intersubjectively testable, replicable only by those who live and work into its particular fields with achievements adequate for judgments there. This is a matter not only of access to a field but also of ongoing functioning within it, of depending on a community that can understand and criticize one's work. Only occasionally can a scientist work individually; usually she needs collegial interaction and even closely coupled teamwork. In science and in religion this dependence on a community differs in ways we soon will specify (there are confessional elements in religion as well as professional ones), but there remains in common an element of personal qualification derived from a corporate education.

The criteria commonly given for assessing a theory are the extent of its agreement with experience, its internal consistency, its simplicity, its elegance, its deployability or interconnectedness with allied fields, its fruitfulness or productivity, its testability or predictive power, and the degree to which it provides a satisfying sense of explanation. In the application of these criteria a considerable degree of argument is appropriate, and in the metric sciences this may be of a computational kind. Some theories are eventually rather well settled into, but appraisal of even the simplest laws and theories always includes a scientist's larger judgment of what can safely be left out in evaluating a particular natural phenomenon. On the cutting edge of science when assessing rival theories, as well as at the philosophical frontier as these theories become more cosmic, perceptive judgment is required no less than conceptual clarity and factual accuracy.

Just how good is the evidence for natural selection as the sole editing factor in evolution? Is the big-bang theory of the origin of the universe now more credible than the steady-state theory? Why did dinosaurs become extinct? *Stegosaurus* had enormous bony plates along its back, and the theory demands some survival value for these. But were they for defense, to make the animal appear more fierce, or to protect against actual attack, or cooling fins, or used in courtship, or some combination of these, or by serendipity first one and then another? Are no acquired characters ever genetically transmitted, or is there enough evidence that they sometimes are, as Trofim Denisovich Lysenko has maintained? Does the evidence for extrasensory perception warrant further investment of research funds there? Just how simple is behaviorism, and is it too simple to be satisfying? Is Sigmund Freud right that monotheism is a projection of the father figure? Has science found causal connections often enough to demonstrate that the apparent randomness in nature is only apparent? Whatever psychological dimensions may operate in decisions here, there are also logical dimensions involved. Max Weber's thesis on the relationship between Protestantism and the rise of capitalism is a distinguished contribution to social science, and it can be empirically supported, but its overall adequacy is very complex to assess.

The answers come as they do to a judge appraising justice, to an ethicist gauging morality, to a theologian testing a religious creed—by a mingling of argument, of weighting of facts, of notions of plausibility, and even by intuitions that yield an informed judgment. Science is decision-laden; one simply cannot do it without grading. Further, the grading is not algorithmic. It is judgmental, and in this respect often draws near to the sort of grading that goes on in religion, considerably exceeding mere observational checking and computing therefrom.

Observer Involvement in Science and Religion

Against the older notions of the researcher as a mere spectator, recent science recognizes the observer's active contribution. The scientist selects what to study and how to study it. Whenever he builds laboratories, sets up experiments, isolates phenomena, or brings theories to direct observations, he is constructing the nets with which he fishes, and his catch is partly a function of his net. In every controlled experiment we tamper with what we observe, even the controls. These factors can partly be compensated for, as with the rat's differing responses to stimuli in the cage and in the field, but they are never eliminated. The more rigorously we probe nature, the more we increase this distorting manipulation. Are we creating the bizarre array of elementary particles that come out of our high-energy accelerators? Do they exist naturally or only as artifacts? Organic molecules have been artificially synthesized in the laboratory by passing electric discharges through selected gases, but how nearly does this reproduce the environment on Earth under which life might have originated? This is how a certain biochemical reaction proceeds in vitro, but is this the whole story in vivo? Existing reptiles can be shown to be less intelligent than mammals, but were the dinosaurs less intelligent than those earliest mammals that replaced them?

To some it seems that astronomers only watch, but even the optical telescope selects the visible range of light as befits the eye or film, and so astronomers have invented instruments to discover other ranges of radiation. What they choose to investigate and how they interpret it in terms of their cosmological theories depends heavily on the manipulations accomplished in the laboratories of physics. What we "see" is a red shift of a spectral line (the Doppler effect), a sight already arranged for by a theory. What we conclude, on the basis of much further theory, is that the universe is expanding. Observation is always a relationship, and this does not cease to be true as science vastly enlarges the range of sentient experience. Whatever is known must come through to us in terms of perceptual equipment that we naturally own as this can be aided by an apparatus that we have created, in terms of cognitive capacities that we may think up within the parameters of our neurological structures. What we are looking for hooks around, often unbeknown to us, to become what we are looking with. So it is not merely the initial selection of a problem in which the scientist is involved; she remains involved in the reception of all information and in the ongoing construction of her instruments and theories of attack.

The simplistic notion that science ought to be entirely a neutral discipline, contrasting with the involvements of religion, distorts what we must now try to see more accurately—how the two fields can be brought into contrast at vital points, which have to do with their essential paradigms, their particular logic, and their extremes. But at intermediate points and in underlying rationality these contrasts can dissolve into similarities, though never without some insoluble residues of difference. The natural sciences deal with a dimension of experience that can be characterized as empirical, while religion, beyond any account of the phenomenal world, deals further with a dimension that can be characterized as existential, moral, spiritual. Natural science can treat things as objects, while religion must reckon as well with subjectivity where it is present.

In this respect the human sciences are problematic. The controversies in psychology that surround a science of the mind, beyond behavioral science, show this tendency of science toward outwardness, and Émile Durkheim's first rule of sociological method is that we must consider social facts as things to be observed like natural facts. So far as these sciences are empirical, they can continue to treat human phenomena objectively. When a psychologist, sociologist, or anthropologist wishes to get into the mind-sets of those studied—for indeed humans cannot be otherwise fully understood—what is inward has somehow, by empathic appreciation, to be laid out for public access. Still the social scientist is an onlooker even where he is an in-looker; he converts his subjects into an object of study, but he does not, qua social scientist, ask of himself the inner questions of subjective experience.

Religion asks about good and evil, about guilt and redemption, about love, justice, and holiness, about the values of the subject in its objective world, and it judges these to be the ultimate or deepest ranges of experience, beside which the empirical explanations of the sciences are penultimate or even superficial. In the natural processes that the physical and biological sciences investigate, most of these issues do not ordinarily appear. So far as they do, as for instance when an evolutionist asks whether the elimination of the less fit is bad, the question cannot be solved with those tools with which the scientist does her empirical work, and so proves to be a nonscientific question. These issues can become, descriptively, subject matter for the social sciences, but when they do science becomes more participatory. The instruments of observation are empathetically constructed; a subject appraises another subject, whom she treats as an object; and the results of even a supposedly descriptive inquiry are increasingly loaded with interpretive categories that demand introspective feeling for their appreciation.

Here the scientist is a member of the class that she observes, at least generically. Yet a scientist proposes to work on an observable other, and to eliminate the consequences of her work for her own experiencing "I." This is sound methodology so far as it achieves universal intent by suppressing personal stakes. Though perhaps selecting for study what promises to be relevant and beneficial, the chemist can neutrally study covalent bonding, or the meteorologist cold fronts, for these are at much distance. But no human scientist has such remoteness from his object, for in the last analysis he is experimenting on himself. Psychology has psychiatry as its cousin; all the human sciences become helping sciences, broadly therapeutic. When physiology and psychology describe normal human conduct or functioning, they are not far from normative conduct, ethics. The line between an "is" and "ought" can be drawn and ought to be drawn, but it proves semipermeable. Ethnology is never far from ethics; nor is human ecology. Like physicians, human scientists mix scientific detachment with a deep concern for the patient. So far as there comes with them any concern for human welfare, they not only describe the normal but also expect to have relevance in normative prescriptions. Thus, science may be value-free at the physical end of the spectrum, but at the humanistic end it courts values it cannot itself provide.

On any occasions where prescriptions are offered, some values must be superadded to empirical data, and science has moved over to the participatory level of religion. Reformatory elements begin to appear, and in religion reformation of the person is a primary goal. Religion is accordingly less hypothetical and more categorical than is some science, since the latter can have less nearness, be less imperative, and therefore be more negotiable. Religion is thus to be trusted in, while science is sometimes more lightly believed. But some sciences operate in areas where convictions are not tentatively held, as those find who investigate whether some races are lower in intelligence or who propose genetic experiments to enhance human genius.

Science stays hard, that is, objective, proportionately as it stays empirical, and in all the sciences—physical, biological, and social—one can isolate out elements of strict science. Are there special brain centers for the emotions, others for reasoning? But so far as the issues are observational, they are proportionately superficial, piecemeal bits of analysis that form technical science, technical because it is experimental and manipulatory. It does not demean this impressive technical element to see it for no more than it is and as but instrumental to deeper human concerns. In this technical domain scientific assertions can be put outwardly and impersonally, contrasting with the way that religious assertions require also inwardness and personality for their comprehension.

Science becomes soft, that is, participatory and subjective, as it nears the experiential beyond experimental dimensions. Are humans more emotionally driven than rationally guided? At length, religion is participatory in ways that science never reaches. How best do we use our emotional capacity to love? To what world view does it seem most worthwhile and reasonable to give my allegiance? Here science *has* a way of truth; religion *is* a way of truth. In science one knows "about" the object; religion removes that "about" to know with more intimacy.

Here the judge must be up to what she judges; that is, the character conditions are more demanding. Aesthetic achievement or sensitivity is required of the music or art critic in a way not needed in the chemist or even the sociologist. Moral experience is required in the counselor, a sense of justice of the judge. Spiritual qualification is required of the theologian, involving talent at levels not demanded of the physicist qua physicist. Only the pure in heart can see God. Every discipline requires its relevant sensitivity; and learning and thinking in the biophysical and social sciences, so far as they operate empirically, are simpler morally, aesthetically, and spiritually, however complex a causal logic may be used, than these are in religion. Proportionately as truths become more significant, combining cosmic with personal importance, they require more sensitivity for their reception. One cannot verify merely by painstaking observation or imaginative construction what has been discovered and confirmed by passion, sacrifice, faith, and suffering. This relative restriction of science to empirical levels and to descriptive, technical logic partly explains why, among those competent to judge, there can be broader intersubjective agreement in science than in religion. Sometimes it even seems that the elusiveness of an answer is in proportion to the importance of the question!

The danger of such sensitive involvement lies in slipping into a no-lose setup, where negative results prove the observer is out of tune with his object of study, while positive results prove that the observer is in tune with it, and right! One must carefully allow place for being both in tune and critical enough to hear falsity. Meanwhile, no one can judge with competence any enterprise with which she is not competently and seriously engaged, because the absolutely crucial thing about any scientific or theological inquiry is that it be controlled by the reality that it intends to study, and this demands adequate engagement with it, adequate receptivity and sensitivity. Religious thinkers too attempt to be "scientific," that is, systematically to scrutinize their beliefs for their consistency, simplicity, deployability, for adequately explanatory accounts, for practicality in and congruence with experience. Needless to say, this requires a specific method adequate to the subject matter, and this cannot always be scientific in the positivistic sense but may demand instead considerable existential involvement.

To be objective is not in most cases to be neutral or indifferent; nor does it prohibit the holding of previously gained, presently owned, presumed beliefs. Objectivity requires only that one be willing and anxious to test convictions against experience and logic, and to reform them accordingly. Those who are prepared to accept such criticisms do not hold convictions subjectively, that is, only from within a private subjectivity. They own no bliks, but they look to be informed and reformed from without; they seek external involvement in correcting their judgments. We began this section tracing how both science and religion are, in one sense, subjective knowledge; we can close it by noticing that they intend also to be, in a further sense, objective knowledge.

4. SCIENTIFIC AND RELIGIOUS LOGIC

Causes and Meanings

Science and religion share the conviction that the world is intelligible, susceptible to being logically understood, but they delineate this under different paradigms. In the cleanest cases we can say that science operates with the presumption that there are causes to things, religion with the presumption that there are meanings to things. Meanings and causes have in common a concept of order, but the type of order differs. "Cause" has proved a difficult notion to explicate. Some scientists have tried to reduce it to, or to substitute for it, bare functions between variables. But most scientists find it difficult to escape the conviction that the variables are efficaciously connected. In a stretched sense, or in loose everyday use, cause refers to any contributing factor in an explanation (as with Aristotle's four causes), and it may include deliberations, reasons, and even meanings. But in science cause is restricted to outward, empirically observable constant conjunctions, attended by an elusive notion of necessary production of consequent results by the preceding spatiotemporal events. Where causes are known, prediction is possible, and an effect is commonly thought explained if its causes are known, especially if it is subsumed under a covering law (as with gravitation, thermodynamics, or natural selection), that law giving a certain logic to the process. It does little explanatory work to refer *x* to the class *X*, and to notice that *x* produces *y* because all *x*'s regularly produce *y*'s, when we do not understand those other productions either; we have only gotten used to them. So law alone, although it permits deductive prediction, provides only the beginning of illumination, which further requires some intelligibility past regularity in the relationship between cause and effect.

"Meaning" is the perceived inner significance of something, again a murky but crucial notion. Occasional apprehension of meanings does not constitute a religion, any more than occasional recognition of causes constitutes a science. But where meanings are methodically detected out of a covering model, which is thought to represent an ultimate structure in reality, one has some sort of religion or one of its metaphysical cousins in philosophy. Science holds that causality runs deep in the nature of things; religion holds that what is highest in value runs deepest in the nature of things. It may be objected that one can search for meanings without being religious. This has not often been true historically in any broad sense, for, until the twentieth century, cultures, so far as they were systems of meaning, have been everywhere interwoven with religion.

More recently, under the impact of science some humanists and existentialists have held that meaning is merely a human construct, nonreligiously selected, since the world itself neither offers nor bears any meaning structures. It remains to be seen, in view of the contemporary problem of meaninglessness, how viable these latter accounts are and whether any culture can be sustained on them; but here perhaps one has the anomaly of systematic nonreligious meaning. However, if meaning is thought to be given in the world structure, or to be had in dialectical relationship with the natural order, or to evolve as a sacred cultural emergent, then one has a religion, though perhaps a new immanentist or naturalistic one rather than a classical supernaturalistic or transcendentalist one. Relative to the distinction between cause and meaning, it may be said that science answers how questions and religion answers why questions; but these words, while suggestive, are not reliable indicators of syntax and the kind of explanation sought.

Social scientists and psychologists are disagreed as to whether their sciences are ever sciences of meanings, and the puzzle as to how far human subjects can be causally understood has left the human sciences unsettled. Rigorous behaviorists insist that psychology is entirely a causal science, while humanistic psychologists seek to understand personality as a function of meaning. Social scientists find that causes operate in human affairs; there are causes of inflation, war, revolution, depression, suicide, birth and death rates, environmental crises, etc., and these causes operate comprehensively, including overriding or negating what the members of a society may mean and intend. At the same time no society is entirely understood without appreciating its meaning structures as these interlock with the causal factors that constrain it. Meaning structures too can be understood in terms of a governing model out of which conduct follows. Given a certain meaning model (M), a certain pattern of conduct (O) will be observed (if M, then O), and thus meaning models, no less than causal law, can be embraced under the sort of logical inquiry we have here been tracing. They too have their regular operations and predictable dynamics. We have already maintained that creeds and theologies can be studied in this way.

Although social scientists or psychologists may inquire what meanings other persons have and how these function in their lives, they do not use—the majority will insist—their sciences to discover meanings for themselves. These sciences may describe the meanings that others have, but they do not prescribe what meanings scientists themselves ought to have. The scientist may find meanings in his subjects and make these his object of study, but he does not, with his science, find meanings in the world structure or cultural structure and make these life-orienting. Whenever one undertakes this latter task, one has passed over into the province of religion and its cognate fields—ethics, comparative religion, the humanities, philosophy. Thus, in the human sciences we find an overlap between science and religion, but so far as there is disputed ground this is because we know what the master paradigms in the two fields are—that science is a study of causes and religion is an inquiry into meanings.

Negotiability and Compatibility of Causes and Meanings

Each master paradigm is virtually nonnegotiable, a dogma within that discipline. These paradigms arise out of experience, for the scientist has found many causal connections, while the saint has discovered much of significance. Such realized causality and meaningfulness are universalized into the beliefs that everything is causally sequential and that all events are meaningfully interpretable, and with this they become presumptions brought to experience as well as derivations from it. In modern science this yields a universe of precise law, which persons can successfully study and profitably manipulate. In modern religion this yields a universe that has cumulative meaningfulness, coming to focus in God, the Absolute, or a divinity of the natural whole.

These dispositions to interpret things causally and also meaningfully are built into the deep structures of the mind, and we have to some degree an innate psychological drive to find things intelligible. But neither the causality found by science nor the significance found by religion is to be dismissed as merely psychological, for these also are present as logical structures in the mind. The mind has evolved as a natural fit in response to the environment in which life occurs. What an individual mind brings innately to the world recapitulates the edited genetic experience of this species.

There is some temptation to say here that causal relations are "really there," discovered, objective, but that meanings are invented, subjective, only "in us." A truth in this is that causal relations, after we have recognized how they are subject to our mental structures and constructions, may be outwardly reviewed for their constant conjunctions, while meanings appear as the subject is experientially related to her world. But, again subject to value structures provided by the mind, it would be anomalous if humans had evolved their enormous innate thirst for meaning in life in a world where life is a natural event but where all these investings of life's relations with meaning in the world (for example, those of love and hate, fear and joy, birth and death, of beauty and fruitfulness, of work and parenting) were inappropriate and superficial. In this case all those appearances of language as it seems to lodge meanings in things and relationships would in fact be deceiving and refer in a hidden way only to the psychic state of the user of such language, a state that was disjoined from his biological origins. This might be so, but any argument strong enough to prove it is likely also to carry the implication, with Immanuel Kant, that causes as well as meanings are nothing but compositions of the mind. Until such argument prevails, it is simpler to hold that causes are experienced in the world and that meanings, however self-involving, are sometimes given, often relational, even if on occasion created *ex nihilo* in the mind.

It is perhaps true that disciplined science can abstract out bare causes, devoid of any meaning; but this is a very sophisticated, high-level analysis, only recently accomplished in the intellectual life of humankind. The real world of nature and culture in which we live is one in which we meet facts, values, disvalues in fusion; they come at us together. It seems natural to say that we meet and find both causes and meanings there. The gut nature of living on, surviving, makes the world a field of values and disvalues, never neutral to the pursuit of life; and at this point it becomes artificial to leave by analysis the causes objectively there and wrench the meanings out of it as a subjective appearance or fabrication. What is given, what is protocol, is not naked sense data, not bare constant conjunctions, but a milieu of events, with causes and meanings in-mixed, sought and found, made and coming at us, opportunities, a world we have to move through and to evaluate.

Can either discipline tolerate anomalies? Yes, but both will so minimize the exceptions that their respective gestalts still govern. A pathologist may search without success across decades for the causes of a baffling disease, but she will not conclude that the disease is uncaused; a psychiatrist is likewise likely to insist that every mental disease has in fact some cause. A monotheist may admit frankly that he finds some events meaningless, although he also may believe that even these have some divine purpose, which he cannot now find. Quantum mechanics has come to permit the possibility that there is some genuine indeterminacy in subatomic nature. So far as evil prevents the assignment of meanings, its presence has always troubled theism.

Randomness on the one hand and absurdity on the other do challenge these paradigms but are allowed only when effectively overridden by a statistical causality or a net meaningfulness that does not interrupt a larger intelligibility. By some accounts this reduces these paradigms to regulative maxims. The scientist proceeds in the effort to find all the causes she can; the theologian will pursue meanings as far as he can. Neither must then claim that her or his procedure will in every case be successful. But both are still prone to think of their procedures as appropriate because the world is constructed so as amply, if not universally, to bear relations of causality and of meaningfulness.

The warfare between science and theology is often a struggle to clarify to what extent causal explanations are compatible with or antagonistic to meaning explanations. Particular disputes may result in adjusted claims about the territory occupied by each account. While no one denies that each field commands some territory of its own and that there is partial complementarity, are they always commensurable? Some kinds of causal accounts, for example, the competitive survival of the fittest, do seem to inhibit some kinds of meaning accounts, such as that every species was divinely designed at an initial, sudden creation. Some causal explanations show some meaning explanations to be inaccurate, inadequate, or irrelevant. But if these are really different tracks of explanation, how can they compete as they sometimes do? Science, by redescribing nature, places constraints on what concepts of God are credible, even though science by this redescription prescribes nothing about God's existence. It sets limits within which meaning accounts can work.

Does the presence of sacred meanings in the world require any tearing in the weft of causes and effects, any perforation of the natural by a supernatural order? Does the meaning account sometimes constrain the causal, as when the experience of autonomy and moral responsibility seems to demand that persons be something more than effects predetermined by antecedent causes and stimuli? If there is randomness that proves causally baffling, inexplicable by science, does this imperfection correlate with the absurd in religion? Or can an account be reached whereby such causal looseness provides just that novelty and unfinished openness to nature and to life that religion can enjoy? Experience that is counted puzzling under the one paradigm may prove intelligible under the other.

Differing Kinds of Logic

The causal paradigm favors a computational logic, whether inductive or deductive (at least for routine science, though perhaps not for revolutionary science), while the meaning paradigm involves an intelligibility that is more holistic. Causes go into linear networks, which often permit a quantifying theoretical overlay measuring with numbers such things as wavelengths and stimulus-response correlations, although we should not forget that those numbers, which look so accurate and objective, even with their margins of error, are in the case of scientific measurement always the product of a theoretical overlay on nature and never purely natural computations at all. The validity of such quantifying depends on the quality of the overlay.

Even nonmetric science is prone to taxonomic serial catalogs and phylogenetic chains, the steps of which can be isolated for analysis. This brings a particular occurrence or individual under a covering law or type. Such repeatability and parallelism are not always found or verified by either induction or deduction, and just what counts as patterns similar enough to warrant their inclusion under the same law is always a matter of some discretion. But the causal character presumed here sometimes permits to science a level of rationality and thus of testability different from that in religion, a step-by-step checking that can be summed up into near-compulsory argument.

Religious meanings are not integrative in this scalar way. When set in their gestalt, the particulars give rise to meaning. In detecting more sophisticated patterns, as when, despite her aging, we recognize the face of a friend whom we have not seen for decades, there is a subtle interplay of textural features by which the whole is constituted. This sort of logic is present in science when a geologist recognizes the facies of rock strata, or when a dendrologist notices the differences between the bark of spruce and that of fir. But it looms much larger as one approaches the perception of meaning in a novel, such as *Gone with the Wind,* or in a historical career, as of Abraham Lincoln. One must join earlier and later significances in ways more qualitative than quantitative, more dramatic than linear. The sense of scenic scope is more crucial than that of incremental detail, hence the nonmetric character of religion.

Pattern statements differ from detail statements, alike in science and religion, but in some science it is easier to go from detail statements to pattern statements, owing to the metric-causal character. The holographic character of meaning models is not merely sequential with the chronology of life but requires more cross-play and interweaving, a logical network sometimes said to be more characteristic of the right than of the left cerebral hemisphere, more characteristic of the brain in general than of a computer. But this remains in the if-then mode, for even in the analysis of gestalts one says such things, to recall the reversible drawing, as "If that is a young woman, then this is a necklace and that an ear. But if it is a hag, then this is a mouth and that an eye."

The finding of meanings is not as simple as is identifying unvarying conjunctions. Those unique, nonrepeatable factors present in each occasion can often be integrated into its meaningfulness, while in subsuming an event under causal law these are irrelevant. The *Victory of Samothrace* instances certain universal forms of grace, strength, and flair, found also in other great sculptures. However, its aesthetic value is not constituted in abstracting these but rather just as these are indissolubly particularized in the individual integrity of one historical statue. There are recurrent religious meanings, as when persons rediscover the significance of forgiveness or of sacrificial love, but each occasion instantiating this will be cherished not only for its generality but also for its particularity.

There are various modes of interest of the human mind, not all of them either scientific or religious. Science and religion share a theoretical mode of interest. Both want to operate out of a model or theory, a plot or a pattern, that gives a universal intelligibility to what is observed in particular episodes. But science has little interest in particulars for their particularity after they have been included as instances of a universal type. It has little interest, for instance, in proper names as essential to its content. But religion retains its interest in particulars both for their constitutive power in enriching the universal model and as loci of value. It is thus full of proper names, no less than of creedal models.

Because of this inclusion of particulars in the composition of meanings, religion can tolerate the presence of surprise more than can science. The history of science is beset with surprises, of course. But real surprises are quite upsetting to prevailing theory, for scientific models must be specifically extensible in advance to all forthcoming phenomena, and any incapacity to predict is unnerving. Religion is less inclined to predict, less insistent on similarity of cases; rather, it waits to see, after the fact, whether its paradigm can extend to cover these surprises, whether if the theory is true then a novel observation can be seen by retrodiction to follow from it. Neither the causal flow nor the meaning flow is reversible in fact. Yet causal accounts are projectable in thought symmetrically forward and backward (remembering, however, the logically troublesome status of induction). The admission of the singular existent implies that a meaning account cannot always, on the basis of recalled experience, limit its expectations as to what will and will not be absorbable into its creeds. In this sense a religious theory has an openness beside which a scientific theory is closed.

One does not always have to say in advance exactly what would refute one's theory, for that requires too much prophetic power; but one must be willing to examine each new bit of evidence as it comes along with widening ranges of human experience. A Christian judges Jesus, the Christ, to be the key to meaningful life in the world by perceiving in him the normative expression of a life style of agape. The claim follows that the agape life will always be found meaningful, but one is not able to say, in prospect, just what will count as a context for agape. Such contexts are too idiographic, although one can say, in retrospect, whether those meanings launched in Jesus have been continued as embodied in the historical particularity of each disciple's life. Here one has to judge the cumulative effect of severally inconclusive and partial verifications, which are woven not to prove but to corroborate a creed.

One can deduce only in a looser logic of weak connections. One can know out of his theory something of the possibilities that the future may hold, but he cannot make the watertight predictions that a positivist will insist that the hypothetico-deductive model requires. But to know, out of one's theory, something of the possibilities is already to know something, just as to know probabilities is already to know much, although it is not to know everything. Neither science nor religion arrives at certainties. They at best predict probabilities, but religion is looser here than is science and often can predict only a range of possibilities. Still, there is a logic to it, a model out of which one can derive the oncoming particulars and a symbolic system that functions as a regulatory model, albeit a noncausal one, into which the events of life are fitted (composing and recomposing this creed) and out of which they are interpreted.

Thus, the hypothetico-deductive method in religion does not employ the narrower sense of "deduce" that science sometimes uses. Although a new event cannot be entirely foreseen from the theory, that event, when it does occur, does follow and unfold from the theory, while some other events may not. From the first half of a play we cannot predict just how the second half will proceed, although as it proceeds we have a gathering sense of how the several events fit into an overall plot. We reason back down from the general to the particular, more broadly deriving from the paradigmatic plot what episodes may be allowed to constitute it. Thus, the dramatic plot is testable against unfolding experience. But this testability is not a stringent one. There is no single logically necessary deduction from what has gone before, although certain events can, and others cannot, be significantly emplaced in the scheme. Even in science this may occur, as in evolutionary theory, where later specific developments in their novelty cannot be unequivocably forecast, although after they occur they may be examined as to whether they are consistent with the theory.

Given that science remains causal, leaving off any assignment of meanings, it is a value-free enterprise, while religion is a valuational one. This is not a simple matter, however, because there is a spectrum of meanings attached to value and to neutrality. Science, as we have noticed, shares certain pervasive values with religion, such as those of truth and critical inquiry. In science one makes judgments about good instruments or research. One operates on the presumption that science itself is good, either instrumentally or intrinsically or both. But where science is confined to causal accounts, it never prescribes life values, for these lie in the realm of meanings.

Science is not, as is sometimes thought, merely instrumental to value, for intrinsic science does redescribe the world for us. The descriptions here cannot be ignored, for such discoveries as the age and extent of the universe, the evolution of life and its biochemical nature, the human neurophysiological structures, or the electronic character of matter have forced theology to reform earlier accounts of meanings. Persons always shape their values in some correspondence with what they believe the world to be actually like. But these descriptions never constitute prescriptions, however much they may force a reconstituting of them. In this sense religion is fully operational, completely functional in joining theory with practice, as science is not, for religion has its own value setup, which permits the translation of principles into conduct, while any scientific system is parasitic on some value system before it can become operational in life. Religion, however, is not so operational that it can ignore what science reveals about the character of the world and of life.

An older form of this claim is that science seeks knowledge, but the spiritual quest is for wisdom.[19] Knowledge and wisdom are neither coextensive nor mutually exclusive, but they overlap. In part, but only in part, a person remains naïve and unwise until she has integrated the best available knowledge from the current sciences into her world view. Still, such knowledge is not sufficient for wisdom, for no accumulation of causal explanations can ever produce the significance of a thing. The latter comes at another level of insight. In this sense, science explains but religion reveals; science informs, but religion reforms.

It is often said that science operates in an I-It mode, that of experience, while religion proceeds in dialogical encounter, the I-Thou mode. This distinction is founded on the biting difference, noticed daily, between dealing with persons and things.[20] This dichotomy recognizes the outward objectivity of science, where an "I" describes "things" in their causal relations and manipulates them as a result. In religion this "operational I" is replaced by a "relational I" that answers to the world and constitutes meanings in exchange with it. Demands flow to the "I" as well as proceed from it, for the existential "I" is called forth by that which is known. The subject has gone out to its object, which is no longer bare object, but is itself a subject, that is, a source of prescription to me. Wisdom appears in this intersubjective encounter, while the objective mode can provide only descriptive knowledge. So the notion of subjectivity loses some of its unwanted flavor, and the word "operational," often favorably linked with objectivity, becomes annoying so far as it is manipulative. The unilateral operator is ill fitted to hear the address of another or to respond to its worth.

Monotheism, moreover, detects the divine as a depth presence, an "Eternal Thou" in, with, and beyond the sacramental, superficial objectivity of the phenomenal world. This detection is comprehensively extended from the way in which we detect other minds in the behavior of human bodies. That sense of divine address is more elusively present in Eastern religions, but what is present is a depth engagement of the sacred so gripping as to draw forth the entire person, a meeting of the world at its inciting ground such that the whole self is called to respond, nearer like my relation with a "Thou" than with an "It." Further, though, theism and monism aside, meanings may arise where we attach no "Thou-hood" to this gripping other, as in encounters with nature or in aesthetics.

Some accounts find religion to be less linguistic and thus less logical than science. This may be taken by critics as a vice, but it also may be taken by proponents as a virtue, that religion plunges to deeper levels than the conventional ones of science. This latter position is not without merit, for the religious object, God, if it exists, is incomparably greater than any routine scientific object, such as rocks, fish, or atoms. Logic and language may have evolved, and be evolving, best to fit the mundane, phenomenal world, and they may ill fit the transmundane, noumenal world. Sometimes in the West and often in the East, mystics cultivate noncognitive states supposed to transcend all logic and language.

We do not need entirely to dismiss such claims to recognize that nevertheless logic and language enter steadily and decisively into religion, just as fully as they do into science. Interpersonal Thou relationships are hardly less linguistic than experiences of an It; if anything, they are more so. The discovery of meanings, which humanizes us, requires language no less than the discovery of causes. If meanings are more resistant to language than are causes, if they have nonverbal dimensions, that may indicate that the intelligibility that religion seeks requires a richer logic than the scientific sort. If all created things derive an intelligibility from their Creator, then the phenomenal world is a product of the noumenal Logos and sacramentally points to it. The prescription of values takes more, not less, thinking than does the description of events. Possibly our religion outruns our rational capacities further than does science, but, whatever consequently is the place occupied by mystical moments, these do not constitute the whole of religion; nor can they stand alone. All the classical faiths have their speechless moments, but they all have their supporting scriptures, creeds, arguments, and education.

The immediate experience of God, Brahman, or *nirvāṇa* always proves on examination to be quite as theory-laden as are any of the protocol data of the sciences. This does not disparage the intensity or firsthand directness of such experience; it only insists that there is a logic that leads up to and unfolds out of it. In this sense "God," "Brahman," and "*nirvāṇa*" are postulates, inferential theoretical entities used to explain what underlies the world and certain marvelous encounters had within it. The personalness in religion does not prevent its being logical.

It is logically and empirically possible that religious knowledge would come by occasional interruption of an otherwise regular world order, by fluke and visitation, unprecedented, unrepeatable, not amenable to methodological study of even the theological sort, much less the scientific sort, proposed here. This could be not only in the context of discovery, which could well be nonpredictable, charismatic, mutational, revelational, but even in the context of verification. Revelation, miracles, oracles once confirmed could never again be reconfirmed, but would ever after have to be taken on sheer faith. But this would be an odd sort of knowledge, one that had no carry-forward features, with no way it could be shown to be true, reasonable, probable, repeatable in experience. Such knowledge would be just true, inserted once for all into historical time. Whatever elements of this kind one can find in the classical religions, those faiths have also claimed that their truths could, in some measure, be verified in life, tested out in each new generation, seen to work again and again, despite the once-for-all character of the launching visitation.

Self-implicating Meanings

Meanings are always self-implicating. Values are by definition those things that make a difference. This might be thought to bias a person's capacities for logic in religion. One cannot think clearly about what one is wrapped up in. But the other side of this is that one will not think at all about that for which one does not care, or rightly think about that for which one does not rightly care. This caring becomes more self-reforming as the inquiry passes from the scientific to the humanistic to the religious. The task of religion is to examine that self in its relationships with the world, unmasking illusions and false cares, reforming it from self-centeredness, centering it on that which is of ultimate worth. This is worship, produced out of and returning to reflection. This worship, conceived as the self's disengagement from private concerns and engagement with the absolute, is precisely that universal intent that makes logic possible. Only such enthusiasm, or divine inspiriting, can get the self off-centered enough to reason aright.

The religious judgment is that the self must be reformed in order to eliminate its tendency toward rationalizing, and it is just this positive combination of worship and reflection that makes possible an unbiased rationality. Religion shares with science then a concern for objective rationality, only it knows far better than science that the path to true objectivity lies through subjective reformation. This passion makes for genius. Religion is the science of the spirit, where a rationality suited for objects is inadequate. Here the reflective scientist will not say that he comes to nature without assumptions, despising the theologian as being overcome with them. But he will see that, so far as his selection employs empirical causation as his fishing net, he has a different set of assumptions; and he may even wonder whether just these assumptions might prevent him from receiving the data of religion in an undistorted form.

Perhaps some will complain that the account here has dealt too much with religion as a means of *copying* reality, with correspondence in truth, and too little with religion as a means of *coping* with reality, with its instrumental functions in life. So we readily grant that religion is a means of coping. But that is just as true of science, which is driven by the need to cope with reality not less than to copy it. Like different sorts of maps, both help us to get around in the world (supply a "method") because each in its own way represents that world ("follows after it") more or less faithfully.

NOTES

1. Compare Carl G. Hempel, *Philosophy of Natural Science* (Englewood Cliffs, N.J.: Prentice-Hall, 1966).

2. See Richard Anschütz's account, "August Kekulé," in *Great Chemists,* ed. Eduard Farber (New York: Interscience Publishers, 1961), pp. 697–702.
3. Fred Hoyle, *Encounter with the Future* (New York: Trident Press, 1965), p. 93.
4. Dimitri Marianoff, *Einstein: An Intimate Study of a Great Man* (Garden City, N.Y.: Doubleday, Doran & Co., 1944), p. 68; Gerald Holton, "Constructing a Theory: Einstein's Model," *American Scholar* 48 (1979): 309–40.
5. Lubert Stryer, *Biochemistry* (San Francisco: W. H. Freeman & Co., 1975), pp. 327–28.
6. Stephen Jay Gould, "Darwin's Middle Road," *Natural History* 88, no. 10 (December 1979): 27–31.
7. Karl R. Popper, *The Logic of Scientific Discovery* (New York: Harper and Row, Harper Torchbooks, 1965).
8. Tested and statistically verified in research by Jack D. Jernigan and Steven L. Nock, reported in "Religiosity and Family Stability: Do Families That Pray Together Stay Together?", paper presented at the Society for the Scientific Study of Religion, Knoxville, Tenn., November 6, 1983.
9. Further claims involve the inconsistency and meaninglessness of this world (*saṃsāra*), and such a paradigm of world unintelligibility is difficult to test.
10. Norwood R. Hanson, *Patterns of Discovery* (Cambridge: Cambridge University Press, 1958); Michael Polanyi, *Personal Knowledge* (New York: Harper and Row, Harper Torchbooks, 1964).
11. Thomas S. Kuhn, *The Structure of Scientific Revolutions,* 2nd ed. (Chicago: University of Chicago Press, 1970). For the continuing and often unsettled debates about what constitutes scientific method and its logic, see also Imre Lakatos and Alan Musgrave, eds., *Criticism and the Growth of Knowledge* (Cambridge: Cambridge University Press, 1970); Stephen Toulmin, *Human Understanding,* vol. 1 (Princeton: Princeton University Press, 1972); Larry Laudan, *Progress and Its Problems* (Berkeley: University of California Press, 1977); Paul Feyerabend, *Against Method* (London: NLB Verso Editions, 1975). For the relevance of the debate about scientific explanation to religious explanation, see Ian G. Barbour, *Myths, Models, and Paradigms* (New York: Harper and Row, 1974), and his *Issues in Science and Religion* (New York: Harper and Row, 1971). Critical issues here include whether and how far new paradigms are continuous with or incommensurable with older paradigms, what are the criteria for paradigm choice, the degree to which standards of rationality are controlled by the paradigm, and whether paradigms are to be given a realist or an instrumentalist interpretation.
12. See the story in George Gale, *Theory of Science* (New York: McGraw-Hill, 1979), pp. 169–90, 278–85.
13. R. M. Hare, "Theology and Falsification," in *New Essays in Philosophical Theology,* ed. Antony Flew and Alasdair MacIntyre (New York: Macmillan, 1955), pp. 99–103.
14. Arthur Koestler et al., *The God that Failed,* ed. Richard Crossman (New York: Harper and Row, Harper Colophon Books, 1963), p. 60.
15. Karen Horney, *New Ways in Psychoanalysis* (New York: W. W. Norton & Co., 1939), pp. 7–8.
16. Mark 11:12–25; Matt. 21:18–22.
17. Albert Einstein, "On the Generalized Theory of Gravitation," *Scientific American* 182, no. 4 (April 1950): 13–17, citation on p. 13.
18. Polanyi, *Personal Knowledge,* p. 65.
19. Augustine, *The Trinity,* 12–13.
20. Following Martin Buber, *I and Thou,* trans. Walter Kaufmann (New York: Charles Scribner's Sons, 1970).

Chapter 2

Matter: Religion and the Physical Sciences

Physical science has simultaneously a great distance from, and serious implications for, religious belief. *Life* belongs centrally to both religion and biology; *mind* belongs to religion and psychology; *society* to religion and sociology. But physics excludes life, mind, and society; it restricts its focus to *matter-energy.* Nothing is said of culture or history, much less of the spirit. When inquiring about atomic particles, compounds, and catalysts, or about mass, radiation, and binding forces, physics and chemistry come at reality at too low a level to touch guilt and forgiveness, faith and love, good and evil. Experiential religion is light-years away from the stuff of physical science. Yet the impact of physical science on religion has been as great as has that of the life and human sciences. Its descriptions of the natural world, though dealing with prereligious levels, are so subultimate that everything afterward is colored by its paradigms.

In one sense, physics and chemistry are the most abstractive of the sciences. They leave out all the emergent eventfulness with which the other sciences and religion will want to deal. They drastically oversimplify. But then again, physics and chemistry take no special subset of natural entities for their subject matter, while biology takes organisms, behavioral science takes behaving animals, sociology takes societies, and even the special physical sciences—geomorphology or meteorology—have their restrictions. But, just so, physics and chemistry are the most universal sciences. They apply to everything, everywhere in the universe. Accordingly, what these sciences say about nature will have far-reaching consequences. Thus, we are alerted to the paradox that these universal sciences, which seem to legislate the most for any subsequent studies by any particular sciences or by religion, are also the most abstractive. They achieve their universality by omission, gaining authority not only through what they say, but also through silence.

Here nature has, or at least seems to have, its most evident objectivity. In sheer matter subjectivity is absent. In our knowledge of such matter subjectivity can be increasingly eliminated. So runs the initial promise of physics, the first and most successful science, the hoped-for ultimate science. But here too, with increasing scientific sophistication, we run into paradox. The further we probe into the ultimate constituents of matter, the more our subjectivity returns to dog us. As we move further away from our native ranges in nature, probing very small subatomic particles, or very large black holes, or counterintuitive relativistic effects, the more theory-laden our observations become. Human observers tamper with what they observe, or contribute conceptual schemes that filter what they know. So physics, chemistry, astronomy, these most objective of the sciences, do not escape subjectivity; they rather run more and more into it.

Still, it is no accident that science began several centuries ago in the severe beauty of physics, nor is it surprising that physics has since dictated so much for metaphysics. Its descriptions seemed then and seem now to prescribe what we ought to believe about the nature of things, and thus it constrains that religious belief which it does not properly investigate. It might seem that the sorts of values we choose need have little connection with microphysics or astronomy, but first appearances are deceiving. A material view of the universe will yield a material culture; a spiritual view of the universe will yield a spiritual culture.

1. NEWTONIAN MECHANISM[1]

We will be concerned in this chapter principally with recent physics, but we back up to take a running start by recalling the Newtonian mechanism in which physics was born and spent its juvenile years. Not only does contemporary physics need to be understood against this history, but much of the mood of twentieth-century science in general, and much of the current religious puzzlement before science, is still governed by what such a physics is supposed to decree. We are living in a century of revolution in physical science, and we can begin to understand what scientific revolutions are like by recalling the revolution in which modern science began, that of the sixteenth and seventeenth centuries.

A Revolution in Explanations

The medieval world had inherited its science and metaphysics from the Greeks and its religion from the Hebrews. Thomist philosophy coherently fused all three, employing Aristotle's four "causes," or explanatory factors in understanding a thing: (1) the *material cause,* matter, the stuff of which a thing is made, (2) the *efficient cause,* the effecting, mobile operating force that produces changes, (3) the *formal cause,* the plan or structure inlaid into a thing, (4) the *final cause,* a goal, the end state toward which a thing is drawn. The two earlier causes look backward to ask about generating antecedents. They look for a compositional substrate that has been pushed into its present conformation. Explanation is thus in terms of *physical origins.* The latter pair of "causes" look forward to ask about plans, a will-be, a will-to-be, and a pull onward. The former pair feature the objective side of reality, while the latter suggest a working at ends, rationality, intention, and, at length, subjectivity. In the medieval account, the first pair are secondary, necessary but insufficient for understanding. Explanation must subsequently be completed by setting forth the primary components that account for why things and events are what they are. Explanation is in terms of *significant endings.*

In terms of the complementary paradigms of the preceding chapter, we can say that formal and final "causes" tend to inquire into meanings and belong more to religion than to science, at least to physical science.[2] In contemporary usage in science (though not in history), we would not refer to goals or plans as "causes," except somewhat awkwardly, because in strict science, at least since David Hume, we restrict the term "cause" to material and efficient causes. The scientific revolution programmatically repudiated formal and final categories for understanding and expanded material and efficient categories to do the whole work of explanation. It did not base understanding and prediction on teleological factors, but on material and energetic propellants. Jacques Loeb, a physiologist, finds such explanation the key to all knowledge: "What progress humanity has made, not only in physical welfare but also in the conquest of superstition and hatred, and in the formation of a correct view of life, it owes directly or indirectly to mechanistic science."[3]

This revolution in explanatory strategy can be diagrammed, perhaps oversimply, as in Figure 2.1. Applied to the objective physical world, this move had spectacular results, seen in the successes of physics, chemistry, and astronomy. It was deployed with more uncertainty in the biological realm, for organic life did seem to be regulated by plans (formal causes), and organisms with a subjective life even seemed to select goals (final causes). Nevertheless, the apparent goal-directedness of "organic machines" (as Descartes calls them) rested on physicochemical determinants. Life was a derived phenomenon overlaid on fundamentally mechanistic processes. The word *anima,* "animating vitality," is reducible to the word *vis,* "force." So it has seemed, and even in our own times half of bioscience (but only half) is based on this conviction. Thus, modern science began with a revolution in the sorts of explanation looked for, and while we may be glad for the resulting insights into the nature of things, we may also be wary of a blindness imposed by the governing blinkers. Physics achieves its successes, we remember, by clever decisions about what to leave out. Science involves more careful observations, but not simply so; it comes with decisions about what to look at, and what not to.

In our own century we are in another scientific revolution. The natural world, when looked at more closely, has defeated or shown to be inadequate many Newtonian presumptions. So even in physics we need to worry about what was getting left out in the earlier abstractions. We will try to move through the Newtonian picture and to reach a more adequate account of physical nature, as given in contemporary physics. We will do this, moreover, with an eye to anticipating how,

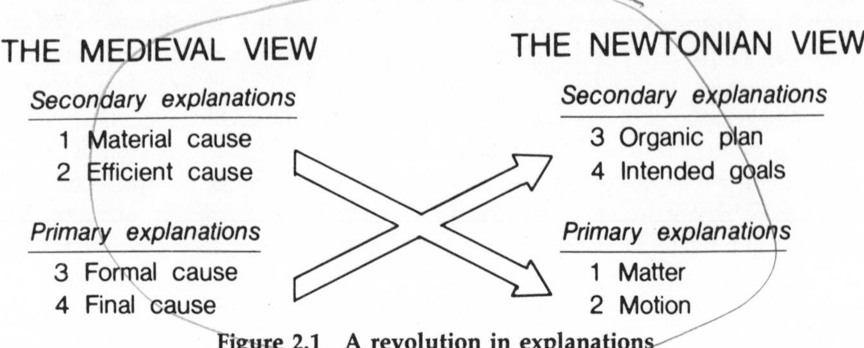

Figure 2.1 A revolution in explanations

beyond physics, the other half of bioscience needs a more open, larger view of nature than the limited Newtonian view, as we find in the chapter to follow. Beyond that still, we must eventually reckon with such phenomena as mind, society, and history, about which physics is silent.

Matter-in-Motion; Primary and Secondary Qualities and Explanations

An allusion in the revolutionary sketch (Figure 2.1) to primary and secondary qualities is not incidental. The Newtonian primary explanations were thought to correspond with what is really there in nature, *matter-in-motion,* while any secondary explanations are observer-dependent, the introduced products of subjectivity, appearing when mind interacts with matter. This switch in explanatory emphasis was to have far-reaching results. It yielded an enormous rise in technical power. But this know-how for manipulating the motions of matter came with a corresponding impoverishment of any sense of purposes, those upper-level explanations that were now said to be secondary, derived, subjective, only apparent. The revolution that gained increasing competence over causes in the world came with a reciprocal loss, a decreasing confidence about meanings there. What was no longer looked for was no longer found.

The celestial decay was an early and dramatic symptom of this. The heavens were classically emblematic of the ethereal and the divine. But the Newtonian view redescribed them as mere matter-in-motion. Before, the planets were thought to have a Platonic formal perfection, typified by circular orbits. Afterward, they are simply other earths, with Earth one among these demoted planets. The cosmic system was anciently a three-decker universe (heaven, earth, hell); it became medievally geocentric. Later, by the scientific revolution it became heliocentric, and finally astronomy depicts it without any center or hierarchy at all. Earth has no special position. The starry heavens were more immense and awesome than before, but they had also become mindless, swirling clockwork.

Nicolaus Copernicus broke up the Earthcentric system; Tycho Brahe destroyed the heavenly spheres; Johann Kepler substituted inferior and approximate ellipses; Galileo Galilei with his telescope confirmed what Copernicus had hinted, that there was a material homogeneity to the whole. Everything was dirt, more or less. Isaac Newton discovered the gravity that kept everything in place. By the time Pierre Simon de Laplace appeared with his nebular hypothesis, material and efficient causes had expanded so as to explain all, both how things are and how they came to be. The classical first and final cause, the God hypothesis, became quite unnecessary in a mechanomorphic world. There was, of course, a small catch, but the catch was hidden by what physics was electing to leave out: the human phenomenon, evident in mind, society, history. Perhaps God is banished from a mechanical world, to be found neither on earth nor in the heavens. But if all teleological explanations are everywhere repudiated, even outside physics too, what are we to make of this paradoxical program of the scientist setting out (purposing!) to show that there is no need for purposive explanations?

This new world is differently perfect, a perfect clock. Here the archparadigm of physical science reigns supreme. Each present event occurs necessarily, completely determined by sufficient causal linkages that run rearward in time. If, at the beginning, one had been given the set of natural laws and initial conditions, and enough mathematical ability, one could have predicted all the motions of the whole, a staggering and really inconceivable task, but seemingly possible in principle. In practice, the heavens proved quite predictable and retrodictable. Celestial mechanics was soon verified in the planetary orbits and stellar courses, and presumed to be true in Laplace's nebular hypothesis. Even gaseous clouds were in fact fluid machines; they too worked like the planetary clocks.[4] Nor do we need to exempt terrestrial motions from this determinism, as mechanical physics, hydraulics, and stoichiometric chemistry were soon to demonstrate, with uniformitarian geology to follow. Whatever moves—stars, planets, mountains, rivers, particles of dirt—moves as and only as physical forces push it so. If any cloudiness remained (as indeed it did), this was in human heads, not in the natural machines, which exactly obey the laws of matter in its motions.

Notice how addiction to this paradigm forces out anomalies. Consider the following graph (Figure 2.2) of functions that we may for the moment leave unspecified. Anyone who has worked in a physics or chemistry laboratory can supply concrete examples. We want to move from the particulars of observation to a general law of nature.

The points are plotted and the curve is drawn, ignoring of course the anomaly. Some extraneous effect or auxiliary hypothesis can account for it, but nature makes no such jumps. It can be eliminated by repeating the experiment (Figure 2.3). We can do this in physics and chemistry, although when we earlier noticed a similar sort of anomaly on the anthropologist's curve of increasing skull sizes (ER-1470), we could not repeat the experiment.[5]

This time the offending point is gone, although another, lesser one has appeared, which is the more easily overlooked because it was not there before. With further reruns the curve is averaged out, and the premise reappears in the conclusion: the phenomena under investigation are precisely determined by natural law. If this does not suffice, we can build better equipment, tighten up the controls, use meters with less margin of error, and by a series of similar but more rigorous (more constraining!) experiments further erase any anomalous empirical cloudiness to our theoretical uniformity. That there are any perfect machines, we have honestly to admit, has

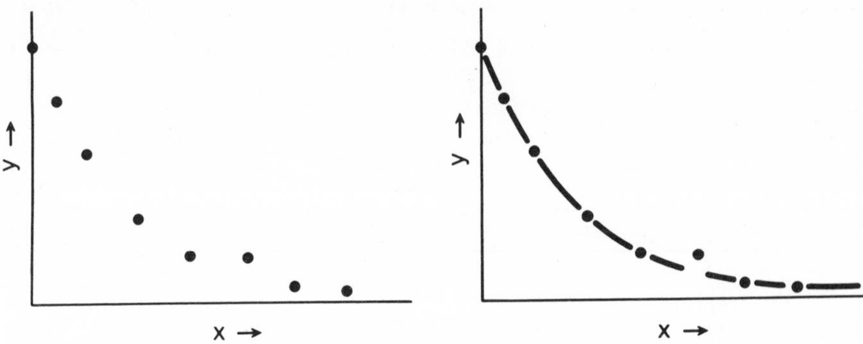

Figure 2.2 An offending point on a physicist's smooth curve

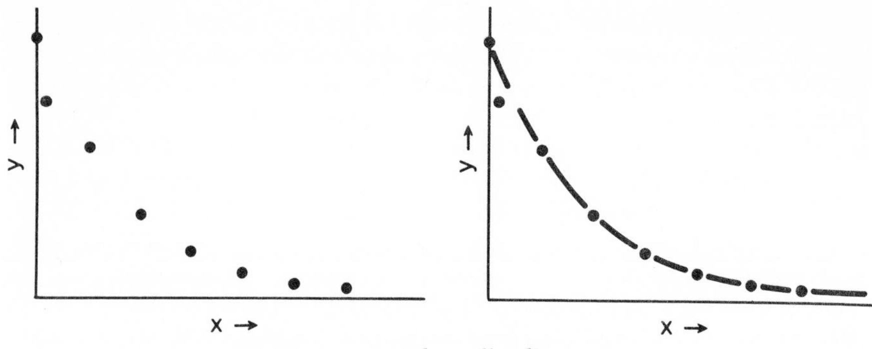

Figure 2.3 Another offending point!

really never been empirically verified. But this can almost be seen in some data that approximate flawless curves and perfect repetitions, where the actual data come very near the ideal form.

We may next erase *secondary* qualities by assigning them to the observer. What we perceive as *heat* is in actual fact a summation of impacts by gaseous and fluid particles, or the result of molecular vibrations within a solid crystal lattice. *Sound* is a wave motion carried by a vibrating medium. *Smells* arise when airborne molecules enter the nostrils and react with receptors there. *Tastes* are molecules affecting the tongue. *Colors* are in fact electromagnetic radiation. With the observer removed, the real world contains only *matter-in-motion,* only the *primary* qualities of mass, length, shape, time, velocity, and their derivatives. These are the objective, absolute, natural facts. All else is modification by the personal subject who somehow aggregates these atomic motions and superadds the relative, introduced, projected qualities. The familiar appearance of things is just that, an illusory front put up over nature, a virtual sham. Mechanism unmasks the deceptions of the common senses.

The primary qualities are essentially metric. They register in quantities read as grams, meters, ergs, seconds, wavelengths. With secondary qualities reduced to primary quantities, the phenomena must, above all else, yield a number. Objective measurements make science possible. Nature is really like a great clock, with the scientifically skilled mind a passive observer. Under this paradigm, it was easy to forget that all but a few of these "natural" numbers were in fact artifacts, gained with ingenious theories that constructed the looking units, schemes composed in the attending mind. Meanwhile, nature, unobserved, contains no ergs, seconds, or meters, and no wavelengths. These were really not objective units, only intersubjective ones, shared by those educated into an agreement about their appropriateness. This seeming objectivity came with conceptual overlays that in their very success had become invisible.

Further, all this made invisible any phenomena that could not by a forced fit be reduced to a mathematical complex of quantities. The secondary qualities (or should we say the tertiary or quaternary qualities?) in their compounding include practically everything that the particular sciences and religion are interested in. Nevertheless they were found by physics to be unreal, epiphenomenal, "ideal," queer subjective additions to the primary facts, which were objective, real, empirical, material, energetic. Now it was all the easier to return to the revolution in explanations and to find it justified. Those Newtonian secondary explanations—goals and plans—were only apparent. The real explanations were the primary ones: matter and its motions.

The Divine Architect

The prescriptions of mechanism and its proscriptions of subjectivity, teleology, and God could seem undefeatable. What began as a dream—the scientifically explainable world—seemed about to become the nightmare hardworld of secular determinism. But religion also managed considerable accommodation and counterattack. The word "machine" only deceptively belongs purely to physical science. More carefully considered, all the machines that we routinely know are psychological or biological products. Sewing or computing machines have functions. If the organic body is machinelike, it also functions. A machine that is not a machine for anything is not really a machine. One can always ask *why* machines work, as well as *how*, and so be led from structure to design, from material and efficient causes back again to formal and final causes, to life and mind.

That is obvious where life and mind are at work, but also the astronomical and geological worlds, though they have been found to work like machines, are not machines for nothing. They have no intrinsic purposes, but they have extrinsic purposes. They are the machines of God. One can in a short-range perspective speak of suns and moons, rivers and volcanoes, eroding rocks or crystallizing minerals as machines without functions. But on a larger scale these are parts in an earthen and cosmic whole that does have both purpose and design. Mechanisms are really arrangements whereby parts cooperate toward a desired result.

In this perspective, the appraising of a great world machine, so far from eliminating God, rather demands a Divine Architect, a Prime Mover. The celestial heavens in their mathematical intelligibility have not lost their beauty; they have gained more. If Earth is one among the other planets, how admirably it is endowed with water and light, with atmosphere and climate, continents and oceans, with elements well proportioned for life, with soils and rivers, minerals and ores! How well populated it is with a largess of life, with every living organism splendidly fitted to its environment! The spacious firmament on high still has its awe, the ecosystemic Earth still has its providential goodness. So there is room for religion, for God.

Indeed, to turn the tables, it was monotheism that launched the coming of physical science, for it premised an intelligible world, sacred but disenchanted, a world with a blueprint, which was therefore open to the searches of the scientists. The great pioneers in physics—Newton, Galileo, Kepler, Copernicus—devoutly believed themselves called to find evidences of God in the physical world. Even Einstein, much later and in a different era, when puzzling over how space and time were made, used to ask himself how God would have arranged the matter. A universe of such beauty, an Earth given over to life and to culture—such phenomena imply a transcending power adequate to account for these productive workings in the world.

At this point, cosmological and teleological arguments for God regained their force. A really perceptive mind could follow the natural signs back, up, and on to God, the only sufficient Cause for this effect. To anticipate an idiom of computer science, every intelligent program points to a Programmer. Especially in those early stages when they found order only approximately, the founders of mechanism often believed in the well-ordered world just because they believed in a theological harmony beyond a mathematical one. They found the order their religion helped them to value and led them to hope for. The smoothing out of their curves (Figures 2.2 and 2.3) was done for theological reasons combined with scientific ones.

As an added boon, *rational religion* can replace *revealed religion* with some sense of relief. Faith, while not eliminated, is no longer dependent upon supernatural phenomena, such as the miracles of a Savior or prophetic oracles. Nature is a Bible of God; the heavens show a divine mathematics; the Earth is a sacrament that recalls him. There is "the book of God's Word," but there is also, said Francis Bacon, "the book of God's Works."[6] Bishop William Paley's watch and his Watchmaker God superbly illustrate the religious response to Newtonian science.[7] But the reasonableness of his argument about the perfection of creation, an ideal that often led to overlooking the actual erratic disharmonies in nature, would greatly trouble Paley's disciples a century later, when Darwin's evolutionary model, with its blind randomness, broke over them.

Notice that the logic of religious thinking here is quite consistent with the logic of science. Every theory (T) is constructed to explain observations (O), adjusted to accommodate new experiences, and corroborated by backtracking. If T (theism), then O (the world as observed). The set of world observations shifted dramatically with the scientific revolution. An observed world mechanism (O_m) replaced a medieval Aristotelian scene. Could classical theism, which had accounted for that earlier-experienced world, still stand in the face of such a redescription? It could, and did; but not without revisions that pushed the medieval monotheism toward deism (T_d), while retaining the basic theism. If T_d, then O_m.

Many of these revisions are positive in tone. There is enlarged size and grandeur to the universe, and its mathematical character enriches God's perfection. The cosmological and teleological arguments offer a more rational approach to God, recalling always that such arguments never offer hard proof, for neither the scientific nor theological logics offer conclusive proof, but only corroboration. Laplace to the contrary, the hypothesis of a deistic God rationally implies the world as observed. Meanwhile, too, we want to notice that the observed world mechanism (O_m) was itself the product of another theory—the theory that material and efficient causes were primary, with its corollary that formal and final causes were secondary. So we have theism adjusting to scientific observations that are not quite the hard facts they seem to be, but are partially and importantly the products of decisions about what to look for and what to overlook.

Some of the theological adjustments are negative, and there are anomalies, some of which can feasibly be erased, with others more recalcitrant. The Divine Mechanic who built this world mechanism—will he not be rather remote? Is he as active in national or personal affairs as was Yahweh of the Hebrews? Of course, there is still some cloudiness in the clockwork. Here the intervention of God can be posited at gaps in the machinery, here perhaps even intercessory prayer is allowed. Again, what is the role of the human spirit in dialogue with the Eternal Thou, surrounded by such a spiritless environment? Above all, what is the place of freedom and sin, of grace and responsibility, in this neutral and amoral machinelike world? Much new wine was being poured into old wineskins. Would they burst?

2. QUANTUM MECHANICS AND INDETERMINACY

That question was never conclusively answered. Before the Newtonian paradigm fully tested the capacity of religion to absorb it, a surprise developed, jolting to those whose mechanism tolerated surprise poorly: the Newtonian paradigm itself burst in the theories and observations of twentieth-century physical science. Determinism has remained at least statistically as a methodological assumption for all science, but no longer with the ironclad certainty of earlier science. This paradigm shift began about the turn of the century with certain anomalies that were to prove revolutionary. These centered at the start around the phenomenon of radiation. Radiation was first a mysterious energy spontaneously delivered out of matter, as when radium emits X-rays. But soon it proved a powerful observational tool, which, like the telescope centuries before, enabled scientists to see what could not before be seen, only this time the peering was not at things large and heavenly, but at things small and material. The earlier physics had found everything in the heavens above to be material and deterministic. It had claimed that all the activities in our Earth environment were matter-in-motion. By the account of a maturing physics, what was matter now to prove like, deep down inside?

Electronic Characteristics of Matter

One can make a diffraction grating for light by drawing lines closely parallel, and with such a spectroscope analyze the radiation emitted from incandescent matter, which registers as line spectra. If one draws another set of parallel lines perpendicular to the first, one will get dots arranged in circular patterns. But can one make a really small diffraction grating suitable for use with X-rays? Fortunately, nature has already done this for us. A crystal is just such a diffraction grating, in three dimensions. We can use it to analyze the radiation—but we can also use the radiation to analyze the crystal; by this technique windows began to open into the structure of matter, in the work of Ernest Rutherford and Lawrence Bragg.

But it was not only molecular and crystalline matter that could so be studied. The atom itself, when bombarded by suitable and more powerful radiation, can be forced to yield its subatomic structures. What theories will account for these arrays of radiation observed as spectral lines and dots, or, in other experiments, as streaks on photographic plates or clicks in Geiger counters? It looked in the beginning as though the atom were somewhat like a little solar system, after a model proposed by Niels Bohr (Figure 2.4). To take the simplest atom, hydrogen, the electron was in perhaps circular orbit around a central nucleus. Atoms become excited under bombardment; they go into higher-energy states as they absorb some of the incident radiation, and they may later emit radiation. The raised energy levels have larger

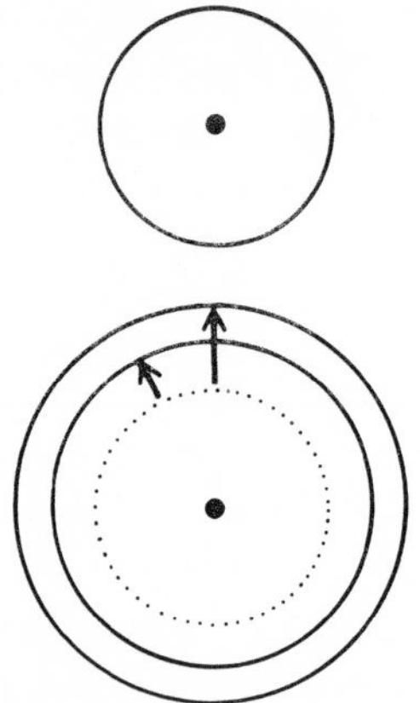

Figure 2.4 The Bohr atom

radii, or orbits, with the atom returning after emission to its ground state. So far so good. Physics had already learned in Newton's classical mechanics how to calculate orbits.

But a problem arises. The moon has an apparently fixed orbit around Earth, though it was in ancient times much nearer and revolved faster in the intermediate positions. But in the planetary model of the atom, for causes initially anomalous to Newtonian theory, we must posit privileged orbits. At these minute levels, energy is quantized, perforated with jumps (quanta), hence the discrete spectral lines that always show up in our analyses. In macroscopic use we can continuously vary the frequency from a tunable emitter, as with relatively long waves in radio. But this is not so in atomic work, with the relatively short wavelengths of electron radiation from a specific atomic emitter. A violin will yield any frequency, but a piano yields notes only by steps, without intermediates. The excited atom proves more like the piano. In watching a child swinging, we see her continuously move; but in watching a motion picture of her swinging, if we look closely at the slowed action, we see her jump through an arc at a foot per frame, without intermediate stages. The orbiting electron seems more like the film, jumping orbits, with the resulting spectral lines nowhere letting us see electrons in between. The jumps are always multiples of what has come to be a celebrated number in microphysics, Planck's constant (h), named after Max Planck, who discovered it. But one would think the real world would be more like the child in the swing than the movie of her. A little cloudiness begins to appear in the clockwork atom.

Albert Einstein and Arthur Compton had noticed that light, usually thought of as a wave, could on occasion, as with the photoelectric effect, be better understood if considered a particle. Louis de Broglie ventured the reverse suggestion that an electron, hitherto thought of as an orbiting particle, might be thought of as a wave. When Erwin Schrödinger worked out the necessary equations, adapting standard wave equations from classical mechanics, supposing that the electron formed a sort of standing wave about the nucleus, the orbital lengths of the circumference had somehow to fit the integral multiples of the wavelengths needed to form various

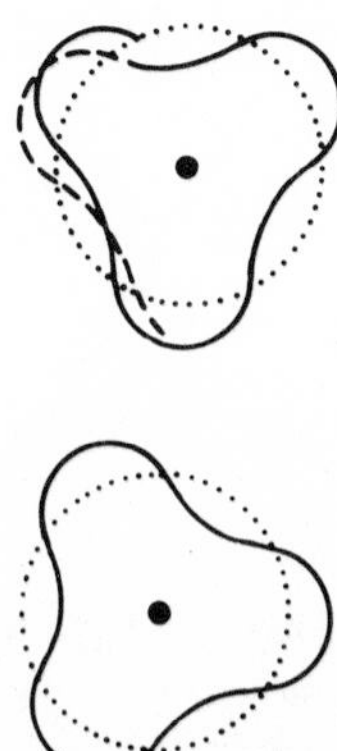

Figure 2.5 The electron as a stable and unstable wave[8]

standing waves, revealed in the discrete spectral lines. It was no longer surprising that there were no intermediate orbits, since an odd length would prove self-destructive for any standing wave (Figure 2.5).

But this was not the end. Already the electron, once a particle, was being "smeared out" (in a phrase of Schrödinger's) as a wave about the nucleus, and this not only in one plane but entirely surrounding it. Some physicists have tried to call it a *wavicle* (!), but that cloudy notion suggests an ontological unity about elementary particles that complementarity argues against. The electron was to prove, by yet a further account, a sort of cloud, alike a charge cloud and a sort of little clock. The notion of circular orbits and discrete radii proved impossible. Circular orbits are not permitted in later quantum theory. Electrons are more or less everywhere about the nucleus; sometimes even inside a large nucleus (though not at its center), sometimes very far away. The standing waves and specific orbital lengths were only one way of looking at it. We portray the electron in still another way as a charge cloud enveloping the nucleus, without the hard edges of a precise radius (Figure 2.6).

If we still want to think of the electron as having a precise location somewhere about the nucleus, we can only draw a probability pattern, showing that the nearer we look at a certain distance (the radius of the old Bohr orbit), the likelier we are to find it there, and the farther away from the nucleus we go, or, alternatively, the closer in to it, the less chance we have of finding it there (Figure 2.7). We have a

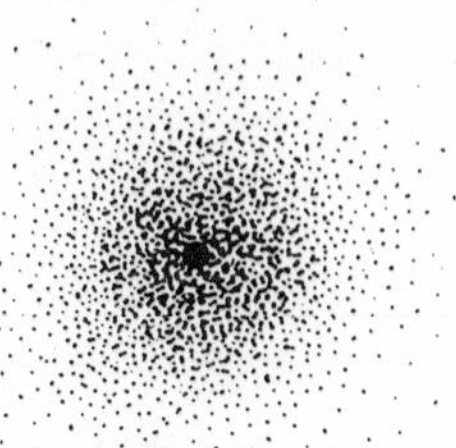

Figure 2.6 The charge cloud in a simple atom

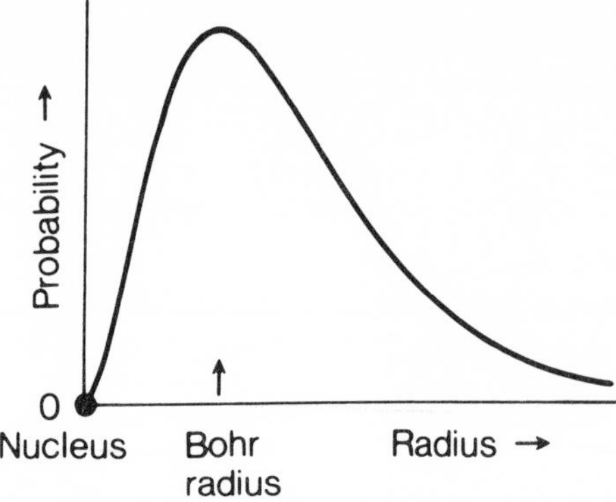

Figure 2.7 The probability of finding the electron at a definite radius in a simple atom

smooth curve this time (unlike Figures 2.2 and 2.3), only because we are graphing possibilities. What any particular electron is doing at any moment may be something else. This is partly suggested, partly hidden in the fact that Schrödinger's waves, like ocean waves and acoustical waves in classical mechanics, have an amplitude (related to what he called ψ functions), but just what this amplitude means is by no means clear. It has something to do with the probability of an electron's being more or less distant from the nucleus. The electron does seem to be moving, but it is no longer clear that the notion of an orbiting electron was the right one. By now it is less clear that we are giving, or ought to give, the electron an ontological interpretation at all, as though we were describing something real. We may be just reporting methods we are discovering for relating to whatever is there.

So far we have tried to describe only the simplest atom. The common elements on Earth have dozens of electrons, which go into multiple shells occupying more or less the same space as the first one, since the atom is, for all the seeming density of the charge cloud representing the probabilities of electron location, mostly empty space. These electrons, as charge clouds and waves, have increasingly complex interactions with each other and with a nucleus that also becomes more complex as it is built larger. The nucleus itself, in fact, proves even more complex than the electron shell, and here too the particles, mostly protons and neutrons, are simultaneously to be thought of as waves, with protons charged although neutrons are not, all of them having probability locations. They are bound together by dramatic new forces, the nuclear forces, which are so short-range as never to be manifest macroscopically. Analogously to the electrons in their shells, the nucleus too emits and absorbs radiation and is subject to excitation.

Uncertainty and Indeterminacy

At all atomic ranges of operation, we must add the uncertainty principle discovered by Werner Heisenberg. If electrons are converted into a beam and directed through a single narrow slit, they behave about like little Newtonian bullets and produce a single spot on a photographic plate beyond. But if the beam is directed at two slits, which form a diffraction grating, the beam produces spectral bands. The electrons reveal themselves as waves, which by now is not surprising. What is surprising is experiments that attempt to make this experiment more precise, as by narrowing the distance between the slits, yielding more precise information on an electron's location, or by arranging for sharper spectral lines, yielding better information about the frequency, energy, velocity, and momentum of the electron. We are here forced into a trade-off in which, for the case of electron position and momentum, we can have only limited information about each, or complete information about the one with complete loss of information about the other. We cannot (to put the point in ordinary language) know with accuracy both where it is and where it is going.

Heisenberg concluded that the uncertainty in the position multiplied by the uncertainty in the momentum could not be less than a function of Planck's constant. The result is that the clearer an electron's position is, the fuzzier its momentum; the clearer its momentum, the fuzzier its position. A similar uncertainty appears if we try to get a precise hold on energy and time. These considerations apply not only to electrons but to all the atomic particles (indeed, to all macroscopic particles, where effects are vanishingly small), with the result that we cannot get a close access, a tight hold on subatomic nature. The cloudiness somehow lies hidden by that quantum character, a jumpiness masking intermediates, symbolized by Planck's constant. What in the Newtonian account would be termed a margin of error, and expected later to be eliminated by closer observation, now becomes an observational limit, a barrier beyond which we can peer only with our capacity for resolution limited. From behind the limit nature emits information only in jumps.

The phenomena of physical science must above all else, we have said, yield a number. But at this point atomic particles will not yield such numbers, not both simultaneously and with enough precision to allow a Newtonian to calculate in detail the affairs of these bits of matter-in-motion. Worse yet, just what the state of affairs is within these wavicle clouds is becoming increasingly uncertain, despite the new wealth of information, since so many of the answers come in terms of belts and bands, in functions reciprocally related to something else, in statistical averages, in interrelated probability distributions, and all with a certain looseness.

Imagine, in a thought experiment, a shrinking blue ball.[9] We can watch it shrink in a conventional way until it becomes invisible to the naked eye. Then we can use a microscope until it has shrunk a thousand times smaller, when it has become so small that Earth's gravity ceases to have much effect on it, and some surprises begin. Blue light has a wavelength of 470 nanometers, so the ball, when shrunk to the range of viruses (100–200 nanometers) has become too small to have color. Or temperature. That is discomfiting, for one hardly knows how to visualize an entirely colorless object, but it confirms the claim that color and temperature are only secondary, not properties of balls large or small.

Although the ball has become too small to reach with any visible light, such light being too long-waved, we can watch it shrink further to megamolecular ranges with an electron microscope, since electrons can be made very short-waved. So we follow its size and shape, while trying to shake off the feeling that the once-blue ball is now gray, the color of its supposed "photograph" as supplied by the electron microscope.

When the ball has shrunk to the size of a few thousand atoms, we can still get a picture of it, and maybe even (with the latest electron microscopes) a representation of sorts at atomic size. But are we getting a picture, or something more like shadows of it, an aura cast by the electron cloud? The electron is very much smaller than an atom, and this might mean that one day we could image the parts of an atom, although these images are likely to remain cloudy, since we are dealing with wave clouds. Even where atoms cannot be seen, their effects can be seen, as with the impacts of atoms causing Brownian motion, the haphazard drifting of tiny particles suspended in a liquid.

When the shrinking ball has become roughly electron-sized, how shall we reach it? The injected beam of electrons dislocates and disrupts the investigated "ball," as if one were to investigate Ping-Pong balls by shooting bullets at them. Shorter-wavelength radiation yet is available, gamma rays, only the recoil problems grow worse. The once-blue ball has shrunk smaller than we can "see" by any means, in principle or in practice, and all subsequent evidence of its existence is circumstantial or indirect. The data coming out of it halfway suggest that it is not a ball, no longer just or always a particle, but equally and alternately a wave, or a probability cloud. With this last shrinking, almost all its remaining qualities become diffused.

We lose touch with its shape and velocity, qualities once thought primary and objectively there. Perhaps they are still there, only so minute that we cannot get at them? But that supposition is only partially consistent with the observations, since we do not any longer have observations, or much hope of later attaining them, that yield these qualities precisely. Electron microscopes, like all other instruments, are subject to Heisenberg's uncertainties. We can believe that such qualities are there only by extrapolating our experience at upper levels down to that level. But perhaps, just as the shrinking blue ball has lost color and temperature, it has lost the capacity to have size and shape, velocity and position, except as these are observer-introduced?

Observer Involvement and Loss of Picturability

The upshot of these observational difficulties is a puzzling blend of the *loss of picturability* with *observer involvement*, each owing to and inseparable from the other. Perhaps paleontologists can remain naïve realists when they reconstruct a dinosaur from fossil fragments. They believe their model directly pictures, replicates, the extinct animal that they postulate. (Even so, they have to guess about its color.) But physical scientists must here become, at best, critical realists, since our images of the submicroscopic world are no more than inexact models, symbols. Indeed, still more skeptical accounts and agnostic conclusions are possible. Our pictures are not so much of nature as of our interactions with nature. Like rainbows, electrons are a product of an interrelationship between the observed and the observer. The microphenomenal electrons can be made to "come through" in terms of our macroscopic, Newtonian concepts—a wave, a particle, position, velocity. But every inquiry becomes something of an inquisition, and torments what we wish to observe. Every inquiry becomes something of a translation, and interprets, as well as interrupts, what we observe.

We are unsure that "spin" as applied to these particles, sometimes thought of as a turning on their own axis, has any macroscopic analogues at all. We have no picture whatsoever of what is meant, in more recent quantum theory, by "strangeness conservation," or by the "quarks" almost comically said to have their "flavors"—"up, down, sideways, and charmed"! Quarks are supposed to be subconstituents of the particles, but the theory expects no free quarks to be detected (though researchers sometimes think they have found free quarks), because energy enough to blast a particle apart into its quarks is energy enough to make new, additional particles that at once combine with any would-be-free quarks. So quarks are always hidden, incorporated into something else. So far from demanding their observability, this theory predicts their nonobservability, and hence expects little in the way of their isolated description.

It looks as though an electron may not have precise location or momentum; indeed, it may not have any location or momentum at all in macroscopic senses of these terms, at least not until we give it one or the other of these, or some indeterminate mixture of the two, by the conditions of our experiment. When our observations are not taking place, these quantities are not always or usually fixed, and even observations do not fix simultaneously all the qualities present. The electron, and all other atomic entities, are really what are called *superpositions of quantum states,* nested sets of possibilities as regards their forthcoming world lines, partly indeterminate matrices, partly determined this way and not that way by, among other things, the demands of observation. In the involvements of these observations, they are at once coagulated or translated over into what we call forth, and known symbolically through our models that make an analogical connection between the microscopic and the macroscopic worlds. We are not exactly mapping the microworld, but somewhat making it. We make the glasses through which we darkly see, but also partly make the events we see. Sometimes, too, this knowing shows up as a complementarity (as of waves and particles, or of position and momentum, energy and time) in quantized epistemic units, which is really a kind of ignorance and cognitive dissonance.

One is reminded of how theologians suppose that God "comes through" in terms of our human concepts—a person, a father, a judge, a spirit, analogically applied. Agnostics at that level also doubt that these categories are genuinely informative, but instead are, like rainbows, mostly projections. The problem that began as an *instru-mental* one, that of building more precise measuring instruments to gain access to formerly inaccessible data, has become also a *mental* one, the failure of any macroscopic models nonsymbolically to describe the microscopic world. What started as an empirical cloudiness is now a theoretical epistemic indeterminacy. The lack of our capacity experimentally to interact without modifying what we study has shown how our theories are linked with our observations rather more indirectly and rather more inseparably than we had hoped.

Status of Determinism

Coupled with the loss of picturability and with observer involvement, we have also to worry about the *status of determinism.*[10] How far Heisenberg's uncertainty challenges determinism—the logical backbone of all science—is itself uncertain! At a minimum it revises the status of that assumption. If determinism is true, then ever more precise observations might be expected to follow. But Heisenberg has observed limits to this precision. We cannot refine our observations below the grain of the quantum, and what theory best predicts these revised observations? There are three possibilities, and deciding among them provides an excellent example of the difficulty of scientific judgment on the frontiers of knowledge. The case is tougher than that described in Chapter 1, of knowing whether to hold on to Newton's theory before anomalies in the behaviors of Uranus and Mercury. We can here refine our observations no further. Notice that there is really no disagreement on the data, only on how it should be interpreted.

(1) *Subjective indeterminacy.* Perhaps the old paradigm holds. Determinism is true, and the uncertain observations will prove a matter of temporary human ignorance. It is difficult in the present state of our knowledge to imagine, much less predict, how the observational barrier will eventually be broken, but the history of science is full of the later doing of what was earlier thought impossible. So, extending the old paradigm by coupling it with inductive prediction based on past experience, one can plausibly hold that physics will eventually be able to see where it cannot now see with accuracy. It will later attain the now-denied precision. Determinism is to be reaffirmed, despite the new, anomalous evidence, on the strength of its previous appeal. Einstein, Planck, and de Broglie so believed. There have been a number of efforts to show that there are "hidden variables," inaccessible but nevertheless exactly defined, yet with some indirect evidence of this fact. These proposals and their experimental verifications have remained inconclusive. David Bohm is a contemporary advocate of objective determinism in nature, though this is decidedly a minority position in contemporary physics.

(2) *Experimental and/or conceptual indeterminacy.* Perhaps our incapacity to overcome the observational uncertainty is permanent, and accuracy in the world of the very small will be forever denied us. Indeterminism is a kind of veil, which may be doubly woven. (a) It is in part an experimental veil. We disturb what we measure. While at macroscopic levels that distortion can be allowed for, it cannot here be inferred past, not entirely. Like a blind person who seeks to examine a snowflake, which melts when touched, we transform these microevents through our touching of them. We cannot see what they are like unobserved. They might or might not be determined; here we must be agnostic. (b) Perhaps the veil is also conceptual, not merely a matter of instrumental reach, but of the structures of the mind, which has evolved for life at the gross level, and is unable to conceptualize, except analogically, and hence imprecisely, what reality is like at its finest grain. What was first observed experimentally has in fact turned out to be implied from within our theories about waves and particles, and we may be hitting logical limits not less than empirical ones. Bohr held a position like this.

(3) *Objective indeterminacy.* Perhaps the uncertainty observed is a result of objective indeterminacy in nature. Determinism is thus ultimately false, at this level, although of course it is an approximation of the truth and mostly true at gross levels. Scientists are still well advised to seek all the determinacies that they can find. (We shall presently see that, in relativity theory, temporal simultaneity is ultimately false,

although it is approximately true in common life.) This does not mean that no causes operate in such microscopic events, only there is no set of causes that uniquely determines the outcome. Some genuine spontaneity and plasticity appear within fixed parameters, and this is not so much hidden as revealed by these quantum jumps. The system may in fact be open, partially undefined until it later closes, or until our participatory probing forces it to close this way or that.

Uncertainty is not a veil; it is an observation of the way things are. Strict determinism is not demonstrable here, and those who continue to believe in it do so without evidence from that realm itself; rather, they merely extend there an old paradigm. Here indeterminism is equally as warranted, indeed more so, since the data are in fact indeterminate. Einstein particularly disliked this possibility. "I . . . am convinced that *He* [God] is not playing at dice."[11] It left his universe too unsimple, too irrational. Nevertheless, by far the majority of interpreters consider an unqualified determinism no longer supported by the data and not requisite for science. Heisenberg, Henry Margenau, and Milič Čapek are among the many who have adopted this theory, known as the Copenhagen interpretation.

Not only do we lack this precise predictive ability when we disturb an atomic system, as Heisenberg did, finding that we cannot "spy" on it at closer range than is permitted by these quantum jumps; we lack predictive ability when we do not disturb the system, as, for instance, in our inability to specify in emission or absorption exactly when or in what way the orbital jumps of a particular electron within a naturally excited atom will take place. In radioactivity, the Geiger counter merely records spontaneously emitted radioactivity, and such radioactive decay is not known to occur under specific determinacy conditions, although it occurs with probability. Indeed, we find it virtually impossible to interfere with radioactive decay.

Returning now to a data plot left unspecified earlier, and describing it as a plot of radioactive decay across time (Figure 2.8), if the sample is small and the decay occurs in microseconds, we become rather reluctant to smooth out the curve, erasing anomalies. These observations may in fact represent genuine indeterminacies in nature, and thus they will not recur just so if the experiment is repeated, although an average will smooth out the curve. As the shrinking blue ball lost its color, temperature, and contact with gravity, and maybe even its shape and velocity, so

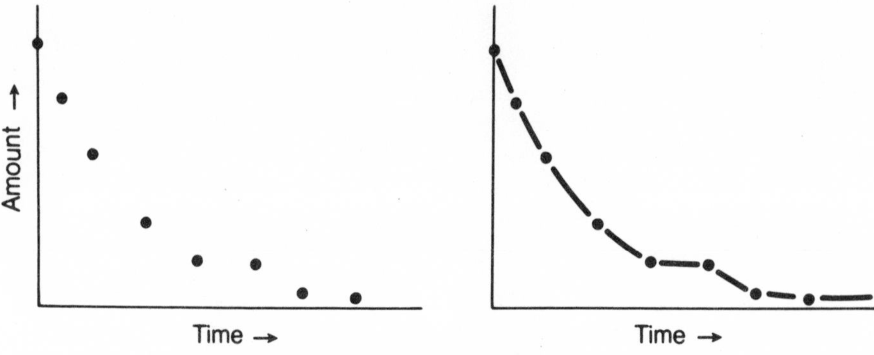

Figure 2.8 Radioactive decay

the subatomic "particle" is too small to have that exact predictability that larger objects can bear, and our graph with its increasing detail needs to show these fluctuations in the averages, the more so as we resolve nature in the finer grain.

Indeed, what we have been saying involves more. Imagine for a moment that the data dots (Figure 2.8) are the flashes from a firefly on a summer night, and that we are trying to plot the "world line" of the bug. We hardly have reason to prefer one set of routes over another (Figure 2.9) within the general constraints of the dots, and really no reason at all to smooth out the curve. But of course we think that the erratic bug did have an exact position and velocity during all the intermediate moments, when it emitted no light, and we might have taken a flashlight and spotted it in between. But if these dots were the world line of an electron, just this in-between information about position and momentum is what we cannot get, owing to the quantum problem. Indeed, we are now saying that the information is not there to be had in between, not precisely, for the electron does not have an exact position until an investigation or some other interaction demands one of it. It is a mistake to draw any sharp line between the dots, for in between are only nested sets of possibilities, superimposed quantum states, which at the dotted occasions of observation coagulate into these locations.

Randomness and Interaction

Has all this any macroscopic implications? Two seemingly opposite but related questions meet here. One is of *randomness*, the other of purpose or informed control, involving *interaction* phenomena. They both tie back to that revolutionary emphasis on material and efficient causes with which science began. So far as determinism in world events is concerned, indeterminism in the atomic world might have no import for our native ranges of experience. Any uncertainty will always be statistically masked out. Some single macroscopic causal sequence of events ($C_{mac} \rightarrow E_{mac}$) can proceed unaffected by differing microscopic compositions ($A_{mic\text{-}1}$, $A_{mic\text{-}2}$, $A_{mic\text{-}3}$, . . .). The grains can permutate while the gross form remains. Thus, a macrodeterminism remains, despite a microindeterminism. We could have a whole state that is completely defined, although there is a looseness of the parts. It

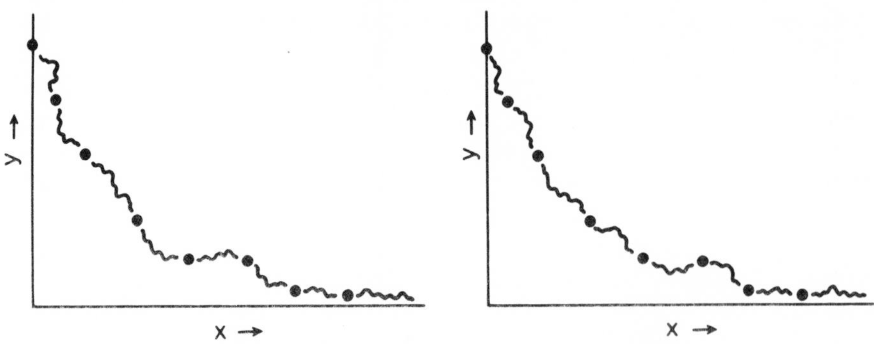

Figure 2.9 A firefly on a summer night

might first seem that one cannot have an exactly specified whole if one has indefinite parts, but on further analysis perhaps the composite levels completely wash out the minute atomic fluctuations. Despite the atomic uncertainties, we can still have clocks accurate to millionths of a second, because the averages are that reliable. Stochastic processes at lower levels are compatible with determinate processes at upper levels. The atomic indeterminacies imply nothing for human affairs or for a broad-scope world view. Statistico-determinism remains, having the same cash value as absolute determinism at everyday levels.

But then again, perhaps there are sometimes gross random effects. The macrodeterminism may be only an approximation, not only overlaid on a microindeterminism, but sometimes affected by it, similar to the way in which Newtonian classical concepts have often proved to be good approximations, but only approximations, of the larger truths of nature. No doubt a sort of statistico-determinism is usually the case, and fortunately so. But not always. In fact, we have not far to seek for evidence that molecular and even atomic phenomena are often amplified.

In biochemistry and genetics, events at the phenotypic level are profoundly affected by events launched at the genotypic level. Such events may sometimes be affected by quantum events, as when random radiation affects point mutations or genetic crossing over. This may affect enzyme functions or regulatory molecules, as when allosteric enzymes, which amplify processes a million times, are in turn regulated by modifier molecules, of which there may be only a few copies in a cell, copies made from a short stretch of DNA, where a few atomic changes can have a dramatic real-life effect. A single base pair altered can shift a whole reading frame. Indeed, by the usual evolutionary account, the entire biological tale is an amplification of increments, where microscopic mutations are edited by macroscopic selective processes. These increments are most finely resolved into molecular evolutions.

"If Cleopatra's nose had been longer, the history of the world might have been different!"[12] Might a single random mutation, and a consequent shift in reading frames, have affected the length of her nose? That supposition is extreme, but it illustrates how history is not without contingencies; nor is it unthinkable that some of these may reach down to the molecular and atomic levels.

In the case of the molecular basis of thought and mind—and hence what we notice and neglect, what pops into our mind or catches our attention, what inspirational ideas are launched, and what we remember out of our past to combine with new ideas—we may well think that everyday life can be affected by events at the molecular level. There will be some selection pressures to pack cognitive processing abilities and to store information in memory in as economic, lightweight, and miniaturized a form as possible, and this will drive thought and memory down to the microscopic levels (just as computers have been made smaller and more efficient by going down to microscopic operational levels). This will have the serendipitous effect of putting thought in contact with the open, experimental possibilities derived from quantum effects triggering thought mutations, though a brain will also need long-term memory stability and reliability. The brain has evolved as an extremely sensitive instrument and a powerful amplifier, in which microscopic physicochemical changes down to the molecular and atomic levels cannot be dismissed as unimportant. Especially near thresholds, microscopic variations can have great macroscopic results.

In the case of the triggering of molecular diseases, including sometimes cancer and diabetes, we simply do not know whether microscopic quirks can play parts in the triggering or the timing of macroscopic affairs, as with the death of a leader. At least the Newtonian paradigm has lost its iron grip. What sort of openness remains we cannot as yet establish; but sooner or later, and beginning in physics, a little cloudiness appears in all the clockwork.

Rigorous determinism is also being questioned even in large-scale inanimate natural systems, for instance, in what is called irreversible thermodynamics or in climatic trends.[13] In the case of large-scale astronomical systems, it may be (considering the finite speed of light, black holes, etc.) that we can never get the information we need to settle the question, and we thus run up against a sort of cosmic indeterminacy to match the quantum one. In cosmology there is a good deal of talk of so-called astronomical singularities, where the laws of physics run into such extreme conditions that nothing is predictable. Indeed, in the speculations of cosmologists about quantum effects in the initial big bang, when the macrocosm was a microcosm (the so-called Planck era, 10^{-43} seconds after time 0), quantum effects triggered the density perturbations that have resulted in galaxies later forming when and where they did![14] On the global scale, we must further notice that some major processes in ecosystems and in meteorological systems have been historically altered by genetic mutations, such as the oxygen content of the atmosphere with the evolution of photosynthesis, affecting the ozone layer and radiation levels.

From the foregoing it might seem only that world affairs are occasionally or even dramatically altered at the whim of random quantum events. That much would be puzzling, perhaps dismaying, although we should wonder that it is from what first seems a kind of microscopic background noise and static that all the creativity in the world is emitted. But if we turn from the *random* element of indeterminacy to the *interaction* concept also present, we gain a complementary picture. We are given a nature that is not just indeterminate in random ways but is plastic enough for an organism to work its program on, for a mind to work its will on. Indeterminacy does not in any straightforward way yield either function, purpose, or freedom, as critics of too swiftly drawn conclusions here are right to observe. Yet physics is, as it were, leaving room in nature for what biology, psychology, social science, and religion may want to insert, those emergent levels of structure and experience that operate despite the quantum indeterminacies and even because of them. We gain space for the higher phenomena that physics had elected to leave out.

Organisms in the Microphysical World

Consider the phenomenon of organism. A laboratory apparatus that humans have fabricated can constitute the conditions under which some phenomena appear, and within those conditions we can further coagulate this and not that one from among the superposed quantum states. So the actual phenomena that come to pass are interaction phenomena, as well as being, in other ways, random phenomena. Likewise, an organismic "apparatus," though it has naturally evolved, has evolved to the point where it can constitute the conditions under which the phenomena with which it interacts appear. Within this interaction, it can coagulate affairs this way and not that way, in accord with its cellular and genetic programs. The macromolecular system of the living cell, like the physicist's apparatus, is influencing by its interaction patterns the behavior of the atomic systems.

As a matter of fact, this is probably going on in a much more sophisticated way than it does in the relatively crude physicist's machinery that converts the atomic events into a photographic trace or a Geiger counter click. The organism converts the phenomena into life. This is taking place with instrumental control much closer to the atomic level in a pervasive, systematically integrated way in the organism, while in the bulky physicist's apparatus we can manipulate processes and fabricate the materials of our instruments directly at the gross macroscopic levels only, very indirectly at the molecular levels. But the organism is fine-tuned at the molecular level to nurse its way through the quantum states by electron transport, proton pumping, selective ion permeability, DNA encoding, and the like. In physics, the concepts and instruments of observation participate in forming the course of the world that is observed. Physicists are responsible, in part, for the microevents they observe. But this is likewise true in biology in an even more startling way, for there the organism via its information and biochemistries participates in forming the course of the microevents that constitute its passage through the world. The organism is responsible, in part, for the microevents, and not the other way around.

Both physicists and organisms have executive power as they face future phenomena. The phenomena appear within the frame of reference by which they are called forth or selected. We are getting a different kind of natural selection over a different order of "mutations." To some extent we face just the random bubbling up of indeterminacies, but we find also the drawing forth from an indeterminate substrate of just those determinations that serve the organism. The organism has to flow through the quantum states, but the organism selects the quantum states that achieve for it an informed flow-through. The information within the organism enables it to act as a preference sieve through the quantum states, by interaction sometimes causing quantum events, sometimes catching individual chance events that serve its program, and thereby the organism maintains its life course.

Now we have again, fortunately and in a different way, a whole that persists through the permutations of the parts. It is not now a whole that is statistico-determined, but one that is program-laden, a whole that executes its life style in dependence on this looseness in its parts.[15] There is a kind of downward causation that complements an upward causation, and both feed on the openness, if also the order, in the atomic substructures. We do not have (what the Newtonians might have looked for) a macrodeterminism through microdeterminism, nor do we have (what the statistical determinists supposed) a macrodeterminism despite a microindeterminism, with averages invariant despite the fluctuations of the parts. We have a macrodeterminism, but of an organically "determined" (programmed) kind just because the microindeterminism permits it, providing a looseness through which the organism can steer itself by taking advantage of the fluctuations at the micro levels.

Life makes matter count. It loads the dice. The throttling and interrupting of events is not by physical processes that preset or break up biological events; the throttling and interrupting is much the other way around. Biological events are superintending physical ones. The organism is "telling nature where to go." Biological nature takes advantage of physical nature. Physics can elect to leave out this upstairs control, and it will see everything as diverse energy transformations, which can be brought under a minute yet quantized laboratory analysis. But there is ample space for the specialized sciences to fit in their living, mental, or social phenomena, and for the ordinary course of human affairs to proceed. We may well suspect that mind at, say, the synaptic junctions of the brain, including the memory circuits and molecules, whatever they are, is doing again at yet higher levels what the organism at nonneural levels does merely biologically.

We do not have these processes now described by science, and it may be that, within the limits of the quantum principle, they are inherently inaccessible and indeterminate. Quantum biology, quantum brain physiology, and biophysics may have to remain blurred sciences. But the point is that, under present conceptions of physical nature, this is a likely story, a believable one. Or, if one holds any account to the contrary, as with an old-fashioned determinism that hopes for a complete account of mechanistic organic and neural processes, this too has to be believed beyond the observations, and often contrary to them.

3. RELATIVITY AND MATTER-ENERGY

The quantum and Planck's constant led to lower limits on our resolving powers in observation, but this is only half the story. Another constant, the speed of light (c), soon led to upper limits on our capacities for observation. These effects were first explored by Albert Einstein, and we notice at once that even in Einstein's universe (although Einstein disliked the indeterminacies imposed by Planck's constant) we have an indeterminacy of a different kind—the impossibility of getting current information across long distances. Further, just as quantum mechanics had incorporated observers into their observations, noticing the inescapable interaction between the submicroscopic wave cloud and its interpreters, now again, only more so, the relativity theory, which ensues from the constancy of the speed of light, inseparably relates observers to any astrophysical or microphysical world that we can observe. These most startling innovations ever advanced in science have thus paradoxically brought us closer to what must be true about physical nature—they better approximate in nonanthropocentric terms what must absolutely be the case in nature—yet they simultaneously bar us from knowing just what is the state of nature objectively, absolutely, independently of the knowing mind.

The Relativity of Time, Mass, Energy, Simultaneity

There are two parts to relativity theory: special relativity and general relativity. Special relativity deals with measurements of distance, time, velocity, mass, energy, relating these to reference frames. General relativity incorporates special relativity but goes further to interpret gravitation as the distortion or curvature of space, requiring non-Euclidean geometry for this description. In what follows we can get a taste of what special relativity is like, and move from this into the philosophical implications of both special and general relativity, although any explanation of the technical detail of these theories is beyond our scope.

Once again, as with Planck, an upsetting constant is discovered in radiation. No matter whether light is a wave or a particle, by Newtonian mechanics we predict that its speed will appear differently to observers at differing velocities (Figure 2.10). A first observer at one velocity (O_1) and a second observer at another (O_2) will differently observe the velocity of either a photon or a wave, since their own differing velocities are subtracted from the velocity of the light ray–particle. A third observer traveling oppositely (O_3) will see the light's velocity increased. But this prediction proves incorrect. The speed of light in a vacuum is a constant (3×10^8 meters/second) that is indifferent to the velocities of observers. The Michelson-Morley experiment (1887) is the most celebrated experiment confirming this.[16]

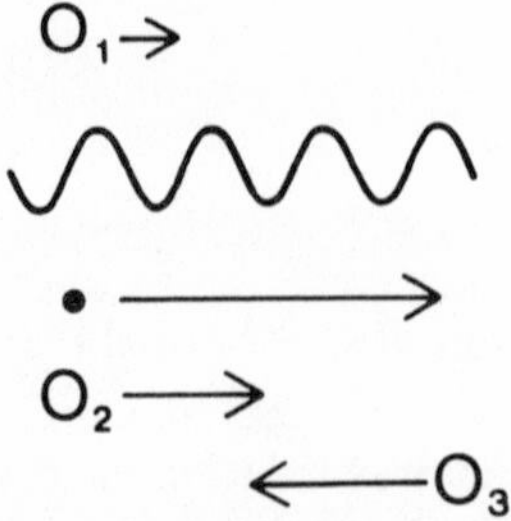

Figure 2.10 Observers of light

If the speed of light is not relative to the observer, then a great deal else is. The detail of such implication is complex, but it is not difficult to get some sense of how this must be so. Consider a cigarette smoker on a moving railroad car, his fellow passenger, and a bystander (Figure 2.11).[17] If we imagine a light ray from the smoker (S) passing to a traveler (T) across the car, the traveler will see the ray coming straight across the car, its velocity computed as the distance (d_T) divided by the time (t_T). A bystander (B) will observe the same path differently, since the movement of the train, of which the passengers take no account, has meanwhile removed the traveling observer from him and thus has accordingly "stretched" the length of the path of the ray (d_B). But traveler and bystander do not differently observe the velocity of the ray. What then of the bystander's observation of the time required for the transit (t_B)?

For the traveler *For the bystander*

$$\frac{d_T}{t_T} = c = \frac{d_B}{t_B}$$

If the speed of light (c) will not give, something else must. The distance (d_T) observed by the traveler is shorter than that observed by the bystander (d_B), and his time (t_T) then must also shorten proportionately to time as observed by the bystander (t_B)! Thus, for the traveler the passage of time is slowed, relative to the

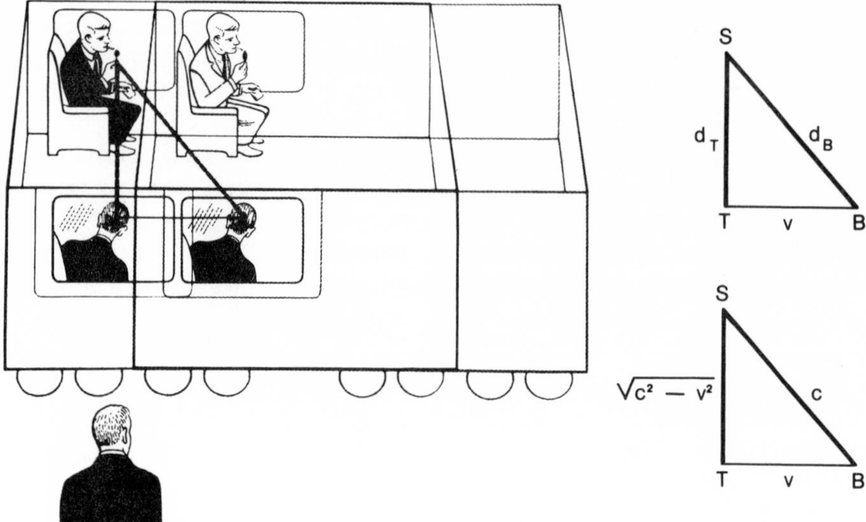

Figure 2.11 Time runs differently for two observers when one is moving relative to the other [18]

bystander. Otherwise, the observed speed of light would not remain the same for all observers.

The ratio involved is easily established from the Pythagorean theorem. If we next choose a unit time and, from the bystander's perspective, let the hypotenuse be the velocity of light (c) and the base be the velocity of the train (v), then the length of the altitude is a function of these ($\sqrt{c^2 - v^2}$). In a unit time these velocities represent the distances involved for the respective observers, as the bystander judges them. These distances are in the ratio $\sqrt{c^2 - v^2}$:c. Hence, the time dilation will have the same ratio. In everyday life the train velocity is so low as to be nearly zero, and the ratio becomes 1:1, which means that no time retardation is observed. If the traveler were in a rocket instead, approaching the speed of light, his time slowdown, as judged by the bystander, would become significant. The first bold and disquieting conclusion of relativity is that time does not run at the same pace for every observer; indeed, it does not run at precisely the same pace for any two observers who are moving relative to each other. While in practice our everyday motions are here inconsequential, it is also true that we find no two observers who are not often moving relative to each other.

But much more is to come. Who is traveling and who is not? Relativity theory also assumes that, at constant velocities in a straight line, we cannot distinguish the traveler from the bystander. From each frame of reference the observer may interpret himself as stationary and the other observer as traveling. At some moment of contact two such observers might synchronize their clocks, but afterward to each the other's clock would appear to be running slowly. Nor can there be any rechecking of simultaneity if the observers will not again meet, and they can communicate with each other only by sending signals, none of which have instant transit time. The fixed velocity (c) of the fastest of these means that the transit time of each such signal will be differently interpreted. If they do meet again, then one must decelerate and reaccelerate to make a round trip, and by this the traveler and the bystander can be distinguished. But even then the traveler will have aged less than his relatively stationary friend, and they must afterward decide whose time frame to use, that of the traveler or that of the bystander to whom he returns. Time is a predicate of the experiencer.

For persons who live on the equator, time passes more slowly than it does for those on the Arctic Circle; indeed, time passes more slowly on the south side of town than it does on the north! The electrons and nucleons in the atoms of which I am composed, if we think of them as moving particles, are traveling at appreciable fractions of the speed of light. An electron typically revolves 10^{16} times per second around the nucleus. Nucleons move within the nucleus at nearly one-quarter the speed of light. My particles, each with its own time frame, are aging less than the macroentity that I am. I am getting old, but I am composed, relatively speaking, of young electrons and nucleons! Simultaneity can be established when observers meet, but only relatively and at the moment of contact. The concept of an absolute simultaneity must be abandoned. There is no universal "now." Rather, every observer carries about with herself not only a flux of time peculiar to herself but also a peculiar "now," however largely or approximately her flux and her "now" may be shared with some others.

Due to the lack of synchronized clocks, certain types of measurements become impossible, since the precise measurement of great distances depends on signals from the end points, and differing observers will get different such signals. By the same effect, velocity (distance/time) will likewise be relative to the observer, as is, of course, distance. The Stanford Linear Accelerator is two miles long. If one were riding in a rocket alongside the electrons that are accelerated in it at their exit speed, the length of the accelerator would contract to 4 cm.[19] We would view the rocket riders as flattened to nearly nothing.

In an outcome of the theory that we can here only report, any measurement of mass also becomes relative. From the viewpoint of a (relatively) stationary observer, if a rest mass (m_r) is given a velocity (v), its traveling mass (m_v) will increase proportionately by the reciprocal ratio of the time dilation.

$$\frac{m_v}{m_r} = \frac{c}{\sqrt{c^2 - v^2}}$$

So the train traveler, if aging less, weighed more than his friend who stayed home, within the bystander's frame of reference. In everyday cases, any velocity here approaches zero, and the ratio becomes one, and no increase in mass is observed. But if the velocity is large, mass increases significantly, and at the speed of light the mass becomes infinite!

But this is to say that any body *with mass* would offer infinite resistance to such motion, and hence that no material entity can be accelerated to the speed of light. An electron, considered as a material particle, cannot be accelerated to this speed.

In accelerators enormous energies have been used in the attempt to drive electrons to higher speeds, but in the upper ranges these accelerate them very little, contrary to the Newtonian predictions, by which, for instance, the Stanford Accelerator would need to be less than an inch long, instead of two miles, to achieve similar speeds.[20] But photons do not have any rest mass at all. A photon has only energy and no rest mass; it is pure energy, nonmassive, immaterial, though it can perhaps be converted into some particle that does have rest mass, if it is slowed or stopped relative to the observer.

Relativity theory further discovers the celebrated interconvertibility of energy (E) and mass (m), related by the speed (c) of light ($E = mc^2$). Where quantum mechanics gives us what some tried (rather vaguely) to call *wavicles* instead of waves and particles, relativity theory gives us what some have tried to call *mattergy* instead of matter and energy, stumbling again for a term that fuses solidity and materiality with dynamism and energy. Nature (sometimes called matter) has two interchangeable aspects, mass and energy, and the real agreement is that everything is *process.* We can conceive of an observer only below the speed of light, where all observers see the speed of light as a constant, at which speed nature is (as it were) in the pure energy mode. When nature takes other forms at lesser speeds, it assumes mass, and can be interpreted as resting material without velocity or with variously intermediate motions and masses, relative to the differing velocities of observers. Energy and mass are dimensions of one process.

Mass is an energy knot. When nature is, so to speak, in the material mode, its observation always includes a relationship to the observer, but here not a single one of the former primary qualities survives. Mass, velocity, length, time—all those qualities that the Newtonians thought to measure objectively, independent of the observer—have become secondary, relative to the observer, altered by one's frame of reference, and hence not absolutely quantifiable. They are still natural facts, but not pure ones. They take the form and numbers they do at the demand of the observer. Once again, we cannot "zero in on" the absolute numbers we need to compute a maximum specification of what is the case, independent of the observer.

Objectivity in Relativity Theory

But we must not be overwhelmed with this relativity of everything, for many things are not observer-dependent. The equations, the basic laws are invariant; they do not depend upon a reference frame, although this means that the detailed observations vary from reference frame to reference frame. Some of the physical constants of nature—the speed of light, the charge on an electron, or the number of atomic shells, the atomic table, chemical reactions, and so on—will presumably be the same for all observers, as will ordinarily be the order of succession of causally related events. Thus, there is considerable objectivity in relativity theory. Although measurements of space and time are relative to the observers, many space-time measurements, fusing the particulars of space and time in local places, are invariant between observers. There is also considerable unity: mass is unified with energy, space with time, gravity with acceleration. Their very relativity shows their interrelationships and thus unifies them, objectively so. These features of the world also remain quite real as phenomena, although they are now correlated with other phenomena; they do not exist intrinsically but only interdependently.

Above all, there is enormous intelligibility and rationality invariant from observer to observer, and this is intersubjectively verifiable no matter what one's reference frame, because it is objective to reference frames and observers, even though this is all discovered and filtered through theories that variously and approximately grasp it. There is a world out there forcing its own objective order upon us, submitting to our tests and sometimes agreeing with, sometimes refuting our theories, even at the cost of overthrowing some of our most treasured intuitions and subjective preferences. This world remains an intelligible, mathematical system independent of, and even when doing violence to, our perceptions and conceptions of it. This is true, however much it is also true that we have no access to that world except wearing the eyeglasses of our subjectively fashioned theories.

Matter and Energy in the Space-Time Field

A new problem now arises when we try to examine what a microparticle might itself be composed of. We hit it with an accelerated particle probe, but what it breaks up into depends on how hard we hit it. It takes some of the striking energy and converts it into a display of derived particles with various energetic and material dimensions. The particle is annihilated and others are created out of the total energy-mass on hand. But what shall we say the original particle was "composed of"?

Further, though, we also have to worry whether these once-primary qualities are primary even in a qualitative, nonmetric way—that is, whether, independently of the macroparticipant, micronatural things are in fact matter-in-motion in space and time, as we conventionally conceive and experience these things. We have even to wonder whether, at the velocity of light, where the time dilation is infinite, there is any such thing as the passage of time at all. Even at lesser speeds we now have little reason to think that there is any flow of time except locally to an observer. There is no absolute time. Simultaneity has vanished. If distance can be had only in contracted lengths relative to the observer's speed, and never absolutely, does even distance exist absolutely and unobserved? Events at any other speeds and places do not occur in "our" space and time, though of course other such space-time frames coexist with ours and they will "come through" by signals in our space-time.

At this point we can reconnect with the question of quantum mechanics, whether those microwave clouds really have precise position or momentum, before these are demanded by the observer. Perhaps events are in space and time not only relatively, but also only "in the aggregate" at the molar level, and not at the quantum range, as a blue ball at that range is not solid, liquid, or gaseous. It has color and temperature only if large enough, while these qualities disappear in the very small. Notice, however, that by this account space and time are real, though relative, constituents of the world at least at our native ranges. They are properties, relative properties of all things that exist at the macrostructural level, whether such things (plants, volcanoes, persons) are mentally endowed or not. The "observer" is not a mind, but a macroentity, such as a clock. (This position differs from that of Immanuel Kant, in which space and time are subjective and introduced by the rational mind.)

Perhaps the most fundamental notion of all is not matter or motion, space or time, but *energetic and evolutionary process,* not being but becoming. There are, absolutely, no things, no substances, but only events in a space-time something, not bodies that move in empty space over time, but a series of moving changes with continuity, forming all the "identity" there is, a relative rather than an absolute identity in an incurably successive world. Matter and motion, space and time, as well as size and shape, color and temperature, wave and particle, light and form—indeed, all the interpenetrating and mutable textures of things in life, mind, culture, and history—are various dimensions of this process.

It is difficult to know what space and time mean at the "bottom" levels, but so far as they do survive there, they are joined into one matrix, space-time. Matter is absorbed into it. The most frequent account, based on general relativity, makes each cloudy wave a kind of wrinkle, bubble, or hill in an omnipresent trans-space-time field, which coagulates relative to each disturbance, to each entity. A particle is not some one substance; it is a traveling concavity in a "plasma,"[21] rather as (to use a crude analogy) a dent travels over the surface of a partially deflated basketball. Matter is, so to speak, "freeze-dried energy." In the Newtonian view, space and time provided a passive and empty container, there independent of any contents, regardless of the matter-in-motion within it. But in Einstein's view, while some kind of plenum remains, evidenced grossly as space-time, it is not passive but is the generator and carrier of all the particle play. C. W. Misner, a theoretical physicist, calls space-time an impressively creative kind of ether. "A vacuum so rich . . . in potentialities cannot properly be called a void; it is really an ether. The entire spacetime fabric . . . from beginning to end" is "a library of unused designs," which are creatively "enacted into existence."[22]

Matter is a crinkle in the matrix, an energetic warp in the great plasma-ether. The phenomena come and go; the particles do their trips and identity flips, taking on the spatiotemporal aspects they yield to observers. But absolutely what, if anything, is there? John A. Wheeler, another theoretical physicist, says, "A much more drastic conclusion emerges out of quantum geometrodynamics and displays itself before our eyes in the machinery of superspace: *there is no such thing as spacetime in the real world of quantum physics.* Spacetime is a classical concept. It is incompatible with the quantum principle. It has to be discarded in any deep-going analysis of the foundations of physics. It is an approximation idea, an extremely good approximation under most circumstances, but always only an approximation."[23] Ultimately, there is only a kind of gauzy foam through which quantized pulses run.

The Newtonian world was a great clock, entirely describable in principle as matter-in-motion. These things traveled absolute distances in absolute time, and were absolutely predictable. But now, by quantum mechanics, there are no perfect clocks. The macroscopic ones always were subject to margins of error, and the microscopic ones, remarkably accurate as they are, are subject eventually to indeterminacies, cumulatively disturbing any exact timing of phenomena over the eons of time or at microscopic ranges. Then again, by relativity theory, there really is no one absolute time to keep. There is no one great clock, no absolute "now" to set. There are only local clocks. Everybody carries about their own time zones.

Twice within four centuries we have had radically to revise our concepts of time and distance. Time expanded from a few thousand years to almost forever across the astronomical reaches of evolutionary time. The universe expanded from a few thousand miles wide to staggering, inconceivable distances across intergalactic space. But then time and distance were, by a second move, made relative to the participant, so that the enormity of space and time is from our local, "slow" frame of reference. A proton with the energy of some observed cosmic ray particles, traveling at nearly the speed of light, would, viewed by Earthlings, take a thousand centuries to cross our galaxy. But in its own time frame this crossing would take but thirty seconds.[24] The cosmic drama has surely been more accurately described by recent physical science, but that description has remained interlocked with the human describer. We are still hard pressed to give an objective account of the elementary nature of nature, whether of micronature or of cosmic nature.

Relativity theory relates things—energy and mass, time and motion, space and time—and this unifying of the world process is welcome. It should be noted, however, that relativity theory also separates things. No information or causal influence can travel faster than light, not at least if borne on physical processes as we know them.[25] Since light does not really travel very fast on our cosmic scales, we are effectively separated from any meaningful contact with present states in the rest of the universe. Travelers carry about with them their own reference frames, their own "nows," which may crisscross but not coincide. From portions of the universe, black holes, where no light can escape the gravitational pull, we can get no information at all.

Even within our own earthen system and reference frame, when we add quantum theory, we could not have predicted on the basis of its past states the present states. Nor can we predict the future with exact specification. Once upon a time astronomers might have supposed that, given all the initial conditions and laws of nature, everything could in principle have been predicted in the great world clock. But almost every feature of this supposition has collapsed. There are only probability laws. No set of initial conditions could ever have been simultaneously assembled in any one place. All predictions would have been nonobjective, theory-laden and relative to the calculating astronomer. No predictions could ever have been simultaneously verified at separated points in the system. The whole dream was a colossal misconception.

Nevertheless, there are remarkable powers in the human minds that attempt to do these things and that succeed as far as we have. From Darwinian theory—the other great scientific revolution of modern times—we should expect the human brain to have been naturally selected as a pragmatic, gutsy survival instrument, clever at gaining food, at emotional satisfactions, at passing on the human genes. However that may be, we here find the human brain to be marvelously perceptive at ranges that are, so to speak, light-years away from pragmatic survival quests. The mind is adept at probing micronature and cosmology, at high-level abstract mathematics, at revising its own intuitive conceptions of space and time with bizarre yet compelling logics. Despite the problem of nonpicturability, we know more than we might

reasonably expect to know. The mind is doing an order of thinking that could never have been naturally selected for, that indeed might almost have been selected against.

Only some of the skills needed for life in the African savannas are transferable to our research institutes for astronomy and particle physics, where such skills are of secondary importance compared to the capacity for, and inclination to, abstractive and speculative thought, which would be useless in the jungle. On the one hand, modern physics presents a number of limits to what the human mind can know. On the other hand, that same physics demonstrates the almost endless frontiers of the mind. Physics impresses us with the minds of its physicists, even though physics elects to leave out the study of the mind. To notice this casts us back again to the question of nature in a new light. What kind of micronature and cosmic nature do we have when the human mind, with its startling powers, can evolve out of it to ask these questions?

God and Relativity

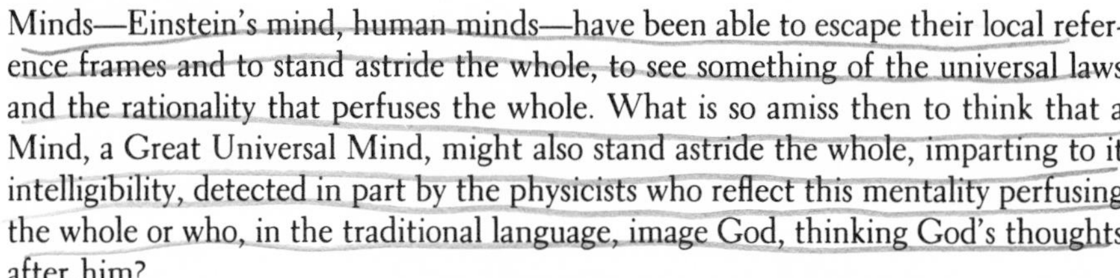

Minds—Einstein's mind, human minds—have been able to escape their local reference frames and to stand astride the whole, to see something of the universal laws and the rationality that perfuses the whole. What is so amiss then to think that a Mind, a Great Universal Mind, might also stand astride the whole, imparting to it intelligibility, detected in part by the physicists who reflect this mentality perfusing the whole or who, in the traditional language, image God, thinking God's thoughts after him?

Some might suppose that any notion of a religious Absolute would collapse under the shocks of relativity and quantum indeterminacy. It is certainly true that, after relativity and quantum mechanics, no theologian would be so bold as to assert that humans can know God absolutely, without observer contribution, without the use of relative models and symbols. But then the wiser ones did not say that even before science. If God is anywhere known, it will be as God "comes through" in our space-time, relative to our local existence, as God is, so to speak, locally incarnate. The divine relativity need not mean that God is not absolutely there at all, but only that God is known in relationship. *God* and *nature* now seem to share this much in common—that each must somehow exist (if at all) absolutely and with real, objective attributes. But we have no access to either except relatively as each is translated over to our own terms, as these local terms can be stretched, extrapolated to grasp something more ultimate than we find at our native ranges. We may have the conviction that nature exists, suspecting that God does not. But at least the

epistemic problem is of the same kind in both inquiries—that of knowing something that transcends our home ranges. That we have done so in physics, however partially, should not prevent, but rather might even encourage, the belief that we can do so in religion.

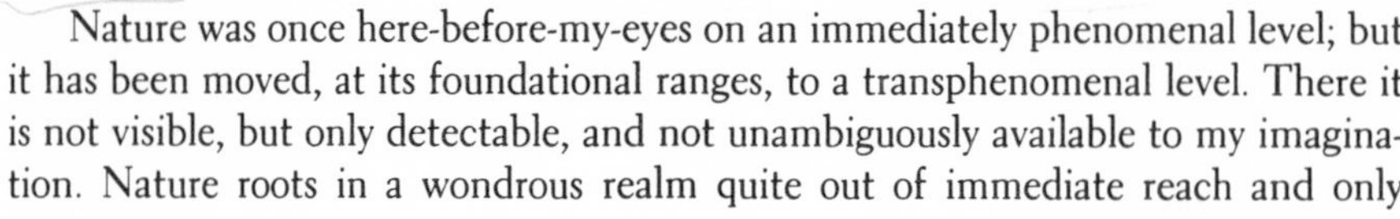

Nature was once here-before-my-eyes on an immediately phenomenal level; but it has been moved, at its foundational ranges, to a transphenomenal level. There it is not visible, but only detectable, and not unambiguously available to my imagination. Nature roots in a wondrous realm quite out of immediate reach and only half-translated into my phenomenal experience, a region into which we gain access by groping out of familiar ranges. But who will then protest, since nature is already transphenomenal, when theologians come along to speak of God (or some Ultimate) in a supraphenomenal way? We can stipulate only that they work back from relevant experiences on the phenomenal level, and ask what hypothetical entity might constructively explain these experiences. By both scientific and religious accounts, what we can see stretches away into what we dimly see and on into what we cannot picture at all. 

But is the hypothesis of God now needed, or permitted? Physics is nontheological, of course; and in that sense one does not need the hypothesis of God to do strict physics. But is physics antitheological? When we pass from physics to metaphysics, after noticing the character of the tale in physics, together with what is left out, in what directions can we travel past these pointers? In one sense, the phenomenal observations from relativity and quantum mechanics have nothing to do with any noumenal divine being. Physical science does not ask or answer this question. In part, its observations even suggest that there are barriers in nature beyond which we cannot go in asking about grounds, beginnings, or fundamental processes, whether subatomic or supracosmic.

Still, perhaps these observations of nature that we have successfully made somehow disprove, or make less credible, the hypothesis of God? The relativistic and indeterminate accounts within physical science might be thought by their character to discourage classical monotheism. They might seem to preclude all possibility of theistic ultimacy, since they allow only randomness, relativity, uncertainty. But perhaps, on the other hand, since nothing absolute, self-explanatory, or eternal has ever been found in nature, the hypothesis of something more primordial still remains an option.

There is certainly no ultimacy in the ultrastructures as now known. We have hit no "rock bottom" in physics, and have few signs that we ever will or can, or would know when we had. At the edges of what we can see, however, we even have a suggestive sort of superspatial model of what God might be like. God too is like the ether in which events bubble up and take place. God is not to be identified with the space-time "plasma" that physics now finds to be fundamental, but lies one or more orders of being beneath it, though analogous with it, since space-time is the divine creation. If we may extend an already comically crude image, God is like the basketball over which the dents travel. Particles, waves, matter-in-motion, stars, planets, persons—all are warps in space-time, but wrinkles in God, too, God's "creations." God—or Brahman, or *śūnyatā,* or the Tao (for we cannot at this level distinguish among the Absolutes)—does not have the individuality of a thing in the world, but is the comprehensive, pervasive, maternal matrix. God is the substrate of the world. Only now we have learned never to think of a stuff, but rather of energetic process and potential.

If we may mix the languages of sacramental religion and recent physics, God is "in, with, and under" the energy pit out of which all comes, the prime mover lurking beneath the scenes. God is in, with, and under the background "noise" out of which all creativity comes, masked by this noise but detectable in what results from it. God is in, with, and under the superposed quantum states, out of which potential all that is actual comes. God is behind the tectonic principles that produce and conserve all the nature that forms over this noise and potential, the nature arising with constancy and variation above the gauzy plasma.

God supplies the infinite and random potential. The world emerges as God not only plays dice, but perhaps loads the dice from below. This does not prevent, but rather makes possible that emergent freedom in the organismic creatures, who through interaction draw this potential into their own programs. So we expect not only creations that form as a result of, and at the whim of, the divine ground below. We posit creatures who achieve a relative independence within their dependence. They move in significant autonomy over the divine subsurface, composing and maintaining their own integrity. Nor should we think of the divine activity merely as a providing ground from below. We are here dealing only with the first-level findings of physics. When the other, later-coming levels (which physics has elected to leave out) are also included, a more robust account may find the divine presence, seen first and minimally as the Ur-ground, later quite omnileveled, superintending each of the emergent stages—life, mind, society, history. God is not the Architect, not the Watchmaker, but the Plasma and the Process—though of course this is only at the physical level, and we have yet to explore conceptions of God consistent with biology, psychology, society, history, and with the sorts of entities that exist at higher structural ranges.

The nature we do know has grown soft. There is something hazy that we can reach with our formulas but hardly imagine. There is a subsurface inaccessibility, plasticity, and mysteriousness that allows us more easily to be religious now than in the hardworld of earlier physics. The old themes of materialism—atomic matter in absolute motion, sensory and pictorial substance, total specifiability, mechanics, predictability, finished logical analysis—have every one an antithesis in recent physics. It is hard to know what synthesis to make, but certainly a religious synthesis is not precluded. Nature is now less material, less absolutely spatiotemporal, more astounding, more open, an energetic, developmental process. John D. Barrow, a theoretical physicist, says that the principal result of recent physics is that "nature has revealed a deep, hidden flexibility, previously unsuspected."[26] If in one sense this nature is still secular, in another sense it is a suitable arena for the operation of a sacred, creative Spirit.

God is not a spatiotemporal entity. God is pure spirit. Having no velocity or mass, God has no time. After relativity, we can even begin to glimpse more of what was formerly meant when eternity was claimed for God and space and time were assigned to the finite creation. If there is God at all, it must be clearer than ever that, being omnipresent and not local, God has no space and time, scan these processes of past-present-future though God may to gather them into a whole, or break through them where God chooses in numinous presence. Nothing in relativity theory prevents there being faster-than-light signals, even instantaneous ones, provided only that these supraluminal processes are not physical as we know them. (Physicists have looked for, though not found, faster-than-light particles, tachyons.) Nothing prevents a divine reality that might, through omnipresence, have contacts perfusing the whole, establishing the universe as a hologram, despite our lack of any capacity to know this. There is ample room for a panentheism, the finding of

God-in-all and all-in-God, for permitting the divine to materialize and energize at its leisure. The basic scientific motifs in physics are *dynamism in power,* and in nature viewed as process there is nothing inimical to the concept of God. To the contrary, when we last lose sight of the shape of nature on the frontiers of physics, there are a good many signs pointing in that direction.

4. MICROPHYSICAL AND ASTROPHYSICAL NATURE

Humans on the Scale of Complexity

The human world stands about midway between the infinitesimal and the immense on the natural scale. The size of a planet is near the geometric mean of the size of the known universe and the size of the atom. The mass of a human being is the geometric mean of the mass of Earth and the mass of a proton. A person contains about 10^{28} atoms, more atoms than there are stars in the universe. Such considerations yield perhaps only a relative location resulting from the fact that we can see only so far up, out, or down. Hence, we not unexpectedly see at similar ranges of the great and the small. We do not know whether we can indefinitely keep on mining ever more tiny things out of micronature, or indefinitely keep on seeing at greater distances into celestial nature. It looks as though observational limits and involvements will become steadily more constraining in both directions. Whether humans are near central range absolutely we cannot say, and perhaps can never say. Relativity theory makes us wary of special locations, as does most science since Copernicus.

It may be that living things have to be about the size they are in order to stay within reach of microphysical indeterminacies, which they need to contact for plasticity as they chart a world course, and also simultaneously to build the complex structures they have evolved. Other engineering requirements that constrain our size are not hard to find. Still, questions of proportion do arise. Perhaps physical science has reached so far beyond the everyday level as to make our human ranges of experience only conventionally significant, and ultimately insignificant. Perhaps by demonstrating some anomalous discontinuities of humankind with physical nature, or by relativizing or shrinking humans from above, celestially, or by deflating us from below, microphysically, physical science has shown the person to be less ultimate, more ephemeral. Humans are dwarfed and shown to be trivial on the cosmic scale. They are reduced and shown to be nothing but electronic molecules in motion on the atomic scale.

On the other hand, astronomical nature and micronature, profound as they are, are nature-in-the-simple. At both ends of the spectrum of size, nature lacks the complexity that it demonstrates at the mesolevels, found in the earthen ecosystem, or at the psychological level in the human person. Astronomical nature is incredibly vast and energetic, but primitive. Such a statement will seem odd, on first reading, for the theories and calculations by which the mind probes such nature are among the most sophisticated known to science, as, for example, relativity theory and quantum mechanics. Physics is no simple science, and the stuff of its observations is abundantly mysterious, as the considerations we are about to undertake will further reinforce. But that stuff, compared with life and mind, is as primitive as it is basic. We encounter advanced forms of natural organization only at the middle ranges and in the other sciences. It does not disparage the genius in physics, especially as it reaches away from our native levels, to remark that we are successful here, and not in the sciences of mind and society, because of a relative simplicity in the objects.

So far as we can still distinguish what it means to be "full" and "empty," astronomical nature is mostly empty space, and so is micronature. A galaxy is mostly nothing, as is an atom, although the more we learn about energy, the less emptiness there is in what we once took to be the empty spaces between particulate bits and clumps of matter. Even so, it is where and how this matter "clots" or gathers into pacts that is of consequence. We humans do not live at the range of the infinitely small, nor at that of the infinitely large, but we may well live at the range of the infinitely complex.

There is in a typical handful of humus, which may have ten billion organisms in it, a richness of structure, a volume of information (trillions of "bits"), resulting from evolutionary processes across a billion years of history, greatly advanced over anything in myriads of galaxies, or even, so far as we know, in all of them. The human being is the most sophisticated of known natural products. In our

hundred and fifty pounds of protoplasm, in our three pounds of brain, there may be more operational organization than in the whole of the Andromeda galaxy. The number of possible associations among the trillion neurons of a human brain, where each cell can "talk" to as many as a thousand other cells, may exceed the number of atoms in the universe. The number of possible genetic combinations in the offspring that a man and woman can conceive may exceed the number of atoms in the universe.

In that sense, the most significant thing in the known universe is still immediately behind the eyes of the astronomer. On a gross cosmic scale, humans are minuscule atoms. Yet the brain is so curiously a microcosm of this macrocosm, since the mind can contain so much of nature within thought and thus mirror the world. This knowledge occurs first at a conventional range. Living at our native range, through natural evolution we already have learned something about nearest nature. But then in microphysics and astrophysics we devise ingenious ways to get at the nature that is beyond our everyday reach. Here science continues the earlier epistemic evolution, further to prove how remarkably elastic is our rational equipment. When the mind reflects on cosmic questions, is that only a human event? Or is this nature coming of age, nature coming around to reflect on itself? Is this merely mind that knows about matter, or is this matter coming to know itself in mind? In the end, even astrophysicists are studying their own origins, and they themselves are wired into the circuits of what they see. They are an end of what they are watching the beginnings of. They are one of the consequences of the stellar chemistry, and they themselves are the most remarkable phenomenon known. We humans too are "stars" in the show, quite as impressive as anything in the skies, because of our majestic role on the stage of natural history.

The Anthropic Principle and the Fine-tuned Universe

But what can we say of any connectedness between these levels? In astronomical nature—in nebulae, in stars, in space—nature mostly exists at the low structural ranges of micronature—as particles, electromagnetic radiation, electrons, protons, hydrogen. Yet this nature steadily aggregates and energetically builds. A dramatic tale unfolds. The stars are the furnaces in which all but the very lightest elements are forged. Without such stellar cultures there could be no later evolution of planets, of life, or of mind. The stars run their courses and explode themselves as supernovae to disperse their matter throughout space. The human person is composed of stardust, fossil stardust!

In this perspective, historically though not at present, an astronomical phase in nature is not irrelevant to, or inconsequential for, human life. It is the precondition of the rational self. Beyond that, relativity theory cautions us not to be unduly anthropocentric about what may be the present and future natural histories of events that Earthlings see as we look out and back into the distant reaches of time and space. In some moods, we may want a democracy of these levels and regions, with none taken to be more or less important than the others, with all equally required or fitting for the show.

But neither can we leave ourselves out of the account. The universe is staggeringly lavish in its size, and within it matter is very rarefied. But matter also condenses into fascinating formations, the rarest and most impressive of which are life and mind. The rarity of any environment that can produce life and mind often suggests that nature on the whole is a ridiculous swirl and empty waste that is hostile to life and mind. Humans are puny and transitory epiphenomena having no essential relationships to these vast, dumb processes, which constitute all but the tiniest fraction of nature.

At this point, one cannot do experiments revising the universe to see whether another might be more congenial, but one can do thought experiments to see what another one would be like. The result of such *if-then* experiments is what is often called the *anthropic principle*, an impressive result with a somewhat unfortunate name. The universe is mysteriously right for producing life and mind, demonstrably on Earth and perhaps just as well elsewhere. This could better be called a biogenic or psychogenic principle. The point is not that the whole universe is suitable to produce Earth and *Homo sapiens*. That would be myopic pride. Nor need it follow that only mind has intrinsic value and every other phase of the story only instrumental value. Value is produced at various levels of natural formation and achievement, not lessened when these are also contributory to the evolution of life and mind. The issue is richness of potential, not anthropocentrism. There is no need to insist that everything else in the universe, in all its remotest corners, has some relevance to our human being here. That is a distastefully arrogant concept of creation. God may have overdone the creation in pure exuberance, and why should the parts irrelevant to us trouble us? Nor is there any need to cram the universe with other forms of life and mind. Life and mind need only to be among its valuable, interesting products.

We might even be a bit sorry if the sublime universe all turned out to be needed for our human arrival here, or even for the scattering of life and mind here and there within it. The sheer extravagance of the universe, the vertigo one gets contemplating space (often interpreted as an overwhelming "waste") is aesthetically stimulating. It would seem rather prosaic should all cosmic history turn out to be necessary for our being here. We should rather enjoy a lavish universe, and creation would seem impoverished if nothing but a container for persons and minds. Just as it would be a pity on Earth to think that all terrestrial and marine creatures had to be of some use to humans, it would be a pity to discover that all the astronomical worlds had to exist merely as a preparation for mind.

But meanwhile, it is impressive to find that the construction of the universe in its fundamental constants, ranges of size, and forms of forces, is not irrelevant to our being here. When we come, we are not accidentally related to the form of the universe, although vast stretches of the universe have nothing to do with our particular destiny. We seem to have a cosmic ecology, a habitat fittedness, to be products of our environment.

If we made a substantial reduction in the number of particles in the universe, or in its total size, *then* what would be the consequence?[27] No mechanism for life has ever been conceived that does not require elements produced by thermonuclear combustion, which requires several billion years of stellar cooking time. But no universe can provide several billion years of time, according to the theory of general relativity, unless it is several billion light-years across. *If* we cut the size of the universe by a huge reduction (from 10^{22} to 10^{11} stars), *then* that much smaller but still galaxy-sized universe might first seem roomy enough, but it would run through its entire cycle of expansion and recontraction in about one year!

If the matter of the universe were not so relatively homogeneous as it is, *then* large portions of the universe would be so dense that they would already have undergone gravitational collapse. Other portions would be so thin that they would not be able to give birth to galaxies and stars. On the other hand, if it were entirely homogeneous, then the chunks of matter that make all development possible could not assemble. So the distribution of matter has to be something like the way it is.

If the universe were not expanding, *then* it would be too hot to support life. Indeed, *if* the expansion rate of the universe had been a little faster or slower (especially since small differences at the start result in big differences later), *then* the connections shift so that the universe would already have recollapsed or so that galaxies, stars, and planets could not have formed. The extent and age of the universe are not obviously an outlandish extravagance, if it is to be a habitat for the life and mind that we know at its middle ranges. Indeed, this may be the most economical universe in which mind can flower on Earth, and perhaps elsewhere—so far as we can cast that question into a testable form and judge it by present physical science.

There are, further, many other physical constants and processes of natural operation, both at the microphysical and at the astronomical levels, that strikingly fit together to result in what has happened. Change slightly the strength of any of the four forces that hold the world together (the strong nuclear force, the weak force, electromagnetism, gravitation—forces ranging over forty orders of magnitude), or change various particle masses and charges, and the stars burn too fast or too slowly, or atoms and molecules, including water, carbon, oxygen, do not form, or do not remain stable, or other checks, balances, cooperations are interrupted. B. J. Carr and M. J. Rees, cosmologists, conclude, "The basic features of galaxies, stars, planets and the everyday world are essentially determined by a few microphysical constants and by the effects of gravitation. Many interrelations between different scales that at first sight seem surprising are straightforward consequences of simple physical arguments. But several aspects of our Universe—some of which seem to be prerequisites for the evolution of any form of life—depend rather delicately on apparent 'coincidences' among the physical constants. . . . The Universe must be as big and diffuse as it is to last long enough to give rise to life."[28]

If one undertakes thought experiments revising the ratios, constants, atomic sizes, and dynamics in the laws that govern these operations, then one runs into similar impossibilities, surprises, and unknowns. When we consider the first few seconds of the big bang, writes Bernard Lovell, an astronomer, "it is an astonishing reflection that at this critical early moment in the history of the universe, all of the hydrogen would have turned into helium if the force of attraction between protons —that is, the nuclei of the hydrogen atoms—had been only a few percent stronger. In the earliest stages of the expansion of the universe, the primeval condensate would have turned into helium. No galaxies, no stars, no life would have emerged. It would have been a universe forever unknowable by living creatures. A remarkable and intimate relationship between man, the fundamental constants of nature and the initial moments of space and time seems to be an inescapable condition of our existence. . . . Human existence is itself entwined with the primeval state of the universe."[29] Concluding a study of energy processes on cosmic scales, Freeman J. Dyson, a physicist, writes, "Nature has been kinder to us than we had any right to expect. As we look out into the universe and identify the many accidents of physics and astronomy that have worked together to our benefit, it almost seems as if the universe must in some sense have known that we were coming."[30]

Fred Hoyle, an astronomer, reports that his atheism was shaken by his discovery of critical levels involved in the stellar formation of carbon into oxygen. Carbon only just manages to form and then only just avoids complete conversion into oxygen. If one level had varied by a half a percent, the ratio of carbon to oxygen would have shifted so as to make life impossible. "Would you not say to yourself, . . . 'Some supercalculating intellect must have designed the properties of the carbon atom, otherwise the chance of my finding such an atom through the blind forces of nature would be utterly minuscule'? Of course you would. . . . You would conclude that the carbon atom is a fix. . . . A common sense interpretation of the facts suggests that a superintellect has monkeyed with the physics, as well as with chemistry and biology, and that there are no blind forces worth speaking about in nature. The numbers one calculates from the facts seem to me so overwhelming as to put this conclusion almost beyond question."[31] "Somebody had to tune it very precisely," concludes Marek Demianski, a Polish cosmologist and astrophysicist, reflecting on the big bang, and Stephen Hawking, said by many to be the Einstein of the second half of the twentieth century, agrees, "The odds against a universe like ours emerging out of something like the Big Bang are enormous. I think there are clearly religious implications."[32]

How the various physical processes are "fine-tuned to such stunning accuracy is surely one of the great mysteries of the cosmology," remarks P. C. W. Davies, a theoretical physicist. "Had this exceedingly delicate tuning of values been even slightly upset, the subsequent structure of the universe would have been totally different." "Extraordinary physical coincidences and apparently accidental cooperation . . . offer compelling evidence that something is 'going on.' . . . A hidden principle seems to be at work, organizing the universe in a coherent way."[33]

Mike Corwin, a physicist, concludes, "This 20-billion-year journey seems at first glance tortuous and convoluted, and our very existence appears to be the merest happenstance. On closer examination, however, we will see that quite the opposite is true—intelligent life seems predestined from the very beginning. . . . Life as we conceive it demands severe constraints on the initial conditions of the universe. . . . It is not that changes in the initial conditions would have changed the character of life, but rather that any significant change in the initial conditions would have ruled out the possibility of life evolving later. . . . The universe would have evolved as a lifeless, unconscious entity. Yet here we are, alive and aware, in a universe with just the right ingredients for our existence."[34]

There seem to be in fact all kinds of connections between cosmology on the grandest scale and atomic theory on the minutest scale, and we may well suppose that we humans, who lie in between, stand on the spectrum of these connections. The way the universe is built and the way micronature is built are of a piece with the way humans are built. The shape of the rest of the universe, of all the levels above and below, is crucial to what is now taking place close at hand. In its own haunting way, the physical structure of the astronomical and microphysical world is as prolife as anything we later find in the prolife biological urges. Prelife events can have, and have had, prolife consequences. The universe is a biocosmos. George Wald, an evolutionary biochemist, says, "Life . . . involves universal aspects. It is a precarious development wherever it occurs. This universe is fit for it; we can image others that would not be. Indeed this universe is only *just* fit for it. . . . Sometimes it is as though Nature were trying to tell us something, almost to shake us into listening."[35]

Cosmic Necessities and Contingencies

Sometimes it is the cosmic and atomic contingencies, and sometimes again it is the cosmic and atomic necessities that are so impressive. Overall it is not just the one or the other but the mixing of the two that yields the impressive natural providences. Sometimes we marvel that it had to be that way, and so the story unfolded. Sometimes we marvel that it could have been otherwise but was not so. Sometimes it is not too clear whether these striking interconnections are necessary or contingent, and we do not know what deepening theory in the next decades and centuries will do by way of revising the necessities and contingencies of these connections. But in the end it hardly matters whether these connections, as they change with shifting theory, are improbable or necessary. So far as they are improbable, we seem to need a guiding hand in the makeup of things by way of ongoing superintendence; so far as they are necessary, the invisible hand seems to have been there from the start. Either way we have a striking result where life and mind are absolutely dependent on these necessary and/or contingent precedents in astrophysical nature.

Some world selection principle is at work in physics, perhaps a precursor of what is later reached as natural selection in biology. There may be dice-throwing, but the dice are loaded.

Nor is it that we can imagine no other world in which intelligence or life might exist; we are hardly intelligent enough to say what the possibilities are and are not. It is rather that in this world in which life and mind do exist, any of a hundred small shifts this way or that would have rendered everything blank, lifeless. For example, John D. Barrow and Joseph Silk, another theoretical astrophysicist, calculate that "small changes in the electric charge of the electron would block any kind of chemistry."[36] Or a 2 percent increase or decrease in the strong interaction coupling constant and there would be no life. The charges on the light electron and on the vastly more massive proton are exactly equal numerically. "Heaven knows why they are equal," wondered George Wald, "but if they weren't there would be no galaxies, no stars, no planets—and, worst of all, no physicists."[37] A fractional difference and there would have been nothing. Yet here we are. It would be so easy to miss, and there are no hits in the revised universes we can imagine, and yet this universe is a hit, a delicate, intricate hit!

The physics that gave us indeterminacy at the microlevel, affecting so surprisingly the macrolevels of common life, now at the astronomical level, coupling back to the microlevel again, takes away the indeterminacy and gives us a fine-tuned universe, one open but pregnant for life, one perhaps not predictable but one that portends storied development. In the plasma we begin to see a plan, or, better, a plot. It is still possible to posit an explanation that retains the randomness; this universe is only one of a run of universes: big bang, big squeeze, big bang, big squeeze, and this one at random has the right characteristics for life—or the many-worlds theory, where the universe is constantly splitting into many worlds, some of which will be right for life. But these are strange explanations indeed—to invent myriads of other worlds existing sequentially or simultaneously with ours, in order to explain how this one can be a random one from an ensemble of universes. That is real addiction to randomness in one's explanatory scheme! It seems a more economical explanation (remembering that science often urges simplicity in explanation) to posit only the one universe we know and some constraints on it that make it right for life. 

What we are marveling at through it all is the hint that human affairs and astronomical and microphysical affairs are not irrelevant to each other, with even the further hint that there must be some great Cause adequate to this great effect, something that infuses meaning across the whole. The macrophysics and the microphysics are affecting our metaphysics.

Sometimes we dismiss the puzzle, noticing that in no other kind of universe could humans have evolved to worry about these things. The skeptic will say, "We hardly know how life got here in the universe we have, although we are sure that it is the kind of universe in which life can appear. It is hazardous to argue that in no other kinds of universe, or with no other arrangements in this one, could there have been life at all. We are here and it really isn't surprising that the universe is of such kind as has produced us. We knew before we started our search that the universe has all the prerequisites for our being here." It is hardly more than a tautology to say that observers will find themselves in a universe where observation is possible. If one is an observer, observation has happened, and one knows before he looks that in his universe observation is possible. Any puzzlement is something like being surprised, after a bomb blast, that one is around to wonder why he survived. Nonsurvivors never wonder.

But those who want fuller explanation will say that this dismissal is more like saying "If the thousand men of the firing squad hadn't all missed me, then I shouldn't be here to discuss the fact, so I've no reason to find it curious."[38] It is no tautology, but rather an impressive empirical finding, to discover that what seem to be widely varied facts really cannot vary widely, indeed, that many of them can hardly vary at all, and have the universe develop the matter, life, and mind it has generated. What we have is a bomb blast (the big bang) that is fine-tuned to produce a world that produces us, when almost any other imaginable blast would have yielded nothing. What we have is friends in (a Friend behind) what would otherwise be a firing squad, a chaotic blast.

Sometimes, in a quest that ventures into religion past physics, these necessities and contingencies, which join to make the world go round, can also by tandem turns on their respective upstrokes feed a governing gestalt that detects Something, Someone behind the scenes arranging for the show. God is "in, with, and under" these arrangements. The world is (shades of Bishop Paley!) a fine-tuned watch again, and this time there are many quantitative calculations to support the argument. Even though it is also a fertile gauzy plasma, this seems to demand a watchmaker God—at least the astronomical and microphysical worlds in these arrangements for life seem to demand a universe-maker, although the biological world, after life arrives, is going to prove more problematic. It almost seems that we need a revolution in explanations all over again (recalling the Newtonian one at the start of the scientific revolution, Figure 2.1), now that quantum physics, relativity physics, and astrophysics have plumbed the material and efficient causes down to the bottom and back to the beginning. The form that matter and energy takes seems strangely suited to its destiny.

Victor Weisskopf, a theoretical physicist, after a review of what is known about the origin of the universe, concludes, "The origin of the universe can be talked about not only in scientific terms, but also in poetic and spiritual language, an approach that is complementary to the scientific one. Indeed, the Judeo-Christian tradition describes the beginning of the world in a way that is surprisingly similar to the scientific model."[39]

Mutability and Permanence in Nature

Whatever we make of God out there, down under, or within it all, we have still on our hands the problem of how to evaluate humanity's place in nature. We commit a sort of astronomical fallacy if we suppose that larger things are more important and real than smaller ones, oblivious to considerations of complexity and capacity. Galaxies and stars are not more real or significant than persons. At the other end of the scale, one commits an atomistic fallacy if one reasons that the smallest entities out of which a macroscopic thing is composed, being more elementary, more elemental, must therefore be more fundamental, metaphysically prior, more real. One is lured to such a conclusion by the long-enduring character of those atoms now embodied within humans, so anciently formed in the stars. Such an analysis is related to the genetic fallacy, which supposes that a thing is to be understood solely on the basis of its chronological and material composition.

The physicist may be tempted still to suppose that micronature has independent primary qualities that are somehow more ultimate than the less real secondary qualities that appear in macronature. The emphasis in physical science on material and efficient causes suggests that the cause must not only be the full explanation of the effect, it must be more real, more significant than the effect. But it does not always follow that the fine-structure of a processive entity, such as a living organism, is less mutable or more significant than the molar structure. Physics gives us nothing that was not once built, nothing that has always been there. Everything is a fossil of something else. Things without a life can still have a "half-life," beginnings and endings, and if some of the microparticles (protons, photons, electrons) have half-lives that are almost forever, even they are inherently unstable. Other particles (muons, pions) have only momentary lives, and all are subject to collisions that annihilate them and create something else.

It has become difficult in recent science to make judgments about mutability, as this may affect judgments about ontological status and ultimate reality. At first impression, the individual seems momentary, its elements eternal. But looked at more carefully, mutability shows up at every structural level. One cannot fully observe a microparticle once, let alone twice to check successive identity. In some sense the superposed quantum states represent potential as much as reality; some of them become what they become under the interactions of the organism. What we can observe of the spontaneous microbits hardly seems primordial; they seem as much energetic waves as static particles. The atoms of which we are composed are in no less flux than we are; electrons orbit at enormous velocities in multiple shells. Radiation produces constant reorganizations within the nucleus. Molecules, once thought to be the most rigid of the structural levels, incessantly vibrate, undergo internal atomic exchanges, outside atomic replacements, and configurational changes. Some are compact, but others are open and flexible; the floppiness of the phosphate molecule in the energy transfers involving ATP and ADP is a key to life. And all these units that whir and jostle carry with them their own time frames.

There can be more stability at upper levels than lower ones. Amino acids form polypeptide chains, and the chain sequences determine secondary helixes and further foldings into tertiary and quaternary structures, which provide the life functions. But often these acid sequences mutate over time, conserved at some key positions, altered at most others, while the macromolecular structures remain stable (as in myoglobin and hemoglobin). The living individual as a sizable whole greatly outlasts most of its smaller parts. It persists across time over changes of materials involving a steady input and degradation of energy. When the individual is set in an evolutionary life stream, that stream mutates over populations, species, and genera, but the stream conserves basic chemistries and maintains an information flow for millions and even billions of years. The cytochrome *c* molecule, used for respiration by plants, animals, and eucaryotic microorganisms, has remained essentially constant in shape for one and a half billion years.

During this time span, even the atoms of micronature decompose and recompose, and the stars in their cosmic courses flux no less than the streaming course of life. In a gestalt, the paradigmatic pattern may remain while the substructures permeate. The figure is more fundamental than its constituent ground. Then again, motifs may persist and recur throughout ever-revising manifestations of them. Given all this, when we survey the whole of natural history as a processive historical scene, do we want to say that some one dimension is less fluxing and thus more real, more significant, than the rest? Nature most fundamentally is not something large or small but a complicated historical tissue of events.

Some might want to say that atoms are more basic than organisms, because one could have a world (as once there was) with atoms but without life, while one cannot have a world with life in it but no atoms. Yet where we do have life amidst its atoms, it is not entirely true that the life states depend on the atom states, while the atom states do not depend on the life states. The precise quantum events that materialize may be called forth to suit the organism. The life states, while they require some atom states, find any local sets of atom states often dispensable. Life moves on with its steady flow of atomic inputs and outputs. A particle too, under the field theory of matter, is not different in this respect. It likewise is not some one fixed and durable substance; it is a persisting pattern traveling across an ether.

Life is a kind of fire that outlasts the sticks that feed it, except that with ordinary fire the flame is merely a chemical product of the fuel burned, devoid of heredity, while the characteristics of the life "fire" are coded in an information flow from the parental fires that light it. The food that fuels the fire is taken over and "informed" into this life form. The information persists and increases over time, is more or less as long-enduring as the particles it employs, and is no less real or significant. In this world in which the atoms present early on have organized themselves into life and mind, the total tale of the pattern states of these atoms hardly seems told until these later levels have been given their place. What a thing basically is should include its last as well as its first stages. This is true no matter whether on the astronomical or microphysical scales, so far as the anthropic principle reveals anything; the atom states are and always have been interlocked with the possibility of life states.

Microscopic and Macroscopic Levels

If one becomes interested in the detail of a newspaper photograph, and examines the photograph under a lens hoping to get more information out of it, one discovers that the fine structure is black and white dots—just dots! One cannot analyze the gross picture by a closer look at the dots of which it is composed, although it can be resolved into these dots. That is true of any photograph. From a distance one may think that there is no limit to the amount of information contained in the detail, but at nearer range one comes up against a grain, which limits one's resolving power. The chemical structure of that grain can be analyzed, but this analysis does not yield any information about the gross picture. In part, the examination of micronature is like this. The information we seek is not always laid into the granular ultrastructures, but it may be overlaid on macrorelationships, on functions in nature that appear at tertiary or quaternary levels of structure and are not present at primary or secondary levels. One ought not to use too high a magnification. We may not need to solve these questions by more resolving power at the microscopic level. We may need more scenic scope at the native ranges, more power to see across the past and to gain a vision of the future.

This is not to deny that life and mind have been greatly illuminated by the study of biophysical structures. It is complicated by the fact that in an evolutionary perspective these atomic "dots" have undergone self-composing into the larger natural picture, as did not happen in the photograph. Along that self-composing route there sometimes seem to come out of micronature effects that are amplified so as to bias some macrostructures (as in genetic mutations). But also, there emerge upper levels that come to govern the lower ones. Thereafter, if one is to understand what is going on, one has to rest explanations at the appropriate level of informational control.

Physical science abstracts for itself a lifeless and mindless nature and therefore comes at nature at too simple a level to provide an entire explanation of natural history. From this point onward, the very strange revelations of physics may not teach us as much about nature as is often supposed. We must, in the succeeding chapters and the later and more specialized sciences, back away from any closer look at the fine grain and look more megascopically—at life in biology, at mind and behavior in psychology, at society and culture in the social sciences, and in the end at history. Ultimacy may lie in nonphysical directions.

God and Physics

Still, the fine structures of matter, the operations of materials and energies there, as these may be found by physical science employing its causal paradigm, must be such that they can support and permit those meanings in which religion is so paradigmatically interested. Whatever meanings may be found in the religious life need to be connectible with and deployable into the character of nature at the micro levels and cosmic levels. If God exists, God may be dominantly encountered in the social and cultural or in the personal and subjective life. But God cannot be absent from physical nature, which forms the support of all life and mind. God may be differently present at different ranges of nature, but God must be present in atomic or cosmic nature in objective ways that are commensurable with the divine presence also in the subjective life. That will be only a partial account. It does not follow that meanings will be given at all, so long as one focuses on at the fine structural, reductive level, or even on astronomical but simplistic levels. If one makes a judgment simply on these levels, one does not consider all the evidence. Perhaps where nature is least elementary, there its most profound dimensions are revealed, and this will be in life, mind, and history.

We have not here been intending to erect a natural theology out of physics. Physics (P) does not of itself imply that there is God (G) down below or up above, as cosmic ground or receptacle. The logic is not: P implies G (if P, then G), certainly not by any sort of causal implication. But neither does physics gainsay God, and the

supposition that God is there can be quite in harmony with anything now being said in physics. This is not only in the sense that there is no logical contradiction, but also in the sense that the sort of world that physics observes is quite "derivable" (in the looser sense, appropriate to religion) from the hypothesis of God. If G, then P. That is, the world observed in physics quite follows in course if God exists. The theory (T) implies observations (O) such as these, under the "if T, then O" pattern of our opening chapter.

The world of physics leaves a space into which God may be inserted. But this is permissive or congenial evidence, nothing more. This inference will not likely be made merely on the strength of causal need. It will be done on a meaning quest, by those thrust on a search that wants God not so much for the First Cause as for the Ground of Meaning. These meanings are not met in physics alone but even more impressively in other chapters of life, in mind, society, culture, and history. Such a Ground is not forbidden by any causal blockage from contemporary physics. To the contrary, physics can readily be interpreted to point in that direction, toward the gestalt of God.

Mathematics and Historical Experience

The impressive rigor of physics and chemistry, as seen in their metric character, with accompanying predictability and testability, is related to their simplicity. Physical science permits great mathematical sophistication just because it chooses to deal with matter so simply. Thought is simpler than reality, language is simpler than thought, and mathematics is simpler than language. We cannot think of all that nature is. It outgoes us and we can only approximate it. We cannot verbally express all that we think, experience, and perceive. Words remain only approximate tokens of thought, however much they in turn enrich our perceptions and facilitate thought. Mathematics is a special form of language, with nonverbal elements; but it is powerless to describe or interpret the world without a text. It is adapted for reliability, speed, analysis, and penetration, which could never be accomplished by prose alone.

It is puzzling to say how mathematics comes to fit the world at all, spun out of our heads as pure mathematics is. Eugene P. Wigner, a physicist and mathematician, contends "that the enormous usefulness of mathematics in the natural sciences is something bordering on the mysterious and that there is no rational explanation for it. . . . The miracle of the appropriateness of the language of mathematics for the formulation of the laws of physics is a wonderful gift which we neither understand nor deserve."[40] We have noticed that most of the numbers of science are artifacts and relative. Indeed, all its numbers, including such seemingly natural measures as dimensionless constants and the count of the moons of Jupiter, are the product of theoretical overlays on nature, and powerless to describe nature except in the context of such theory. All the natural *units* are assigned; fit they must, but they are not bare, naked units. They are as contrived as is the notion of a center of gravity, or momentum, or half-life.

When mathematics does describe the world, it does so by abstractive simplicity. Sometimes this can run ahead of our verbal capacities, as when in relativity theory and quantum mechanics scientists first calculate using mathematical models and then, unless their work is to remain merely operational and nondescriptive, wonder what their calculations signify in nature. The mathematical character of high-level physics, even after we can no longer picture what is going on, suggests that the intelligibility of the universe vastly outruns our sensory capacities for perception or our local capacities for intuitive imagination. As far as our capacities for thought reach, in the finest of our mathematics, the universe is still intelligible, despite the fact that we can no longer visually represent, verbally model, or perceptively sense it. The math still works even in realms in which sense and intuition do not easily serve. There seems to be a realm of exquisite, supersensory rationality that transcends but supports sense, space, and time. Mathematics is, above all, mental; it is the logical creation of the human mind, and the fact that mathematics repeatedly helps us to understand the structure of the physical world corroborates the belief that the world we inhabit is the creation of mind.

Sometimes in recent physics mathematical models have been applied to the real world with very little sense of any physical model that is quasi-representational or even analogical. But often here the sense of explanation drops out in the same proportion. We only *reach* a world we cannot *represent.* We can compute outcomes, but have no inkling of what is going on, hardly better off in our own way than were the ancient astronomers who could compute eclipses but were wholly in the dark about their nature.

It is permissible enough, and even mandatory in its place, to be impressed with the exquisite mathematical nature of reality. P. C. W. Davies concludes, "Mathematics and beauty are the foundation stones of the universe. No one who has studied the forces of nature can doubt that the world about us is a manifestation of something very, very clever indeed."[41] But we do not from this conclude that all the world's cleverness and beauty lie in its mathematics. Even if we were to lay aside the upper levels that metricize less well, or not at all, here at the quantum levels, our metricizing capacities, profound as they are, run to an end zone. We cannot completely metricize the individual quantum event; it defies mathematical specification in its concreteness. At this point, curiously, one of the most impressive of our mathematical theories tells us that nature permits no further mathematical specifiability.

Throughout it all, mathematics remains powerless to appreciate a world until it adds a narrative of events. Perhaps sometimes there is no picture, but the mathematics is useless without a text, without words—no matter how much it is also true that the mathematics accomplishes what words cannot. The space-time diagrams must have a caption, the equations an interpretation. Past this, complex nature is never fully described by mathematical models. To the contrary, in these very much is left out, and mathematics is to that extent stylized and crude as a description of rich natural processes. Its precision is bought with its incompleteness. Eventually, the human phenomenon has to be included in any full science of nature, for humans are the most sophisticated known natural products. Neither mathematics nor other  forms of physics anywhere know the category of "experience," and surely experience is among the phenomena that most cry out to be explained. Doubtless nature is misdescribed with too anthropocentric a model, and mathematical models may offset this. Nature has mathematical dimensions at every structural level, as a musical symphony has mathematical dimensions throughout.

But one runs to the other extreme if nature is held to be only "mathemorphic." John A. Wheeler claims, "This is a world of pure mathematics and when we penetrate to the bottom of it, that's all it will be."[42] But mathematics is not the only mode of thought competent for judging multidimensional nature, or symphonic music, although there is mathematics "at the bottom" of both. Nor have we any reason to think that, being abstractive and sparse, this kind of science is less interpretive and more fundamentally descriptive. The failure of the other sciences to metricize well, the failure of history to metricize its narratives, the failure of theology to metricize at all, should not be interpreted as meaning that nature is mathematical and that these nonmetric inquiries are illusory or crude. It rather means that nature is qualitative as well as quantitative, and that the metric sciences are sketchy. If the scientist is theologically inclined, she must resist the temptation to say that, while God cannot be a person, God might be a mathematician. On the other hand, one has nothing to fear from God as a mathematician; indeed, we may be needing encouragement from that realm of consummate rationality as we next plunge into the seeming chaos of biology.

NOTES

1. We here characterize the paradigms Isaac Newton and Aristotle came to represent, which may not entirely coincide with the views that Newton or Aristotle actually held.
2. Chapter 1, p. 22.
3. Jacques Loeb, "Mechanistic Science and Metaphysical Romance," *Yale Review* 4 (1914–1915): 766–85, citation on p. 785.
4. Adapting a metaphor from Karl Popper, "Of Clouds and Clocks," in *Objective Knowledge* (London: Oxford University Press, 1972), pp. 206–55.
5. Chapter 1, p. 15.
6. Francis Bacon, *Essays, Advancement of Learning, New Atlantis, and Other Pieces,* ed. Richard Foster Jones (New York: Odyssey Press, 1937) pp. 179, 222.
7. William Paley, *Natural Theology: or Evidences of the Existence and Attributes of the Deity Collected from the Appearances of Nature* (1802).
8. Figures 2.5 and 2.6 are adapted from James E. Brady and Gerard E. Humiston, *General Chemistry: Principles and Structure* (New York: John Wiley and Sons, 1975), pp. 61, 75.
9. Adapted from Henry Margenau, "The Method of Science and the Meaning of Reality," *Main Currents in Modern Thought* 29 (1973): 163–71.
10. See Ian G. Barbour, *Issues in Science and Religion* (New York: Harper and Row, 1971), pp. 298–305. Barbour is especially good on the philosophical and religious implications of quantum mechanics and indeterminacy.
11. Albert Einstein in *The Born-Einstein Letters,* trans. Irene Born (New York: Walker and Company, 1971), p. 91.
12. Blaise Pascal, *Pensées,* no. 180, in *Oeuvres Complètes* (Bruges: Bibliothèque de la Pléiade, 1954), p. 1133.
13. For irreversible thermodynamics, see Ilya Prigogine, *From Being to Becoming* (San Francisco: W. H. Freeman and Co., 1980), and Ilya Prigogine and Isabelle Stengers, *Order Out of Chaos: Man's New Dialogue with Nature* (New York: Bantam Books, 1984).

About climate, see Edward N. Lorenz, "Climatic Determinism," *Meteorological Monographs* 8, no. 30 (1968):1–3.

14. Dietrick E. Thomsen, "In the Beginning Was Quantum Gravity," *Science News* 124, no. 10 (September 3, 1983): 152–53, 157.
15. For this kind of account see Alexander W. Stern, "Quantum Physics and Biological Systems," *Journal of Theoretical Biology* 7 (1964): 318–28; Richard Schlegel, "Quantum Physics and Human Purpose," *Zygon* 8 (1973): 200–20.
16. The effect of this particular experiment on Einstein's thought is debated. Einstein was puzzled over the transformation properties between the reference frames of differing observers in James Clerk Maxwell's theory of electromagnetism as this fits the laws of motion, which prompted him to modify the structure of space and time. Nevertheless, the constancy of the speed of light justified relativity, even if the theory was proposed independently of it.
17. J. Bronowski, "The Clock Paradox," *Scientific American* 208, no. 2 (February 1963): 134–44.
18. Figure 2.11 adapted from *Scientific American* 208, no. 2 (February 1963).
19. Edwin F. Taylor and John A. Wheeler, *Spacetime Physics* (San Francisco: W. H. Freeman and Co., 1966), p. 141, p. 27 of "Answers."
20. Ibid., pp. 16–17.
21. Plasma in the etymological sense of something "plastic" that can be formed or molded into particulars. Plasma is not used here in the technical sense found in physics.
22. C. W. Misner, "Cosmology and Theology," in Wolfgang Yourgrau and Allen D. Breck, eds., *Cosmology, History, and Theology* (New York: Plenum Press, 1977), pp. 75–100, citation on p. 95.
23. John A. Wheeler, "From Relativity to Mutability," in Jagdish Mehra, ed., *The Physicist's Conception of Nature* (Dordrecht, Netherlands: D. Reidel Publishing Co., 1973), pp. 202–247, citation on p. 227.
24. Taylor and Wheeler, *Spacetime Physics,* p. 141, p. 27 of "Answers."
25. In some presently inexplicable phenomena information seems to travel faster than light, for instance, John S. Bell's theorem and the Einstein-Podolsky-Rosen effect, where one particle knows "telepathically" (Einstein) what spin a mate will take.
26. John D. Barrow, "Anthropic Definitions," *Quarterly Journal of the Royal Astronomical Society* 24 (1983): 146–153, citation on p. 151.
27. See John A. Wheeler, "The Universe as Home for Man," in Owen Gingerich, ed., *The Nature of Scientific Discovery* (Washington: Smithsonian Institution Press, 1975), pp. 261–96, and his "Genesis and Observership" in Robert E. Butts and Jaakko Hintikka, eds., *Foundational Problems in the Special Sciences* (Dordrecht, Netherlands: D. Reidel Publishing Co., 1977), pp. 3–33.
28. B. J. Carr and M. J. Rees, "The Anthropic Principle and the Structure of the Physical World," *Nature* 278 (April 12, 1979): 605–612, citations on pp. 605, 609. See also George Gale, "The Anthropic Principle," *Scientific American* 245, no. 6 (December 1981): 154–71, and B. Carter, "Large Number Coincidences and the Anthropic Principle in Cosmology," in M. S. Longair, ed., *Confrontation of Cosmological Theories with Observational Data* (Dordrecht, Netherlands: D. Reidel Publishing Co., 1974), pp. 291–98.
29. Bernard Lovell, "Whence?", *New York Times Magazine,* November 16, 1975, p. 27, pp. 72–95, citation on pp. 88, 95. See also Bernard Lovell, *In the Center of Immensities* (New York: Harper and Row, 1978), pp. 123–26; see also Chapter 4, note 21. On the other hand, if the same force (the strong nuclear force) were a few percent weaker, only hydrogen could exist.

    One never knows how seriously to take the talk, common nowadays among cosmologists, about what happened in the first few microseconds of the universe. Astronomers are unsure whether the universe is ten, fifteen, or twenty billion years old, but they nevertheless speculate, on the strength of projected equations that retroactively trace the expanding universe, about what happened in the first microseconds of the universe, when the entire universe was packed into a tiny space.

30. Freeman J. Dyson, "Energy in the Universe," *Scientific American* 225, no. 3 (September 1971): 50–59, citation on p. 59.
31. Fred Hoyle, "The Universe: Past and Present Reflections," *Engineering and Science* 45, no. 2 (November 1981): 8–12, citation on p. 12. See also Owen Gingerich, "Let There Be Light: Modern Cosmogony and Biblical Creation," in Roland Mushat Frye, ed., *Is God a Creationist?* (New York: Charles Scribner's Sons, 1983), pp. 119–37. See Hoyle's "On Nuclear Reactions Occurring in Very Hot Stars. I. The Synthesis of Elements from Carbon to Nickel," *Astrophysical Journal,* Supplement Series, 1 (1954): 121–46.
32. Marek Demianski, quoted in Dietrick E. Thomsen, "In the Beginning Was Quantum Gravity," *Science News* 124, no. 10 (September 3, 1983):152–157, citation on p. 152; Stephen Hawking, quoted in John Boslough, *Stephen Hawking's Universe* (New York: William Morrow and Company; 1985), p. 121.
33. P. C. W. Davies, *The Accidental Universe* (New York: Cambridge University Press, 1982), pp. 90, 110.
34. Mike Corwin, "From Chaos to Consciousness," *Astronomy* 11, no. 2 (February 1983): 14–22, citations on pp. 16–17, 19.
35. George Wald, "Fitness in the Universe: Choices and Necessities," in J. Oró et al., eds., *Cosmochemical Evolution and the Origins of Life* (Dordrecht, Netherlands: D. Reidel Publishing Co., 1974), pp. 8–9.
36. John D. Barrow and Joseph Silk, "The Structure of the Early Universe," *Scientific American* 242, no. 4 (April 1980): 118–128, citation on p. 128. See also John D. Barrow and Frank J. Tipler, *The Anthropic Cosmological Principle* (New York: Oxford University Press, 1986).
37. George Wald, quoted in *New Scientist* 60, no. 871 (November 8, 1983): 427. See further his "Fitness in the Universe."
38. John Leslie, "Anthropic Principle, World Ensemble, Design," *American Philosophical Quarterly* 19 (1982): 141–151, 380, citation on p. 150. See also his "Observership in Cosmology: the Anthropic Principle," *Mind* 92 (1983): 573–79.
39. Victor F. Weisskopf, "The Origin of the Universe," *American Scientist* 71 (1983): 473–480, citation on p. 480.
40. Eugene P. Wigner, "The Unreasonable Effectiveness of Mathematics in the Natural Sciences," *Communications on Pure and Applied Mathematics* 13 (1960): 1–14, citation on pp. 2, 14.
41. P. C. W. Davies, *The Forces of Nature* (Cambridge: Cambridge University Press, 1979), p. 231.
42. John A. Wheeler, interviewed in Florence Helitzer, "The Princeton Galaxy," *Intellectual Digest* 3, no. 10 (July 1973): 25–32, citation on p. 27.

Chapter 3

Life: Religion and the Biological Sciences

Science transforms our experience of life, as it does of matter, but here moves closer to the immediate, vital context of religion. "I am come that they may have life, and have it abundantly," announces Jesus.[1] "Better things for better living through chemistry," advertises Du Pont. These juxtaposed claims first seem incongruous, but their common interest in fuller life soon invites reflection. How far do the explanations and promises of religion and of bioscience overlap and compete? The "secret of life" was once thought hidden in the Spirit of God, but now seems rather lodged in DNA and RNA. When a woman is infertile, or diseased, or neurotic, ought she to cry out to the Son of God or to hope for a more abundant life through biochemistry? Or both? How should we understand the mixture of the scientific and the spiritual in the life process?

Bioscience has developed on two main levels. At the macrolevel its evolutionary paradigm about life's *origin* has been especially disturbing. At the molecular level its reductionist paradigm about life's *basis* has been no less revealing. Both levels fuse to offer explanations about how life comes out of matter, and the miracle of life seems to stand explained, even explained away. The sacred has been found secular, and there is nothing mysterious about it. But, in their turn, neither an evolutionary nor a biochemical account of life proves unproblematic. If merely causal or causal-random accounts, they stumble over what to make of life's teleological dimensions. Given the Newtonian legacy, given that material and efficient causes are the only available categories of explanation, if life is to be explained, then this has to be as a complex material motion.

Both bioevolutionary motion and biofunctional motion may be richer forms of matter-in-motion than physics and chemistry can handle, rich enough to remain distinctively biological sequences, and even to permit the attachment of religious consequences to them. Phrased another way, when bioscience tries to explain the *more*, life, in terms of the *less*, matter, there comes the worry (at either the molecular or evolutionary level) that one may be getting too little explanation, true though that little may be. We worry that a science has abstracted out something less than the whole story that natural history tells. If life proves to be an advanced structural level of abiotic nature, shall we say that our concept of life has been reduced, devalued? Why not say that our concept of material nature has to be elevated and revalued? If the latter, is there still room for complementary meaning accounts, such as may come from religious interpretations?

1. BIOCHEMISTRY AND THE SECRET OF LIFE

It is an understandable first theory that *vitality* is a radically separate phenomenon from nonlife, where only *force* obtains. Living things are self-maintaining systems. They grow; they are irritable in response to stimuli. They are information-processing systems. They reproduce; indeed, the developing embryo is especially impressive. They resist dying. They post a careful (if also semipermeable) boundary between themselves and the rest of nature, and they assimilate environmental materials to their own needs. The very fact that they have *needs* is a crucial point. Whatever has needs, and as a result can be *healthy* or *diseased,* is alive. We may say "My car needs spark plugs," but this is a locution for "I need spark plugs for my car." Machines *serve* purposes, and in the course of this they may have needs, but machines do not *have* purposes, or needs, except derivatively from some biological organism that invests them with purposes and that has needs intrinsically.

Organisms gain and maintain internal order against the disordering tendencies of external nature. They keep winding up, recomposing themselves, while inanimate things run down, erode, and decompose. Life is a local countercurrent to entropy. Organisms suck order out of their environment, an energetic fight uphill in a world that typically moves thermodynamically downhill. They pump out disorder. The constellation of these characteristics is nowhere found outside living things, although some of them can be mimicked or analogically extended to products designed by living systems.

What fundamental theory (T) of nature accords with these observations (O)? *If* what T, *then* these O? It was reasonable at first to suppose that bioscience faced a distinctive natural principle. Such an *entelechy* would be a nonphysical, even immaterial principle, since its properties are so opposed to those of inert, bare matter. Some elementary vital force is resident within an organism, which directs its will-be, its will-to-be, its program. This is something additional to any mechanical propellants as these might be known from chemistry or physics. In such a life principle religion will also be interested.

This mystery of how we tick is one to which there is a very ready religious attachment. If anything at all is to be sacred, it will be this elemental experience of vitality. This comes with a sense of givenness, for we do not bring ourselves into existence, or even do very much about the production of our offspring. Yet it also comes with a sense of the urgent necessity of its being cherished and defended, in that fraction of the life process that is under our conscious power. Accordingly, life, with its respiration and metabolisms, is thought to be sustained in God, given by divine inspiration. Before the rise of science, when all the world religions took their classical positions, what went on inside the body was almost entirely a mystery. Blood circulation and cellular structures were unknown. The seat of emotions was thought to be the stomach. Aristotle believed the brain to be a cooling organ, as shown by how much heat can be radiated if one's hat is removed. In such ignorance, it was easy to conflate causes and meanings naïvely and, in joining diverse types of explanation, to fail to distinguish between science and religion.

Electronic and Molecular Characteristics of Life

Biochemistry and biophysics have altered all this by explaining how life takes place, using—so it first seems—nothing but the materials and forces of chemistry and physics. As the telescope opened up the heavens by making visible great amounts of detail to which the naked eye was blind, so the microscope opened up a hitherto unsuspected world within. A metric shift, again following physics, gave biochemistry increased powers of analysis. Myriads of simpler life forms were discovered, such as protozoans and bacteria, although the simplicity of any living thing has proved something of an illusion. All the life processes have unfolded as chemistries—blood circulation, neurology, enzymes, hormones, vitamins.

Still greater capacities to see, employing the electron microscope, and further metric powers, using electrophoresis, dialysis, gel-filtration chromatography, ion-exchange chromatography, and centrifugation, have enabled us to probe life right down to the molecular levels. It should not go unnoticed, however, that most of the entities of molecular biology, like those in atomic physics, are inferential and no longer observable. They are as much represented symbolically as pictured directly.

At the molecular level, the life process operates something like this. Controlled by a genetic code in which a DNA molecule is unzipped and read, twenty sorts of amino acids are formed into polypeptide chains. Depending on the sequence of amino acids, these chains fold to form secondary, tertiary, and quaternary structures and become the protein molecules of life. Polysaccharides and lipids are synthesized by these proteins. In a fundamental metabolic process, glycolysis, a glucose molecule is sent thermodynamically downhill through about a dozen steps, in the course of which ADP is moved thermodynamically up to ATP. These last molecules serve as a ready energy currency because of their high-energy phosphate bonds. In turn, they drive the mobile processes and syntheses of life.

In the citric acid cycle, which evolved later than and was added to glycolysis, an end product of glycolysis is sent around a ten-stage cycle, reaping energy benefits eighteen times as great as those in glycolysis. In the early days of life, photosynthesis evolved and made possible the chemical storing of solar energies. This also placed oxygen in the air, and afterward oxidative phosphorylation emerged, using the oxygen to provide a ready sink for those surplus electrons that result as energy is stripped off food molecules. Such food molecules are composed in photosynthesis.

Through it all, biochemistry finds nothing but common chemicals and no sort of elementary force that is not likewise present in physics. To the contrary, one is impressed with the electronic character of life. Living is an affair of electron transporting and transferring, of various electronic bondings, of proton and ion pumping, of the whir and buzz of molecules. To this connecting of organic events with inorganic processes we can add the artificial production in the laboratory of amino acids from inorganic materials, as well as the discovery that something like this takes place naturally in the upper atmosphere. We can notice how we have built computers, machines that take on more and more lifelike characteristics. We can puzzle over viruses, hardly knowing whether to call them living. They are at least escaped bits of DNA that reproduce themselves by a parasitic borrowing of life processes from their host organisms.

One might think that these discoveries would provide a welcome unity in nature by integrating the physical with the biological sciences. Nevertheless, bioscience has brought a persistent worry that life has here been shown to be "nothing but" an epiphenomenon of matter. Its molar levels are really under mechanical and molecular control. No separately isolable bioforce, no élan vital has ever been detected in the laboratory. If so, to what can we attach the sacredness and meaningfulness of life? Francis Crick and J. D. Watson, the two key figures in unraveling the secrets of DNA and RNA, have expressed this reductionist hope. "The ultimate aim of the modern movement in biology is in fact to explain *all* biology in terms of physics and chemistry. . . . Eventually one may hope to have the whole of biology 'explained' in terms of the level below it, and so on right down to the atomic level."[2] "Complete certainty now exists among essentially all biochemists that the other characteristics of living organisms (for example, selective permeability across cell membranes, muscle contraction, nerve conduction, and the hearing and memory processes) will all be completely understood in terms of the coordinative interactions of small and large molecules."[3]

If life is nothing but physics and chemistry, nothing but molecular motions, can it still be sacred? It need come as no surprise that humans are made of earthen materials, for even the Genesis account says that God made Adam of dust and mud. Still, God is pictured there as breathing into Adam the breath of life, as though some novel animation or inspiration were involved. But in bioscience not only the matter but even its motions can seem quite mechanical and electronic. Has the mystery of life been dissolved? How then do we conceive life's meaningfulness?

A biological molecule, though it is always physical material, is readily separable from merely physical material *right down to the molecular level.* Does the molecule function? This criterion of *biofunction,* which can also be used to test for life at molar levels and for extrasomatic products of biological systems, nowhere disappears as one descends to the molecular level. If we examine an iron poker microscopically, its functional character is entirely gross and not evident in the microstructures of the iron. But biofunctioning is structurally present down to 10 nanometers and below. Nonbiological molecules have only to *be;* biomolecules have to *work* in order to be. Otherwise, they disintegrate. Indeed, the behavior of the particular atoms and electrons involved, right down to the level of quantum indeterminacies, is not understood until it is understood in terms of a biofunctional analysis. There physicochemical forces and structures are found without prejudice to biofunction. Life is propelled by electric forces, and yet it seems equally appropriate to say that these forces are now put to a vital use, brought under a biofunctional control. Life is indissolubly interlocked with matter.

But what is thereby proved is not so much that life is reduced to matter as that matter here subserves a pulse of life. We have already noticed how organisms by using their biochemistries at the quantum level "decide" events from among the possibilities latent in the superposition states. Rather than the microelectronics predetermining the course of the organism, it is the other way around. The organism by its informational control is taking over the chemistries. Life "brings about" the course of matter, if also sometimes vice versa.

If one is looking for a full explanation, then physics and chemistry can supply only what is taking place, but biological categories will have to supply an account of what is going on. Contrary to the Watson-Crick reductionistic account, it may be that physics and chemistry are getting enriched through these connections with biology, by this showing what dramatic powers lie in those atoms and molecules, assembling themselves as they do into self-replicating organisms and even into scientific and religious persons.

Biological Molecules as Informational Molecules

The secret seems to lie not at the atomic level as such, but in that complexity which, we earlier noted, characterizes middle levels between the atomic and astronomical extremes of size. Now we can identify that complexity more carefully. The complexity includes life *information* with its accompanying biofunctions. What is irreducibly there is not entelechy but vital information. This is the "force" that animates matter so distinctively. The question of an adequate account of the nature, origin, and transfer of this biological information will prove a criterion with which we can test for overly simplistic and reductionist models of life. We shall use it in examining both molecular and evolutionary biology. Notice too that molecular biology and genetics are not fundamentally, in this sense, metric sciences. This vital *information* that is their most essential concept (for example, the code for making hemoglobin) is difficult exhaustively to describe and specify in physicochemical numbers and units, although biologists have discovered the conformations of molecules that do this coding and have used a great many numbers, units, and equations finding this out.

We can trace the vital functioning back to the informational core biomolecules. But what shall we say of them? They have the organism as a product, the outcome of their information. But the genetic material is in turn, of course, generated by the organism. Genes do not preexist the organism, although they may be passed along in reproduction when life is phased through the seed and egg. They coexist with the organism. Now should we say that this genetic part contains the whole, so that the whole is nothing but this part fully unfolded? Or rather should we say that the organism as a whole has the secret of life, which it stores in these parts, the threadlike molecules of heredity? Without a recognition that the DNA is serving the organism, encoding its information, one cannot understand by any conceivable amounts of unadorned physics and chemistry what is happening in the shiftings around of these molecules, or of their constituent atoms and electrons.

The whole organic program is inlaid into nearly every cell, although only a fraction of it is expressed at any time. The whole secret perfuses all the parts, but the secret is a secret of the whole, not of any mere part, even if it is stored in all the parts. There is about one meter of DNA in each human cell, wound up in a fractional millimeter, containing more information than in a large set of encyclopedias. This information is reproduced in each body cell (with exceptions such as red blood cells and bone cells). If the DNA in the human body were uncoiled and stretched out end to end, that slender thread would reach to the sun and back over half a dozen times.[4] That conveys some idea of the astronomical amount of information that is soaked through the body. Nothing like this is found anywhere in physics and chemistry. An active biologic drives events that are superimposed on passive chemistries. It is neither the chemistry nor the physics that makes a gene a gene. Rather, it is this *information,* and the gene makes no sense except as the control center of an organism.

Here is what more there is in life, a "more" never found in physical nature: an informed organization of ordinary chemicals, a morphology and metabolism extraordinary for physics and chemistry, though ordinary to life. Life thus involves no new physical force, no new materials, but it does involve a new process and power, that of informational control. The materials and energies of the substratum are in constant turnover, but the life identity does not derive from them, necessary though they are at any moment in time. It derives from the informational set. On the basis of this, living things can have resources, sources that they take over and redo by incorporating them into a program. In physics and chemistry as such, there can be only sources, never resources.

On the basis of this information, there now arises the possibility of things being important because the organism has interests, a notion that can have no place in mere physical science. By the time we rise to the level of behavior, genetically based though this is, we may wonder if "causes" have not had some "meanings" superimposed on them, so that a "territorial call," "aggressive threatening posture," or a "courtship display" has a meaning even in the animal world. At least something is being "signified."

Perhaps, too, the nature of this vital information will be such that life can continue to have even its sacredness, so that life is not devalued but made the more impressive by its emerging from and its superintending and transcending of physics and chemistry. Thus, the sorts of causes found in bioscience, when seen in the service of these functions that are "meaningful" (or "significant") in an informational context, need not be hostile to a religious interpretation of life.

Historical Continuity in Life

Two complementary facets of organic identity are of interest here. The first is a *historical continuity to the entire life pulse,* especially visible at the molecular level, where kindred metabolisms and structures have been conserved and elaborated upon for several billion years. Life is perhaps four billion years old, one-fourth of the age of the universe. The glycolytic pathway, present in all cells and especially crucial in the blood and the brain, antedates the presence of oxygen in the atmosphere. The genetic coding used for protein synthesis is at least two billion years old, conserved in essentials, modified in details. The three-letter code that is used now seems to have evolved from an ancestral two-letter code, both used to store the information by which, in various life forms, a billion different kinds of protein are keyed. Life overleaps any particular parts and is reincarnated through passing material sequences. Thus, corporate and conformational biological identity persists where individual and physical identity is transient. The germ plasm flows on.

This story is serial and developmental. Small organisms can work by diffusion, but to be of any size, complexity, and mobility, an organism must have a circulatory system and oxygen-carrying molecules. In vertebrates, myoglobins evolved first, probably initially functioning as circulatory molecules, but now conserved within muscles as an oxygen transporter and storer. Myoglobins are compactly folded polypeptide chains that contain additionally a prosthetic group, that is, a functional unit embedded in the folded chains where an iron atom by alternation of valence states carries an oxygen molecule. Later, hemoglobins evolved, consisting of four units now fused, each unit about like the ancestral myoglobin molecule in shape and functional capacity. But differences in the polypeptide sequences make possible the dramatic new capacities of an allosteric protein, that is, a protein that changes its shape to suit the needs of the organism.

There next also arose regulatory sites at which the presence or absence of various control molecules increases or decreases the oxygen-carrying capacity. Further, the loading or unloading of oxygen at one site cooperatively affects the other sites, which allows amplification and fine-tuning of the molecule's oxygen-carrying capacities. The hemoglobin molecule is thus quite active or reactive. Its properties shift in response to environmental input so as to protect the metabolism of the organism. Myoglobin is a fully functional molecule, but hemoglobin is "a macromolecule capable of perceiving information from its environment."[5] Thus, the teleological phenomena associated with life, so far from disappearing on microanalysis, are present in reflexive molecules on molecular scales, and they persist and develop globally across eons of time.

Particular Individuality in Life

But if life is a new kind of *wave,* as it were, which swells over changes in the matter across which it propagates, it is, secondly, a new kind of *particle,* by which we refer to how biological identity is also associated with a *particular individuality.* The living unit has a quantum of centricity, the key to its functioning as an individual. That the organism has a kind of "objective *self*" is evidenced by the barrier it posts between self and nonself, by the maintenance of a program for which it has the cybernetic know-how. An organism can *perceive* and *act* not only appropriately to its kind but appropriately to itself. "Perceive" and "act" are terms that, like the term "self," we know in their richest depths in our human existence, where they are paradigmatically understood as accompanied by subjective inwardness. But they have to be deployed across all of objective life, even in nonneural and relatively immobile forms. Here one may prefer, as in plants, the term "individual" to the term "self." Any living thing can respond to its environment so as functionally to protect its individuality, a capacity that has its foundations right at the molecular level. An individual organism stands off the world by a continuous performance.

Life is a corporate pulse. Yet living things come individually instantiated, bounded wholes in a way that inorganic things are not. This is evidenced at the smallest range by the cellular structure of protozoan life, whether eucaryotic or procaryotic. It is marked by the semipermeable edges, membranes, skins, and bark of metazoans. Colonial forms represent a passing of this individuality from one level to another. All life occurs with support from other organisms, sometimes in a highly social matrix. All organisms *need* their physical environment in order to exist, as inorganic things do not. But they need it as a resource for self. They are not self-contained, but they contain an individual self. An organism must accept and integrate foreign material, bring this within its self-reflexive control, or reject, eliminate, or encyst it. So the organism moves on through time in quasi-independence of its elements, ever overtaking and discharging the physiocochemical materials with which it composes itself.

This capacity for objective individuality, resting on genetic recombinations and mutations, is expressed in protein structure and antibody formations that can be quite specific. There are over three billion humans who share biochemistries with one another and to a great degree with trillions of mammals, reptiles, bacteria, and plant forms. Yet any one human body will recognize as foreign a bit of tissue from any other person or other living individual.[6] Such tissue is not enough like oneself to be one's own, and it will be rejected and destroyed. This is not so with all the fluids and materials of life. Blood can be transfused, food digested, and life materials interchanged owing both to the kinship and powers of capture that we have recognized. But once again biological identity transcends physical identity, now with the addition of an intense individuality to kinship. This is a first encounter with what we shall later recognize as idiographic dimensions in nature, past its nomothetic dimensions.

Some will protest that terms like "self," "recognize," "perceive," and "act" are here inappropriate. All this is quite "blind" or "dumb." But it is just as difficult to know what these latter terms could mean applied to such organisms. We do not suppose that all living things see, have subjective awareness, or have sentient neural capacity. Only some do. But all are demonstrably discriminating in their active capacity to protect their individuality. They are not subjectively cognitive, but they are objectively living, self-protective informational systems, whose workings at the molecular level are not so much "blind" or "dumb" as really rather clever in their active logic. This issue will return with more intensity when we consider the evolution of species.

The ultimate forces and constituents of matter, which physics undertakes to analyze, were initially thought to be rather simply mechanical and material. And relatively speaking this is so. This is partly why physical theory has been able to predict in rather specific form many entities and phenomena that, when looked for, were found, such as the neutrino, Neptune, helium, and time dilation. Still, even the stuff of physics has proved quite clever and sophisticated, full of possibilities for development and defying exact analysis. One striking thing about microphysical and astronomical nature is how it permits the evolution of life. The matter-energy system is suitable for life, although nothing known in physics or chemistry would have enabled anyone to predict its evolution.

When life does come, the fine structures of life and the metabolisms that drive life may once also have looked rather simply mechanical and material. But they have proved even more clever and sophisticated, full of possibilities for development and resisting easy analysis. Unlike physics, biological theory has been able to predict in specific form almost nothing of what it has later found—not cells, not the nucleus, not chromosomes, muscle fibrils, hemoglobin, or any of the specific protein codings. The life processes surpass anything known in physics and chemistry because in life we have an informed organization that preserves an individuality and also enriches itself historically over time.

To use a provocative term, matter here takes on a kind of *spirit.* This spirited, emergent quickening is unprecedented in physical science. We have first located it in biochemistry, calling it *information.* But before we can take full stock of its significance, or set this in still larger perspectives, we have to puzzle over its origins. That moves us next to evolutionary bioscience.

Molecular Biochemistry versus Evolutionary History

Biofunction runs right down to the molecular level, and life is coded to the genes. So it can seem that life has been reduced to molecules in motion. But what determines the shape of these genes? They have been selected for—not at the microscopic level but at the level of organisms in ecosystems. The biomolecules (genes, enzymes, proteins) can seem to determine what the phenotype is, but then again the form of life, the needs, the environmental niche of the organism determines what genotype, what biochemistry is selected and maintained. So the shape of the activity, the conformation, the information at the molecular level is thrown back up to the macroscopic level. Selection operates on the whole organism, not the gene, and so the information stored in the molecular shapes and codings is a story about what is going on at the middle-range level—something like a book with its small print that contains a story of the big world. Molecular and organismic biology tracks big-scale evolutionary biology.

No sooner are we tempted to say that life is nothing but biochemistry than we realize that the biomolecules, selected to provide survival of the better-adapted organisms in ecosystems, are nothing but the recorded and continuing evolutionary story. The reduction can seem complete in biochemistry at the molecular level only in a momentary cross section of what is a dynamic historical process, and the extended process is not merely molecular, but molar, indeed regional and planetary. The ecosystem determines the biochemistry as much as the other way round. The shape that the microscopic molecules take is controlled "from above," as information discovered about how to make a way through the macroscopic, terrestrial-range world is stored in the molecules. Sometimes it is hard to say which level is prior and which is subordinate; perhaps it is better to say that we find storied achievements at multiple levels.

As we turn to the macroscopic evolutionary level, the life processes there may seem interminably slow. So it is perhaps worth noticing, to keep the fullest possible perspective, that from the molecular viewpoint the action is often just as impressively fast. A single *Escherichia coli,* reproducing itself in the reader's intestines, typically synthesizes each second some four thousand molecules of lipid, one thousand protein molecules, each containing about three hundred amino acids, and four molecules of DNA, and thus it will have made ten thousand molecules while you were reading this sentence.[7] In the mitochondrial compartments of your cells, where the citric acid cycle takes place, there are perhaps as few as eighty molecules of oxaloacetate, used in the cyclic process, and the turnover is five times each microsecond. A developing fetal brain can generate over a quarter million neurons in the minute during which the reader is here pausing to consider how much in the life process proceeds at high velocity and has proceeded so for billions of years![8]

2. EVOLUTION AND THE SECRET OF LIFE

Neither speed nor size is the most urgent question. Granting a programmed direction to the individual life course, is this present in the evolutionary wave in which the individual is so transient a biotic particle? Evolution has built a program into the individual, and by reproduction individual programs are passed across generations. But is there any program built into evolution? The *product* has ends, but does the *process* that produced it? That will prove a more difficult question, only partially solvable with the available biological paradigms, despite the fact that evolutionary theory is the most powerful of all those revolutionary paradigms that science has introduced. It is more powerful even than relativity theory, because of the scope of its reach across disciplines and because of the insights it gives into mutability, process, and history. Natural science asks about causes; religion asks about meanings. The question of the origin of life is not the same as the question of its meaning. One could know a good deal about either one in considerable ignorance of the other. Nevertheless, these questions overlap, because some causal accounts inhibit some meaning accounts, and vice versa.

Earlier thought had posited the fixity of species, a plausible first theory. It was held for scientific reasons, since things breed after their own kind and do not interbreed. It was held also for religious reasons, since the historical series of immutable species, traced backward, seems to suppose an original, special creation when species were fitted to their environments by a Designer. Newtonian physics had sought models of creation in a mathematical, architectural form, and the idea of a perfect machine was transferred to that of a perfectly designed organism. God made the creatures as a clockmaker makes clocks. Watches imply watchmakers; rabbits imply a Rabbitmaker. As a result, the then-current paradigm for design persuaded thinkers to overlook not only anomalies and misfits, mutations and monstrosities, but, more importantly, struggle, gamble, and loss.

The Darwinian Revolution

Charles Darwin posited instead small variations in degree within a surplus of offspring, a struggle for survival, and a natural selection by which the more fit survive. Though he had no theory for the variations, genetics was subsequently to supply one, and Darwinism plus genetics is commonly called *neo-Darwinism,* or the synthetic theory. Random mutations and other effects during reproduction introduce differences, mostly worthless or harmful, but rarely beneficial. The best-adapted survive, selected by environmental processes. When such selection proceeds over long periods, the incremental variations accumulate and there arise new species. In place of the divine designing of the fit, we now have a natural selecting of them. Before, one could notice how the woodpecker has feet adapted for grasping bark, two claws in front and two in back, a stiff straight-edged tail that with the feet forms a triangular prop, a sturdy bill for drilling; and the best available theory for these observations was that God had designed it so.

After Darwin, by natural selection alone it could not be any other way. Only well-fitted woodpeckers can survive. Any ill-fitted ones will long since have perished. Evolutionary theory explains the fit on natural principles alone. It describes how the origin of species takes place, going back at least to elementary forms of life. From these few forms has developed all the present biological diversity. Humans had now to view their own origins out of such ancestry. They were not so much children of God as descended from the apes. Michael Ruse, a philosopher of biology, asserts: "The secret of the organic world is evolution caused by natural selection working on small, undirected variation."[9] So now the secret of life, found by Watson and Crick in the DNA, Ruse also finds, following Darwin, in natural selection, and once again this seems to challenge the religious conviction that the secret of life is ultimately laid in God.

The serious trouble here does not lie either in naturally creative processes or in the long time span. Even Genesis reports that God bade Earth bring forth its diverse creatures, and the days of creation can be readily interpreted as eras. Nor, in an extension of the theory soon to be made, is the notion of the spontaneous generation of life a difficult one. Prior to the work of Louis Pasteur, most people believed in a continuing spontaneous generation of some lower life forms. Nor is the notion of an incremental creation itself bothersome.

The troublesome words are these: "chance," "accident," "blind," "struggle," "violent," "ruthless." Darwin exclaimed that the process was "clumsy, wasteful, blundering, low, and horribly cruel."[10] None of these words has any intelligibility in it. They leave the world, and all the life rising out of it, a surd, absurd. The process is ungodly; it only simulates design. Aristotle found a balance of material, efficient, formal, and final causes; and this was congenial to monotheism. Newton's mechanistic nature pushed that theism toward deism. But now, after Darwin, nature is more of a jungle than a paradise, and this forbids any theism at all. True, evolutionary theory, being a science, only explains *how* things happened. But the character of this *how* seems to imply that there is no *why*. Darwin seems antitheological, not merely nontheological.

In the Newtonian centuries, determinism was the troublemaker. It made science possible and was welcome for its ordering of the world. But an unrelieved determinism grew suffocating, for it seemed to leave no place for spontaneity and freedom, none really for autonomous organic life or mind. In Darwin's century, the pendulum has swung to an opposite extreme. Randomness is the troublemaker; it seems to disorder the world. All the higher world structures—most notably those of organic life, including human life—came through accident. They are selected for, of course, but each incremental development is random. Some contingency is welcome enough in a world view. It leaves place for the spontaneity and freedom Newton could not provide. But Darwinian contingency permeates our origins, governs our survival, and becomes oppressive. Life seems devalued by it. Physics in this century has modified its determinism to give us a more open nature, at the same time that it has given us an astrophysical and microphysical nature that is right for life. Is there some way that neo-Darwinism or its successors can derandomize biology by setting this element within a more patterned and constructive nature, a more predictable one?

Einstein and the quantum theorists have given us a startling mathematical world, one of time dilation, wavicles, quantum jumps, energy-mass interconversions, of particle annihilations with new particles created. The astronomers have given us galaxies, black holes, and inconceivable distances and timespans. These worlds are often counterintuitive and remote from the native ranges of life. But Darwin gives us a pragmatic jungle where the rule is eat or be eaten, do or die. The world line of a zebra, grazing and fleeing from its predators, seems almost in another universe from the world line of an electron, orbiting a nucleus and absorbing a photon to rise to an excited state.

The book of nature is not written in the cold beauty and elegant symmetries of mathematics. It is written in contest and ordeal, in blind groping, in waste and survival. Physics gives us a world that is ordered, rational, predictable, at least statistically. But it does so by abstractive simplicity, leaving out these biological goings on of higher interest. When biology attempts to describe that life world, events seem to have too little order, rationality, predictability. They are often inelegant and doubtfully aesthetic. Physics has given us a universe that seems almost "fine-tuned" for life right from the start, and billions of years of impressive cooking up of the precursors of life. But when biology arrives all this seems to vanish into struggle and randomness. Can they both be describing the same world? Does the truth lie somewhere in between? Are there yet further levels to consider?

What are in fact deterministic systems can generate patterns that seem quite random (as with the so-called random number generators used in computers) and, on the other hand, relatively or even absolutely random processes at some levels (the dates of death of individual citizens, the decay of individual radioactive atoms) can quickly generate order at other levels (statistical death rate tables, half-life curves). It may be that randomness at some levels is a precondition of creative, open, flexible order at other levels. Too much symmetry, too much predictability and order can sometimes be an imperfection; disorder can be a principle of dynamic construction, as well as of destruction.

Biological and General Evolution

Before we inquire whether evolutionary theory can be accommodated to a revised religious gestalt, we must look more closely at the theory itself. Darwin's is a *special theory* of the evolution of species by natural selection, but a *general theory* of the evolution of everything soon comes to overarch it. The paradigm was deployed into the abiotic world. Geologists came to speak of the evolution of Earth and its landforms; astronomers report the evolution of stars, galaxies, and even the universe. The concept is extended into the social world. Anthropologists sought to trace the evolution of societies: beginning with chiefdoms, moving through agricultural kingdoms, on to nation-states and industrial societies, with the best-adapted surviving. Evolution can also be stretched into politics, economics, history, even religion.

When it leaves biology for the inorganic sector, the notion of a natural selection of the fittest drops out, for adaptation makes no sense without a survival urge. There are no genes to select; nothing is surviving because nothing is living. Natural selection is replaced, if it is replaced at all, by an analogical and yet importantly different notion of thermodynamic selectivity, the *continuing of the stablest* abiotic configurations. A planar planetary orbit system lasts on, but there is an early collapse of nonplanar, unstable ones. The selection in the abiotic realm is of the probable, of the statistically more likely (at least in classical thermodynamics), whereas the selection in the biotic realm seems rather of the improbable, that is, the rare and less likely mutation but novel and more fit life structure, if random or even statistical factors were alone to govern what persists in time.

When natural selection moves into the cultural realm, what is selected is no longer merely genetic mutations, but, more importantly, selection is of acquired and learning-transmitted traits, a notion more Lamarckian than Darwinian. It is not simply genes that are being selected for, often not genes at all. It is ideas and their resulting behaviors. Culture is not simply a question of organisms reproducing and surviving; there conceptual schemes and world views survive and propagate themselves, or die out and become extinct.

Even confining evolution to the organic arena of its inception, recalling that natural science seeks causal explanations, the notion of natural selection is an *odd sort of causation.* It posits, to begin with, random mutations, which, even when traced to their genetic basis, are not known to be, and need not be, the results of necessary and sufficient causes. These random variations make a difference only as emplaced in the genetic set of an organism with a survival drive, placed in an environment. Natural selection, although a nonconscious selecting, is still a force that picks the few out of many options, picks the best-suited for life, selects for *biofunction.* Nowhere in physics or chemistry do we meet a causal or other force of this kind, not even when we notice how there are physical constants, ratios, and processes on astronomical and microscopic scales that are favorable for life. Indeed, just because we now find a natural pressure that favors biofunctional efficiency, we have a distinctively biological operation, a positively prolife force in this respect, however groping, blind, or unmerciful it may otherwise seem. Physics talks of forces and fields, but biology introduces something brand new: fitness. There is something extraordinary, at least from the viewpoints of physics and chemistry, about a causality that operates so that the system selects A over B because (on the cause) of increased adaptive fit. Or perhaps that is just the extraordinary form that the universe has been taking from the very beginning.

Incremental Evolution and the Fossil Record

We must carefully distinguish between (a) *a description of gradual development* and (b) *natural selection as the key explanatory principle* of this development. Evolution can mean either, and means both in neo-Darwinism. *If* there has been incremental development, *then* several observations will follow. The fossil record will have the simplest forms at the bottom. One will expect kinships between life forms, comparative morphologies showing common yet modified structural themes, nested sets of resemblances, inherited structures branching from ancestral forms, and patterns of biogeographical migration and distribution. All this we do find to an impressive extent, although the fossil record is a bit piecemeal. There are transitional individuals and successive species within the corals, bryozoans, gastropods, pelecypods, ammonoids, and the like. There are some convincing fossils of major transitional lines, for example, *Archaeopteryx*, linking the reptiles and the birds.[11] Sometimes these are fewer than evolutionists could wish, especially at the start of the Cambrian period, when the major phyla emerge. It can be expected, to some extent, that transitional forms will be rare and that the more ancient and softer-bodied forms will be preserved poorly.

A layperson to paleontology will have to puzzle over how some experts think there is and others think there is not ample evidence of gradation. David Raup, a paleontologist and curator of geology at the Chicago Field Museum of Natural History, writes, "The evidence we find in the geologic record is not nearly as compatible with Darwinian natural selection as we would like it to be. . . . We are now about 120 years after Darwin and the knowledge of the fossil record has been greatly expanded. We now have a quarter of a million fossil species but . . . the record of evolution is still surprisingly jerky and, ironically, we have even fewer examples of evolutionary transition than we had in Darwin's time."[12] (Some transitions once thought plausible have been found implausible with new fossil evidence.)

David Kitts, a paleontologist and philosopher of science, concludes, "Paleontology . . . has presented some nasty difficulties for evolutionists, the most notorious of which is the presence of 'gaps' in the fossil record. Evolution requires intermediate forms between species and paleontology does not provide them. . . . Discontinuities are almost always and systematically present at the origin of really high categories."[13] "Phyletic gradualism," say Stephen Jay Gould and Niles Eldredge, paleontologists, "was never 'seen' in the rocks;" it was an assumption that "has now become an empirical fallacy."[14]

Nevertheless, and in general, an evolutionary description of events, together with a number of contributing factors, is an empirical hypothesis that can be tested. Evolution is false unless the serial life process is incremental with heritable variations, and unless there is a surplus of offspring that must compete for survival. Nonblending variations that do not get swamped out but persist make the theory more credible. We can see why large gene pools have a certain genetic inertia and small populations will sometimes speciate more rapidly. The experimental results in genetics, mapping gene flow across generations, provide considerable confirmation, though in a somewhat oblique way, for Darwin's descriptions. Darwin's contention, against Lamarck, that acquired parental traits are not transmitted to the offspring was subject to experimental verification. Darwinism would be falsified if human bones were to be found in Cambrian strata, or if there were other major surprises in the fossil record. Darwin has a place for leftover or vestigial organs, such as the appendix, male nipples, and degenerate wings on now-flightless insects.

Many biologists are impressed with the more sudden, steplike character of phases of evolution.[15] Some species seem to have arisen almost at a stroke, perhaps by chromosomal doubling, hybridization, or other major structural innovations, followed by close inbreeding. It is often difficult to visualize how structures—rudimentary legs or wings—could have been useful during the incipient stages of a very gradual development. These are a liability unless they are working, and one needs to suppose some more rapid construction at the start. So one may need to look for faster innovative principles than Darwin supposed with his minor variations. The jerkiness is not all an illusion created by capricious gaps in the fossil record; the evolutionary history was itself jerky. Yet there is nothing in natural selection theory to explain why, and the jerks were the principal occasions of creativity. There is (relatively) quick speciation that punctuates the evolutionary flow; there are (to adapt a term from physics) quantum leaps here too, with the subsequent explosive diffusion of a new "idea." Events are as revolutionary as they are evolutionary. But the positing of swifter steps in addition to minuscule variations does not seriously affect the main Darwinian description that there has been gradual development, more or less, or the main explanation that chance mutation (or hybridization, etc.) is edited so that the more fit survive.

A Theory Difficult to Refute: Tautology or Comprehensive Theory?

When we turn to *natural selection as the leading principle of evolutionary development,* the issues become more complex. Here the positive empirical record, which if it were missing would falsify natural selection, does not when found to be present suffice to justify the theory. In light of the "if T, then O" logic of science, some further theory might explain the unfolding sequence more adequately. Similarly, some other theory might incorporate as well the results of experimental genetics. The developing process may be richer than that portrayed by Darwinism and neo-Darwinian modifications. But then again, what would it take to unseat natural selection theory, once it has come to prevail? At this point, natural selection takes on a puzzling and ambiguous character. The theory becomes established only with much patient observation, and after long dispute; but, given the outlines of the process, afterward it becomes axiomatic, some kind of constitutive tautology.

One finds it difficult, even absurd, to imagine that over generations the misfits could statistically prevail, so the long-term survivors, whatever they were, just had to be the best-adapted in their ecological niches. On the one hand, there is lots of evidence, direct and indirect, about selection pressures at work in the world. Birds catch more of the moths that are poorly camouflaged; the distribution patterns of insect speciation on clusters of oceanic islands show responses to available food in niches. On the other hand, natural selection becomes a kind of governing gestalt that eats up all the evidence, casts out anomalies, and pays little heed to any counterevidence. The theory is not only difficult to test, it is difficult to doubt.

It now becomes hard to ensure that the theory is not trivial or circular, that is, that the survival of the fittest does not reduce to the survival of those fittest to survive. In evolution in the inorganic realm, the continuing of the stablest systems means just the continuing of those most likely to continue! Since "the fittest survive" has proved a somewhat troublesome phrase owing to its connotation of gladiatorial struggle, many biologists prefer to say that "the better-adapted survive." But this softer language does nothing to reduce the tautology. The better-adapted are, by definition, those who leave the most offspring. Consider, for instance, how saying (as biologists commonly do) that a species became extinct because it lost adaptation (= survival capacity) is no explanation at all, only a tautology.

Biologists and philosophers of science (entirely apart from the interests of theologians) have often worried over this dimension at the core of the theory. C. H. Waddington, a prominent geneticist, complained, "Natural selection, which was at first considered as though it were a hypothesis that was in need of experimental or observational confirmation, turns out on closer inspection to be a tautology, a statement of an inevitable although previously unrecognized relation. It states that the fittest individuals in a population (defined as those which leave most offspring) will leave most offspring."[16] Steven M. Stanley, a leading paleontologist, concludes, "I tend to agree with those who have viewed natural selection as a tautology rather than a true theory. . . . The doctrine of natural selection states that the fittest succeed, but we define the fittest as those that succeed."[17] The theory is circular, though it serves heuristically to help us understand what is happening on short-range scales. It is not illuminating about the long-range macroevolutionary process. Karl Popper, a philosopher of science, has serious misgivings about natural selection: "The trouble about evolutionary *theory* is its tautological, or almost tautological character."[18]

The theory of evolution, says R. H. Peters, a biologist, is actually "a tautology" and, though useful, "cannot be called a scientific theory." That is the reason why "evolutionary theory is sufficiently general to accommodate any observation."[19] Well, not *any* observation, since many observations could defeat the claim that there has been incremental development from simple forms, and even the natural selection part of the theory could be defeated by some observations—the sudden appearance of complex organs that had no imaginable survival value. But natural selection readily accommodates such an enormous variety of observations that we begin to wonder whether this part really is immune from testing because it is some sort of assumption not open to refutation.

The logic of the theory is like saying "May the best team win" and then defining "best" to mean "scores the most points." "Best team" does have to be defined with reference to some kind of game, baseball, and observations might reveal that such a game was not being played at all. But given that baseball is being played, that such a "best" team always wins is not surprising; it is inevitable. This can be restated as a statistical rather than a logical inevitability. Once in a while a generally superior (= usually winning) team has a sloppy day, or hard luck, and the best team (overall) loses that day. But steady losers are bad teams. It won't be surprising to find, at the close of the season, that the best teams are in the play-offs; the way the league operates, it couldn't be any other way. That the best-adapted survive is not surprising; it is inevitable. But neither principle will illuminate us much about the success stories—about the Yankees who won the series with Babe Ruth batting sixty home runs in 1927 or the Dodgers who won the pennant with Jackie Robinson, grandson of a slave, breaking the color barrier in 1947, or about the Precambrian microbes who won because they discovered (in stages) photosynthesis or the Pleistocene primates who won because they coupled deliberate thought with opposable thumbs.

It is difficult to distinguish by any criteria that could falsify the theory between natural selection as it has now degenerated into a presumption brought to experience (a blik) or a necessary relation that is inevitably true in our sort of world (a constitutive tautology) or a true theory that, though it might have been false, cannot be falsified anywhere in the real world simply because it is entirely true (a perfect theory). Have we finished our inquiry about the formative forces in the evolutionary development? Or have we an alluring dogma that is preempting other forms of inquiry because it is blinding us about what else to look for? Are we committing and recommitting some systematic error? In theory, one would hardly expect the less fit to survive, on statistical average. But it might be that from among competing forms that are equally fit some survive while others do not, perhaps only randomly, or perhaps selected by some other force that our present theory is unable to detect. We do not wish to deny that natural selection takes place and that many organic phenomena are impressively explained by it. But is "adaptive significance" the truth, the whole truth, and nothing but the truth?

Within the main factual parameters, without which the theory could at once be shown to be false (development, variation, surplus of offspring, heritability, etc.), we can now insert myriads of different states of affairs, and any particular state of affairs can be replaced by dozens of other states of affairs without tampering with the theory at all. The theory moves at a high level of generality (the best teams win!) and provides a heuristic explanatory rubric into which thousands of specific cases can nicely be fitted. But so too can thousands of different events that never happened but might have. When any troublesome specific event is encountered, the theory brackets it as an anomaly that can later be handled when more is known.

The theory seems to connect ancestors and their descendants in a tight and statistically determinate way (the best-adapted offspring survive), but on closer inspection the particular ancestor-descendant sequences that have taken place, as opposed to myriads of others that might have but did not take place, are not implied at all, by either prediction or retrodiction. All of them are buried in randomness at the same time that they are covered by the umbrella of the survival of the fittest. So we really never generate falsifiable predictions that expose the theory, though we may find ourselves mistaken about this or that particular application of it, or allow trivial phenomena without survival effect (the dark red flower at the center of the umbel in *Daucus,* Queen Anne's lace, or the locations of various spots and bristles on caterpillars).

That the best-adapted survive is (as computer scientists say) a very short "program," it is (as philosophers of science say) a very clear paradigm, and it may seem to be quite powerful in generating understanding of so much that has happened, quite comprehensive—until we realize that it covers so many nonevents with as much power and comprehension, and then we see that the program is too short, the paradigm too empty, really to explain what has actually taken place. What explains too much, explains nothing.

Natural selection is a key that can be twisted to fit many locks. One will use the notion of fitness in a competitive sense, so long as that suits; but then if the notion of fitness calls for cooperative dimensions, these can be introduced. Siblings compete in the nest; each one defends itself and its genes against the others. But the fledglings soon learn to give alarm calls, and the caller risks revealing its own location to the predator in order to warn the flock. But now, group selectionists will say, selection is not for the individual but for group survival. No, others contend, the caller is really defending by proxy its own genes as these are scattered out and shared by cousins. Or perhaps the alarm really protects the caller by making all flee or freeze together so that it can immerse itself in the fleeing or camouflaged group.

Stags fight for the possession of does, and one might expect the evolution of maximally lethal antlers, so that victors will leave the most genes in the next generation. But the antlers branch and curve in such a way that serious injury is rare; natural selection seems rather to have reduced bloodshed and substituted conventional bouts. So one supposes some check on the individual stag that favors the population as a whole. Here group selection trumps individual selection. No, say others, there is some more complex strategy by which the individual stag reduces his chances of loss.

A distasteful chemical on the surface of the skin benefits the individual toad, as may a chemical inside the body, if the toad is spit out and survives attack. But some toads do and others do not have a bitter taste, and both kinds alike survive, so for the edible ones we posit protective resemblance to the inedible ones, or better hiding behavior, or reproductive efficiency, or something else. In the case of bitter-tasting ants it seems that the individual must be half-eaten for the effect to work, and so, like the bee that dies with its sting, we assume only a benefit for others. We can probably suppose, however, that in former times there was some benefit to the individual by which this phenomenon originated. Or the kamikaze bee is really defending copies of its genes scattered elsewhere in the hive. One will adapt natural selection individually or populationally as need be, genetically or socially, defining the "fit" to be those who one way or another balance "just so" these competitive or cooperative, individual or populational, genetic or social elements, if only with a slight incremental advantage. There are numerous models of various minor trends where structures or behaviors—wings or hiding ability—were progressively elaborated owing to survival value. All this can seem quite plausible, and testifies to the explanatory power of natural selection. But meanwhile it becomes difficult to conceive of observations that could defeat the theory. Those alert to the persuasive, blinding powers of a belief system begin to worry that the theory is too good. If we wanted to put the main principle of adaptation to the test, and not just this or that local application of it, how could we?

Sometimes one can show, on engineering principles, that this rabbit was more fleet-footed than was that one using the same energy budget, and thus one might have predicted it would leave more offspring. But in the full round of complex environmental interactions, we seldom know all the trade-offs. The theory did not come into acceptance, and it does not continue to prevail, by much of this kind of evidence. *If* that rabbit had longer legs, *then* it would more likely survive. That seems testable enough. But if our prediction fails we are quick to introduce compensations to salvage the theory. This rabbit's legs were longer and faster, but, alas, it failed to survive because the extra length weakened the cross-tensions in the bone, which thus broke a bit more easily. Or the extra musculature adversely affected live births, or something of that kind. Any form that survives does so by natural selection. Find the adaptation! Any form that becomes extinct dies by natural selection. Find the maladaptation!

If the *Pteranodons* (large flying reptiles) had been better designed, more aerodynamically efficient, then they would not have become extinct. But they did live millions of years, far longer than humans have lived on Earth, and it is hard to say what was wrong with them aerodynamically. Well, their environment must have changed, or something else outcompeted them for their food supply. When a burrowing animal is being pressed by a predator that routinely digs it out, perhaps it will learn to dig faster, or reproduce faster, or climb trees, or evolve spines or sharper teeth, or develop a noxious taste, or become so rare that the predator finds it uneconomical to hunt it and switches to another prey. Perhaps it will go extinct. Almost any outcome will support the theory. The allegedly decisive influence—survival of the best-adapted—would have been just as present had events gone other ways. There is no serious contrast class of events ruled out, although fanciful ones may be (that the burrowing animal grows wings and flies off to another continent).

If anomalies arise, one tends to say that there must be some survival benefit that will later be learned, or that these exceptions can be taken care of by an auxiliary hypothesis. There was some genetic linkage to something else that did convey a survival benefit, a fitting into a larger functional complex, or a hitchhiking on a beneficial gene, so the main body of the theory is conserved. Where rudimentary organs seem to have no utility, there will be the compensating reply that in these cases a steplike inception did convey a survival benefit immediately. Where there is a local maladaptation, this is due to some evolutionary lag, only a temporary fluctuation in the larger system.

One will allow, now and again, that by chance a more fit individual nevertheless was unlucky, and even perhaps that the more fit races became extinct in extraordinary environmental circumstances (the upset caused by a large meteor) that did not affect their less fit cousins separated geographically from them. One will find opposite strategies with the same effect. The bald eagle aggressively defends its nestlings, and of course that must have survival benefit. But its first cousin, the golden eagle, upon similar disturbance, flies away and abandons them, to nest another day. And so? That too must have survival benefit. There is more than one way to skin a cat! Always, some interpretation is found by which the more fit survive.

The theory has a massive explanatory power achieved by a peculiar way of resting on observations yet with a simultaneous capacity to absorb diverse observations and remain relatively isolated from experience. Have we the ultimate truth? Or a hard-core paradigm exempt from falsification? The problem is that a ubiquitous characteristic that characterizes every living thing in general really characterizes no living thing in particular. Since, within these general features of the sort of world we have, natural selection characterizes every reasonably imaginable course that might have been taken, it does not distinguish the courses that were taken from others not taken, and so does nothing to explain the particular narrative of evolution. Natural selection is so all-inclusive that it does nothing to illuminate the exclusive historical course from microbes to men. The particular state of affairs we in fact have on our hands—this evolutionary ascent from matter to life to mind—might or might not have been the case entirely consistently with the theory. The course of events presently occurring cannot falsify the theory; and, worse, neither would any of hundreds of alternative courses of events. Thus, it is no accident that the principle cannot predict anything, covering, as it does, everything that eventuates or might have eventuated.

Natural Selection as Universal Law?

Is this a provincial theory about what has happened on Earth only? It seems to be of that kind, and thus we have in biology a law of limited, nonuniversal scope, such as is not met with in chemistry and physics. Or is it a theory by which exobiologists might predict what life must be like elsewhere? Does life develop by variation and natural selection on all other planets too? Can we be sure of this even before we discover life there, although we cannot predict what turns those random courses might have taken? John Maynard Smith, a principal theorist, writes, "We have as yet no grounds for asserting that if evolution has occurred elsewhere in the universe, it has done so by neo-Darwinist processes, although I would be willing to conjecture that it has."[20] Will the simplest forms there be the earliest ones? Will the brain cases in the lowest strata there too, if they are found at all, have to be the smallest? Could there be some planet on which, on average, the survivors are not the fittest? The theory becomes in these latter respects much like a religious creed, such as that God oversees history, not only in its resilience and accommodating of massive amounts of diverse material, but in its expectations about what the basic operational patterns of the universe are like.

The Darwinism plus genetics that is commonly termed neo-Darwinism, the synthetic theory, becomes puzzling when we imagine life elsewhere. Does anyone suppose that heritability will have to be Mendelian on other planets, with two sexual parents each contributing one of two genes at a locus, with independent assortment, the chances of getting one gene independent of the chances of getting another? Hardly so; it seems that genetics is Earthbound in the form we now know it, though if there is death elsewhere there will have to be some kind of heredity. Still, even after restricting Mendelian genetics to historical Earth, one might insist that wherever there is life, the fittest survive. If so, have we a universal law of nature? Or a tautology?

Anticipating the dialogue with theology, we can already notice that if all that natural selection involves is differential survival of the better-adapted, whatever the heuristic value of such a tautology in biology, it hardly seems alarming to theism that God would arrange a world in which some individuals leave more offspring than others, which means that they are more fit, any more than it disturbs theism to find that God has arranged a world in which some astronomical systems, elements, or microparticles are more stable than others, which means that the stablest survive. We will not be disturbed to learn that there is "descent with modification" (Darwin)[21] or that there is "change in the genotype of a population" (Dobzhansky).[22] Nor is it surprising that God used such processes in the incremental generation of a universe and of life on Earth.

There can be no world without its stabilities, and yet the instabilities of physics, which involve the decay of one thing, can result in the creation of something else. We can hardly envision life in our world without its being carried by individuals of limited life spans, subject to death, and yet even biological death results in the creation of new and more fit individual organisms. Indeed, a system that selects (blindly or otherwise) increased adaptive fit does not in this respect alone seem ungodly at all; to the contrary, it seems a good thing. Only if what it takes to be more fit, perhaps ruthless selfishness, or the supplying of the increments of additional fitness is blind and chancy, nothing more, will the process begin to look ungodly. Meanwhile, natural selection as such is not troublesome, for this is all but inevitable in the sort of world we have. Rather, it is the character and source of the fitness that need closer examination.

Too Much and Too Little Explanation

In contrast to the theories of physics, evolutionary theory is *nonmetric* and *nonprecise,* and hence *nonpredictive,* and seeks neither manipulation nor control. This is not true of the genetics that has been grafted on to Darwinism, but the mathematics and experimentation in the latter must not obscure the lack of both in broad-scale evolutionary theory. In population genetics, we can model gene flow and selection pressures mathematically with some predictive ability on a statistical basis. One can run computer models of how a gene will behave in a population, assuming certain selection pressures; and sometimes the results can be field-checked. But if a particular model fails in the field, the fundamental theory is never revised, never even suspected; rather, a variant computer model is devised. Evolutionary theory was and is accepted independently of all this, and the mathematical models are limited in scope to tightly defined, idealized conditions. The mathematics apply only to tightly bounded situations that can be completely specified, to evolutionary "cages." In any mathematical statement the fittest must be defined as those that leave the most offspring, and so the sheer mathematical form is largely vacuous.

Particular versions of evolutionary theory (kin selection theory, group selection theory) may be revised because what is predicted is not observed. But the really crucial events resist quantification and prediction, such as the nature and direction of mutations or recombinations, what adaptive significance they will have in the complex ecologies in which they occur, or the potentials of the variations on a gene (alleles) resident within a wild population in an environmental crisis. One predicts how fast melanic moths (if such a mutation appears or such an allele is waiting in a population) will drive out the lighter moths on the soot-stained bark of trees across the industrial English countryside. One predicts that resistant strains of bacteria will evolve, making present antibiotics ineffective, perhaps how soon this will likely happen, though if no such mutants appear, the theory is none the worse. One predicts that white forms will flourish in the Arctic, small plants on the tundra, and perhaps quantifies the survival value of whiteness or smallness.

But we cannot begin to put the whole organism in its field environment in any calculable form. Even if we could, far beyond this, there seems little hope of applying any trends, in mathematical form or not, to the limitless context of the robust world across time spans of any length, given the presence of mutations, surprises, and the impingements of unrelated causal lines. The mathematics nowhere touches the big-scale history. There is no mathematical theory that puts wings on dinosaurs and makes them birds, much less one that transforms *Ramapithecus* into *Homo sapiens.*

C. H. Waddington, though a founder of mathematical genetics, protests, "The whole real guts of evolution—which is how you come to have horses and tigers, and things—is outside the mathematical theory."[23] We know nothing of the causes, if such there were, much less how to measure them, of any of the historically crucial mutations upon which the major events of the evolutionary past have turned.

Natural selection's innovative principles—mutation and other variations—are especially unforeseeable and vague, so that just where the cutting edge will turn next is mostly guesswork. It is about as easy for theologians to predict where the Spirit of God will turn up next. Those who are inclined to characterize science, especially in its opposition to religion, as metric, predictive, and manipulative, subject to tight observational refutation, have to ponder how evolutionary theory, which is that particular science most troubling to religion, is not of that character at all. The phenomena it explains are nonrepeatable and largely nonexperimental. Without repeatability, predictive power, metric precision, or definite falsifiability, one has lost many of the marks of hard science. Rather, after natural selection has established itself based on certain data, it shares with religion a tendency to explain a historical sequence after the fact by the application of a pervasive and persuasive interpretive paradigm.

The main challenge to natural selection from within biology has come from a type of theory that, from a philosophical point of view, is equally puzzling. Various *neutral* or *drift theories* suppose that a significant percentage of mutations are functional but neutral to survival, thus permitting the life pulse to drift.[24] Natural selection cannot "see" mutations that are neither beneficial nor detrimental. Many of these mutations are at the molecular level, and make no difference to the functioning of the organism. Others make a difference phenotypically but not one that is germane to survival, and these are likewise untouched by natural selection. Such a theory, though neutral, is sometimes more testable than the thesis that every significant persisting trait has adaptive value. We have found it difficult to prove that a trait has no adaptive value. As we have seen, one can always hold out for positive value as yet unknown. On the other hand, a trait formerly thought to be neutral can later be shown to have adaptive value. Meanwhile, many biologists hold that in both significant and insignificant events the life process drifts.

There are other sorts of drift. In small, isolated populations, which are frequently important in the early stages of a trend, random components in a few founders' genetic sets can shape the subsequent course of a community. Random differences at a threshold during initiation can lead to widely different outcomes. Whether a fire starts when a spark falls into a dry forest depends on coincidences at the start—a few drops of rain, a puff of wind, how a few fallen leaves happen to lie—although once the flame is ignited, spread of the fire becomes nearly certain. Somehow a few finches happened to get storm blown to the Galápagos Islands, or, among several species blown there over the millennia by thousands of storms, two finches once happened to survive, and anciently to ignite the speciation that Darwin subsequently discovered there.

Or important turnings can result from quite disconnected events, if, for instance, a large asteroid struck Earth and cast up enough dust to reduce photosynthesis for decades, which is thought by some to have slain the dinosaurs at the close of the

Cretaceous period.[25] They were killed by something wholly irrelevant to characteristics they had been selected for; and mammals, not really any more fit in the Cretaceous ecology, replaced them. Some paleontologists claim that the mass extinctions at the close of the Permian period left only a few percent of species surviving and that the life processes passed through such a narrow bottleneck that random survivals affected the subsequent course of life on a global scale.[26] Or perhaps sometimes the slightly maladapted, with a bit of luck, escape the statistics and survive anyway, despite their handicap. Natural selection is true, so far as it goes, in that on average the less fit never survive. But this does not go far enough, because many, perhaps most, of the interesting developments have little to do with natural selection.

Now it is as though we were told about a baseball league in which, although the best teams win on average, there are a great many ties, and on these occasions the umpire flips a coin to see which team advances to the finals. So it is not true that the survivors are always the best ones, although they cannot be bad teams (steady losers). They may be there by luck. The fitness of the survivors does not have to be greater than that of the extinct, but it does have at least to be equal to it, on average. If greater, the outcome is (usually) settled by natural selection; if equal, it is settled by drift or randomness.

Such theories, though they deny that natural selection is an effective editing principle at certain turns of events, pose no positive principle with which to complement it. It might be that there are quite a number of kinds that have enough fitness to survive, all such kinds will have crossed a threshold of fitness, and perhaps the competition between them is not all that intense, not intense enough to settle the matter of which ones survive, but that the outcome is governed by other factors that our theory makes invisible (and which, therefore, we are calling drift). But meanwhile there does not seem to be selection by or for anything else. This only adds to the directionless element about which we have next to worry.

If a scientific theory cannot *predict* (or, what is the same thing, if it retrodicts any of very diverse things that can eventuate), then at least it can *explain,* and we shall be satisfied with it. Ernst Mayr, a leading theorist, believes that "the theory of natural selection can describe and explain phenomena with considerable precision, but it cannot make reliable predictions."[27] Natural selection first seems to explain, but *does it offer enough explanation?* Is this cause adequate to produce its effect? The effect, in sum, is the evolution of life and persons, which interests us at three main test points: (a) that simple things have assembled into complicated things, (b) that living things came to exist where before existed only nonliving things, (c) that psychical subjectivity appeared where once there was only physical objectivity. Once there was matter, later life, and then mind. If a fully adequate explanation, natural selection must give a satisfactory account ("with considerable precision") of how the antecedents are adequate to the consequents. But this proves progressively more difficult with each of the three test areas.

One trouble generally lies in the way that natural selection, though it selects the better-adapted forms and wipes out the ill-adapted, has a soft account of the innovative process. *On the supply side,* the innovations are said to be random. In the composing principle that generates what is to be edited, there seems to be a missing logical ingredient, or more going on than mere randomness conceptualizes. Events can be random with respect to the needs of the organism and to benefits that mutations confer, although caused by the impingement of unrelated causal lines (relative randomness). A cosmic ray might cause a mutation. But a random event can be underdetermined in the sense of being without necessary and sufficient causal antecedents (genuine randomness). These two levels of randomness can combine. Within the quantum indeterminacies, there is no complete causal account of why this cosmic ray came when it did, with this electronic result in the gene. While we can statistically show that certain radiations or chemical mutagens step up the mutation rate, we cannot in principle, much less in practice, be precise about this in any evolutionary sequence. Within the probabilities, there is a range of genuine randomness where there can be no further explanation, and it is absurd to expect anything more to be said.

But then again, how can we be sure that such randomness is not sometimes a veil of our ignorance, a veil of some creative power that our fishing net is not catching? Is this a randomness that leaves the supply of innovations "in the dark," a darkness that may later be penetrated by a better theory? Or does this randomness claim here an impenetrable darkness, banishing the possibility of later scientific explanations or complementary accounts from other quarters? Is it perhaps only a confession that, beyond this point, the matter is incapable of being studied scientifically? Nevertheless, we must take many things as given in science. In physics certain mysterious powers and processes are just brute facts. So perhaps here we must take as given this generative randomness. It may not be a bad thing, even though it is an inexplicable thing.

On the retention side, the editing that preserves the better-adapted is only a half-positive principle. Beyond our earlier worries about this theory perhaps being a blik, tautology, or all-inclusive generalization, natural selection, though it retains survivors and is prolife in this sense, does so indifferently to their advancement. Most biologists hold that natural selection does not guarantee, explain, or make probable any overall direction or ascent to life. John Maynard Smith says, "There is nothing in neo-Darwinism which enables us to predict a long-term increase in complexity."[28] Natural selection posits only the survival of the more fit. Evolution takes place wherever there is any change in gene frequency. It has nothing to do with the selection of the advanced. There can be wanderings up or down the ranges of life's complexity.

A Random Walk?

Natural selection can give account of a *random walk.* A random walk is illustrated by a child's penny hike, one where he flips a penny at each corner to see which way he will go. A random walk is the trip that a suspended mote, too small to be under the significant influence of gravity, takes under random impacts from atoms, the phenomenon of Brownian motion. If life is kept moving by a survival urge, with random shifts in direction, and the options selected are those most likely to keep it moving (= surviving), then one can see why life will keep moving but not why it will move directionally toward complexity or sentience. The principle predicts that there will be survivors, but not that there will be any advancement.

There is a kind of macroscopic randomness that results from the microscopic randomness. The evolutionary course, so far from being a directionally ordered whole, or having headings anywhere in its major or minor currents, rather wanders. It wanders in the first instance due to atomic and molecular chance (both relative and absolute), and, given these chancy mutational possibilities provided from the lower levels, it wanders in the second instance due to the nonselection for anything but mere survival, without bias toward progress, improvement, or complexity.

It may happen, of course, that the life pulse constructs "advanced" forms, as it may happen that a boy gets to some interesting place on a penny hike. It is unlikely in the short range, but with enough time and trials even the slimmest possibility converts into a probability. So a random walk, if it goes on long enough, will get us what has happened in the evolutionary sequence. Here the causal selective principle explains the continuing of fit survivors, but any ascent is accidental to the process, despite the fact that it will in time probably happen somewhere, but not everywhere, in the life process. This account appeals to biologists who, surveying the staggering array of fossil and surviving life forms, see it as full of struggling, chance, zigzag, and groping omnidirectionality, some trials happening to work, most failing, a very few of them eventuating in the ascent of neural forms.

Curiously, the astronomers who until a decade or two ago often interpreted the lavish universe as mindless waste have now become prone to argue that the universe is anthropic, formed from the very start with those constants and potentials that are right for life. But when we reach biology (after fifteen billion years of astronomical and geomorphic developments) the biologists think the upslope progress of life, once it arrives, is all random.

Jacques Monod, a geneticist, wrote: "Chance *alone* is at the source of every innovation, of all creation in the biosphere. Pure chance, absolutely free but blind, at the very root of this stupendous edifice of evolution: this central concept of modern biology is no longer one among the other possible or even conceivable hypotheses. It today is the *sole* conceivable hypothesis, the only one that squares with observed and tested fact. And nothing warrants the supposition—or the hope—that on this score our position is likely ever to be revised. . . . When one ponders on the tremendous journey of evolution over the past three billion years or so, the prodigious wealth of structures it has engendered, and the extraordinarily effective telenomic performances of living beings, from bacteria to man, one may well find oneself beginning to doubt again whether all this could conceivably be the product of an enormous lottery presided over by natural selection, blindly picking the rare winners from among numbers drawn at utter random." Nevertheless, "a detailed review of the accumulated modern evidence [shows] that this conception alone is compatible with the facts. . . . The ancient covenant is in pieces; man at last knows that he is alone in the universe's unfeeling immensity, out of which he emerged only by chance."[29]

George Gaylord Simpson, a paleontologist, concludes an evolutionary survey, "Man was certainly not the goal of evolution, which evidently had no goal. He was not planned, in an operation wholly planless. He is not the ultimate in a single constant trend toward higher things."[30] S. E. Luria, a virologist, writes, "The essence of biology is evolution, and the essence of evolution is the absence of motive and purpose."[31] Stephen Jay Gould, another paleontologist, agrees: "We are the accidental result of an unplanned process . . . the fragile result of an enormous concatenation of improbabilities, not the predictable product of any definite process."[32] The process is mindless and mechanistic. Even when biologists are not so outspoken, there is often the supposition, voiced or implied, that evolution is incompatible with any orderly principle that prompts directional achievements, that designs life.

If there remains any doubt, the learned and authoritative *Encyclopedia Britannica,* after telling us that "evolution is accepted by all biologists," concludes the matter: "Darwin did two things: he showed that evolution was a fact contradicting scriptural legends of creation and that its cause, natural selection, was automatic with no room for divine guidance or design. Furthermore, if there had been design, it must have been very maleficent to cause all the suffering and pain that befell animals and men."[33]

In this sense, the biggest events (the coming of mammals and men) not less than the smallest events (the microscopic mutations) are accidental or random with respect to anything the theory can predict or retrospectively explain. It first seems that in one part of the theory, the supply side, we are given randomness, but that in another part of the theory, the retention side, we will have a positive "automatic" selection that explains what has taken place. From among the myriad trials that come momentarily into existence, the fittest are selected to stay. The new events occur at random with respect to their significance, but are preserved for their significance.

But when we look more closely at even the retention side, randomness with reference to any sustained creativity is equally present there. We have a twice-compounded account of what is not there—no explanation of the microevolution, and no explanation of the macroevolution, either. One has no covering law, or trend, enabling one to say that microbes, or mammals, or men could statistically be expected. They just occur as historical events, and the theory is surprised by them, although in retrospect they are consistent with the theory. If something else had happened instead, the theory would have taken care of that too. The invincible theory can never lose, because it permits almost anything to happen.

Though natural selection is true as far as it goes, in that the less fit do not survive on average, *natural selection does not go very far in explaining how far up the scale evolution has gone.* Among the equally fit, some are more complex, some less so, and while survival might have been possible without advancing complexity, there is nevertheless advancing complexity consistent with, but not required by, the principle of natural selection. It is not surprising that many critics complain that this theory, with its nonexplanation of the crucial journey, will not do enough explanatory work with regard to the coherence we observe in natural history. This paradigm treats the most striking feature of all, the ascent of life, as an anomaly, that is, something that cannot be predicted, derived, or given account of out of the theoretical model. Some of the movements here are causal, to be sure, but the movements we most desire to have an explanation for are *casual,* not *causal.* They happen without sufficient cause, in the shuffling of materials. The seminal principle is missing entirely, and randomness stands in the gap. And randomness is noise, not explanation.

The real problem is not ongoing *survival,* but oncoming *arrival.* The problem is not just that evolution is a *partial* account but that it is *impartial* with respect to the main events. There is no theory for why birds developed from reptiles; no evidence, for instance, that the land was crowded, driving some forms to seek the air. There is no theory for why sea creatures moved onto the land, no evidence that the early lungfish, stirring in the drying seas, could not have migrated back into the sea rather than transform their modified fins still further into agile limbs and invade the land. There is no theory from which we can deduce that certain primates will leave the trees and take an erect posture on the savannas. When these things happen, we may guess the increments, and suggest their survival value. We can tell a likely story. But if something else had happened instead, the theory would have been just as well conserved. There is no insight into the succeeding chapters of the life saga.

Notice here that, although one is getting a description of what has happened and a semicausal account of why it should keep moving and vary, one is not getting *any explanation at all* of why it must or did ascend, but rather the assurance that there is not any overall orthoselection, not even in those episodes where simple forms eventuated in complex ones. We do not need, much less have available, any explanation for this. The explanation is that, even on the scientific level (and altogether apart from religious interests), there is no explanation, no adequate cause for this effect. We can say that if life starts out simply, there is nowhere to go but up. So some development of complexity is not surprising. But life does not steadily and irreversibly have to go up. "Nowhere to go but up" is true at the launching, but not thereafter. There is down, stable, and out, and many forms take these routes. The evolutionary process might have achieved a few simple, reliable forms, needing little modification, and stagnated thereafter, as has sometimes happened in little-changing habitats. Nor is there any account of why the life process, if it happens to ascend, will not happen to descend, earlier more complex, later simpler, devolution after evolution, since up or down is immaterial to survival. Life might have gone extinct.

It does not help here to appeal to time to guarantee complexity. Time does nothing to cure randomness, not unless there is some further principle (which natural selection does not supply) that locks in the upstrokes. John Kendrew, a molecular biologist, writes, "It may be surprising that a random process like this can improve a species or even produce a new species, indeed lead eventually to the whole vast diversity of animal and plant life we see around us. But it must be remembered that these processes have operated over an enormous span of time, more than five hundred million years."[34] Natural section, by this account, is a mechanism that, over time, converts improbabilities into probabilities, and thus complex life becomes more likely the longer the process continues.

But not so. Nothing in the theory makes probable a continual ascent, since, at every point in time, the probabilities of descent, stagnation, and ascent are equally great. Nothing says that the better-adapted are more complex. Nothing tracks ascent. The Preacher in Ecclesiastes was nearer the truth: "I saw that under the sun the race is not to the swift, nor the battle to the strong, nor bread to the wise, nor riches to the intelligent, nor favor to the men of skill; but time and chance happen to them all."[35]

We do get, of course, hundreds of weak explanations involving mutation, speciation, adaptive radiation, specialization, filling vacant niches, and the like. But there are always hundreds of nonprogressive and less diverse courses, alternatives equally consistent with the theory. We are not getting even a weak, much less a strong, explanation for the sustained drama from microbes to men. In computer language, we have gotten into the *subroutines* but do not yet have hold of the *executive program,* and are being told that there is no such main program, only the ever-spinning subroutines. "Natural selection is a theory of *local* adaptation to changing environments. It proposes no perfecting principles, no guarantee of general improvement." It provides no reason, continues Stephen Jay Gould, to believe in "innate progress in nature"; none of the local adaptations are "progressive in any cosmic sense."[36]

So we know why hair gets longer and whiter in colder climates. We know why horns evolve repeatedly. But we know nothing at all of the trend toward sentience, toward awareness, why humans come, why things grow more complex. We know all about the microevolution and nothing about the macroevolution. Natural selection reads the *subplots,* but is really powerless to explain the *big story,* the *history* that, over time, nature comes so dramatically to have. It cannot give a causal, much less a meaningful, pattern to the storied achievements in the rise from matter to life to mind. That is not a very impressive resolving power, if we really want to know why the picture in the Pennsylvanian period, or the Pleistocene, is like this and not like that.

Natural selection bears to the course of natural history something like the relationship thermodynamics bears to the weather. The laws of thermodynamics are never violated in any weather that comes to pass, but neither do they explain any particular weather event in contrast to others, nor do they enable us to predict or retrodict the historical sequence of weather. But in evolutionary biology there is a long-term prolife pattern, absent from meteorology, which demands explanation. At this point the most powerful and provocative of scientific theories turns out to be in some ways a tautology or something close to it, something that had to be so in the general sort of world we have, and in some ways a descriptive truth, but a nonexplanation of what we most want explained, the long-term history of this marvelous evolutionary universe. Yet in an introductory text on evolution, after admitting many gaps in the theory, Douglas J. Futuyma can go on to affirm that "the known mechanisms of evolution" provide "both a sufficient and a necessary explanation for the diversity of life."[37]

A cardinal rule of scientific method is that the observations (O) must in some sense, strong or weak, be derivable from the theory (T). If T, then O. But on the theory of natural selection (T_{ns}) have we any implication of the particular long-term course of natural history (O_{nh}) that we do observe in the fossil record, of this general course as opposed to some less striking one? If T_{ns}, then O_{nh}? We really have no such implication at all. Myriads of other eventuations of natural history follow just as well, provided only that there are well-adapted survivors, since all such outcomes are, in fact, held to be random.

The theory explains the events that do occur, but in such way that it does not explain why a great many other events did not occur instead of each, and if, moreover, one of these alternatives had occurred, the very same explanatory theory would have been invoked to explain the alternative. To anticipate distinctions we will draw in Chapter 6, there is a covering law, but there is no historical explanation, and the covering law does not touch the actual historical narrative. Natural selection is commonly thought to be a more economical theory than the grander religious or the modified scientific theories that compete with or add to it. But one wonders if a theory is really economical if it supplies a zero explanation at the critical thrusts. All we are really getting out of the strange picture of omnipresent randomness and survival of the fittest is a nature that is always and everywhere full of indefinite probabilities. Organic nature, to recall what C. W. Misner said of physical nature, is "a library of unused designs" that are creatively "enacted into existence."[38] But the main story, the plot, has completely escaped this theory.

We must insist, in the name of sound methodology, on probing a theory where it is weakest.

An Alternative Paradigm?

Is there not an *alternative paradigm* that can make intelligible this anomaly for which no explanation is possible under the Monod-Simpson-Luria-Gould account? Such an alternative, at a minimum, is going to have to load the dice, and to use natural selection to make improbable complexity probable. Or at a maximum it will hold that life and mind are inevitable consequences of the biochemical logic and evolutionary pressures within the planetary ecosystem. It will posit some constructive forces that give a *slope* to evolution. But it may not at first be noticed that this must be an *upslope,* even if a slower version of ascent. When one modifies neo-Darwinism so, the selection of the fittest begins to pass over into the selection of the advanced, and this is a move commonly made, but is it a scientific one? Is it neo-Darwinism? We are getting a new picture painted over the old one, although many of the old colors still show through.

The physical world overall moves thermodynamically downhill, despite some negentropic eddies (that is, some events moving counter to the statistical increase of disorder with increasing entropy); but now in bioscience we need an overall upslope force, or set of forces, a sort of biogravity that accounts not only for a survival drive but for the assembling and conservation of more advanced forms. Across a slope, of course, one can still gradually wander up or down, but there will be cumulative directionality. Now, with the passage of time and trials, there will, by ever more probability, be ever more salient constructions of life, enormous distances traveled upward. But one does not anymore have an unalloyed selection of the fittest from random mutations. One has inmixed selection upslope. This may seem at first a slight revision, since it retains much randomness, but it admits a radically different logic into evolution.

Thermodynamics need be nowhere violated, because there is a steady "downhill" flow of energy, but some of this energy comes to pump a long route uphill. This would be something like an old-fashioned hydraulic ram, where the main downstream flow is used to pump a domestic water supply a hundred yards uphill through a pipe to a farmhouse—except of course that the ram-pump is deliberately engineered and the "life-pump" spontaneously assembled itself as an open cybernetic system several thousand times more complex and several billion years long. Not only is there energy present, not only have the precursor materials assembled, but some force or forces are present (for which biologists will need a naturalistic name, but in which theologians might detect a Presence) that suck order in superseding steps creatively out of disorder. The energy irradiated over matter is order waiting to happen.

Once we can entertain this gestalt, the Monod-Simpson-Luria-Gould account seems blind to overwhelmingly evident, longstanding evolutionary trends across three to four billion years. Indeed, even Monod and Simpson had to keep dispelling the illusion that there was directionality to evolution. But, retorts Michael Polanyi, a philosopher of science, "there is a cumulative trend of changes tending towards higher levels of organization, among which the deepening of sentience and the rise of thought are the most conspicuous," and "only a prejudice backed by genius can have obscured such elementary facts. . . . From a seed of submicroscopic living particles—and from inanimate beginnings lying beyond these—we see emerging a race of sentient, responsible and creative beings. The spontaneous rise of such incomparably higher forms of being testifies directly to the operations of an orderly innovating principle."[39] But it seems fair to complain that mainstream evolutionary bioscience does not have an orderly innovating principle clearly in hand. There is no agreement how much of such a principle is needed, if any. If natural selection is made over so that it does guarantee or make highly probable orthoselection for advancement, then we have undergone a paradigmatic shift.

When we do that, the randomizing element begins to look a bit different. It does not need to be taken away, at least not all of it, but it can remain as openness and possibility. Again in biology, as before in physics, what we get is a world of infinite possibilities, one in which there is a superposition of possible mutation states over actual ones, but also one where many of the possibilities become briefly actual, real mutants, and then a fractional few stay actual (survive). Once again, but at a higher level, microscopic possibilities are edited "from above" in accord with the needs of the macroscopic organism. Further, organisms are edited so that from many options, the well-adapted survive, and this results, among other things, in advancing ecosystemic and evolutionary creativity. There is an editing on the basis of fitness, which stretches on into advancement. Here we are going to emphasize not the shuffling but the overall sorting.

Chemical Evolution and the Incubation of Life

A law is often best tested at extremes, and its breakdown there may reveal its partial scope, and even suggest what larger theory it is to be subsumed under. The extremes of evolution by natural selection are the genesis of (i) life and (ii) mind. (i) *At the earliest incubating of life,* we must explain not so much the origin of life by natural selection as the origin of natural selection itself. The question is not of the evolution of species but of the evolution of self-replicating metabolism, on which natural selection can begin to work. Is there some sort of natural selection already at work in the start-up of life? Any presently known living system is, in some respects, a closed information loop. (In other senses, it is open, involving mutations and learning.) The information in the genetic set reenacts itself in the next generation; the DNA makes the protein that makes the DNA. We know how eggs come from chickens and chickens from eggs, but not how the chicken-egg-chicken loop originated in the first place. Knowing the secret of life biochemically may still leave the evolution of life a secret, until we know how the life loops get established.

The first stage of this chemical evolution is relatively unproblematic: amino acids are constructed by energy radiated over inorganic materials. These collect in ancient seas into a kind of proto-organic soup. The second stage is much more difficult. Many amino acids must be assembled into long polypeptide chains, with no previous templates or enzymes for their hooking up, with no information to steer the process. One worries that, although some partial sequences would undoubtedly be produced at random, their spontaneous rate of thermodynamic breakdown would be vastly higher than their construction rate.

Longer time spans do not much help to make the improbable probable here, so long as the breakdown rate always overwhelms the construction rate. There is as yet no tendency to conserve these would-be proteins. Here the steps can be speculated upon. Perhaps certain clays served as catalysts. Perhaps there is a "bootstrapping" tendency for chains to pair up, unzip, and rezip in variant ways, patching into longer and longer chains. But the historical pathway from abiological materials to coded self-replicating DNA megamolecules is nowhere near being known. This is not to deny that there was such a pathway and that someday it may be known reasonably well. But when it is known, it is not likely to be one hostile to religious accounts of life on account of randomness in the process.

The third stage is to fold these long polypeptide chains into complex functional structures, and this stage, which might have been feared the hardest to explain, turns out to be the easiest of all. They are self-folding! In the presence of the electric pressures of water, that is, the dipolar water molecules with positive charges on one end and negative charges on the other, creating push-pull effects in interaction with the polypeptide chains, the polypeptide chains fold and form their various cross-linkages because they have the sorts of chemistries they have. Those particular polypeptide sequences that later on will function as proteins turn out to be among the semistable forms, which are thus selected for and conserved. There are some parallels to this in the way lipid bilayers, membranes, form the most stable configurations under the electric pressures of water. The hydra and sponge, if pushed through a fine screen, afterward reassemble by amoeboid movement. It is as though shaking the pieces tends to lock a puzzle together.

In a fourth stage, coincident with this, other molecules form, which, likewise under the electric pressures of water, organize themselves into hollow microspheres, empty prototypes of cells. These spheres come to envelop the newly emerging proteins, further protecting the about-to-be-life chemistries from their degradation by the outside environment and providing a semipermeable membrane over which can pass the necessary nutrient inputs and waste outputs. This segregation (selection?) of the right materials, collected into ensembles, nurses the life-forming process. Thereby life assumes cellular form.

But to have life assemble this way, there must be a sort of push-up, lock-up effect by which inorganic energy input, radiated over matter, can spontaneously happen to synthesize negentropic amino acid subunits, complex but partial protoprotein sequences, which would be degraded by entropy, except that by spiraling and folding they make themselves relatively resistant to degradation. They are metastable, locked uphill by a ratchet effect, so to speak, with such folded chains much upgraded over surrounding environmental entropic levels (Figure 3.1). Once elevated there, they enjoy a thermodynamic niche that conserves them, at least inside a felicitous microspherical environment.

Thermodynamicists have recently been surprised at what happens in certain mathematical and statistico-deterministic systems previously thought to "run down" over time, a surprise coinciding with that of biologists at what happens in evolutionary ecosystems as these (randomly?) build themselves up over time. So there seem to be occasional places on the evolutionary upslope where thermodynamics favors advancement. It is as though these stratified, metastable structure slots are there waiting to be filled until one day random fluctuations happen to construct a molecule pushed up onto such a slot, where it stays until another fluctuation knocks it down, or, more rarely, knocks it up into some still further synthesis occupying a higher metastable slot.

Still, it is a long way up any developmental slope to reach an organism with self-coordinating parts in a metabolic whole, a "going concern" that feeds on energy and materials outside itself in its environment and is linked by feedback loops with that environment so as to protect its own functions, its "life." If we remember the enormous complexity of even the simplest of these biological molecules, involving hundreds of amino acids chain-linked in a precisely suitable sequence and then folded dozens of times, and recall how many such molecules of differing function but equal complexity must be assembled to gain an organism, it is striking that there are no known nonbiological stratified structure-slot sequences that ratchet the upstrokes to lure matter along inorganic pathways of this kind, favoring niche-step by niche-step dramatic structural climbs that would otherwise be utterly improbable. We do find metastable slots that favor the building of atoms, compounds, and crystals, remarkable enough in themselves; but, after that, really complex megamolecules are known only biologically.

Only some kinds of order are biological. A crystal has lots of order, simple, spatial

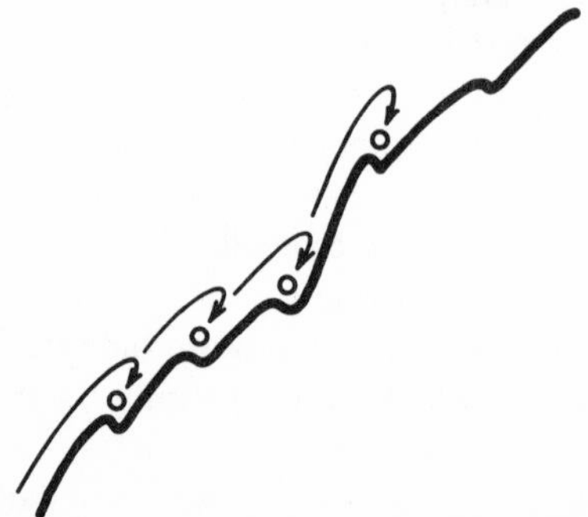

Figure 3.1 Metastable movements thermodynamically uphill

order, sameness, but no life. Biological systems have lots of order too, but a richer kind, functional complexity, organization, diverse parts coordinated in a whole, mobile, autonomous, self-informed process. This kind of high order needs a certain looseness (disorder, randomness) if there is to be novelty, mutation, evolution, creative growth, if there is to be defense of life before a mixedly certain and uncertain environment, a cyclic but also historically developing environment. The complexity is furthered, lubricated by the openness. It seems odd that the natural metastabilities would align only along a tectonic track to favor biologically ordered molecules and no other really complex kinds. Perhaps such complex molecules are sometimes randomly constructed in nature, but nothing seems to conserve them into any noticeable aggregations.

By shaking a tray of printer's type, one can get a few short words, which are destroyed as soon as they are composed. But if sentences begin to appear (an analogue of the long, symbolically coded DNA molecules and the polypeptide chains), and form into a poem or a short story (an analogue of the organism), one can be quite sure there are some formative, even irreversible, constraints on the sorting and shaking that are catching the upthrusts and directionally organizing them. It hardly seems coherent to hold that nonbiological materials are randomly the more and more derandomized across long structural sequences and thus ordered up to life. That is quite as miraculous as walking on water. It seems rather that life is an accident waiting to happen, because it is blueprinted into the chemicals, rather as sodium and chlorine are preset to form salt, only much more startlingly so because of the rich implications for life and because of the openness and information transfer also present in the historical life process. Life is not an accident, whatever place dice-throwing plays in its appearance and maturation. It is something arranged for in the nature of things. The dice are loaded.

When these enormously complex molecules appear, predecessors of DNA and RNA, bearing the possibility of genetic coding and information, they are conserved, writes Melvin Calvin, a biochemist, "not by accident but because of the peculiar chemistries of the various bases and amino acids. . . . There is a kind of selectivity intrinsic in the structures."[40] Peculiar chemistries indeed! With an *intrinsic* selectivity that filters and forces the process upslope, toward ever greater molecular complexity and at length to an informational molecule! If it can be said to exist here at the molecular incubation of life, natural selection is of the fittest (metastablest), but these are just those structures nearer and nearer to biological molecules. Such selection combines with these peculiar chemistries forced toward biochemistries, with the result that the biological consequence, the evolution of life, so far from being random, is "a logical consequence"[41] of natural principles.

Then one day, signals appear! Where once there was matter, energy, and where these remain, there is information, symbolically encoded, and life. There is a new state of matter, neither liquid nor gaseous nor solid, but *vital.* But by this selective account, much of the randomness is gone; indeed, we seem almost to be saying that life is the earthen destiny of these chemicals. Randomness does not rule out purpose; randomness plus something to catch the upstrokes can guarantee purpose—at the same time that it puts adventure, freedom, drama, and surprise into the storied evolutionary course.

But we must not overdo this "selectivity intrinsic in the structures," for there is not much in the physics and chemistry of atoms and molecules, prior to their biological assembling, that suggests that they have any tendencies to order themselves up to life. Even after things have developed as far as the building blocks of life, there is nothing in a "thin soup" of disconnected amino acids to predict that they will connect themselves or be selected along metastable upward courses into proteins, nor that they will arrange for DNA molecules in which to code the various discoveries of structures and metabolisms specific to the diverse forms of life. All these events may come naturally, but they are still quite a surprise.

We do not know that life, if it occurs on some other planet, being there built too of the same atoms, must select these same biochemistries, although the amino acids found on meteorites and the prebiotic molecules guessed to be present in interstellar dust clouds suggest that the potential for life is omnipresent in matter. "This universe breeds life inevitably," concludes George Wald, an evolutionary biochemist.[42] But we hardly know whether to put it quite that strongly. Nothing suggests much coding for life in the microscopic particles as such, although we have noticed an anthropic principle that finds the materials right for life, and puzzled over how in some universes otherwise constructed life would be impossible. To the contrary, quantum physics gives us an open system and nested sets of possibilities; but, while some atoms and molecules take living tracks, called forth as interaction phenomena by the cybernetic organism, most atoms and molecules take nonliving tracks. If there is some "inside order" to matter that makes it prolife, it is in the whole system and not just in the particles. Even there the "selectivity intrinsic in the structures" does not rule out a universe of myriad options, only some of which are realized. Physics and chemistry, unaided, do not get us very near to life and mind.

Meanwhile, we do posit a primitive planetary environment in which the formation of living things had a high probability, or, in other words, a pregnant Earth. Here we may not so much need interference by a supernatural agency as the recognition of a marvelous endowment of matter with a propensity toward life. Yet this latter sort of natural performance can be congenially seen, at a deeper level, as the divine creativity. We may still need something to superintend the possibilities. Once again, it is not just the necessities or the contingencies, but the prolife mixing of the two that impresses us. It is not just the atomic or astronomical physics, found universally, but the middle-range earthen system, found rarely, with its zest for complexity, that is so remarkable.

Here there is a mixture of inevitability and openness, so that one way or another, given the conditions and constants of physics and chemistry, together with the biased earthen environment, life will somehow both surely and surprisingly appear. After a long study of the possibility of the evolution of biological molecules capable of self-organization, Manfred Eigen, a thermodynamicist, concludes "that the evolution of life . . . must be considered an *inevitable* process despite its indeterminate course."[43] W. Ross Ashby, a neurologist and cybernetician, claims, "The development of life on earth must thus *not* be seen as something remarkable. On the contrary, it was inevitable."[44] Life is destined to come as part of the narrative story, yet the exact routes it will take are open and subject to historical vicissitudes. But none of this is much, if at all, explained by natural selection in its usual hard-nosed, rigorously scientific form. Modified forms of the theory deemphasize the randomness, or use it productively in the inevitable climb toward life, and leave us puzzled about the mix of inevitability and indeterminacy in a chemical system that eventuates in historical life.

The Evolution of Mind

(ii) At the other evolutionary extreme, the brain appears and slowly but surely enlarges, not in most forms, but in some and eventually dramatically *in the passage from primates to humans.* (See Figure 1.4, p. 15, though some think bursts of development punctuated the gradual growth.) But why? Is this because the smartest are the fittest, other things being equal? At first this means only those who are more perceptive, with more information-processing capacity, the quickest to respond, since neural circuitry works faster than circulation or diffusion. But later on it means those who can learn, those more aware, and, at length in humans, those who can deliberate and form culture. Now it is easy to believe that increased cerebral power is being selected for, at least in this one crucial strand of a many-stranded process.

Perhaps we cannot say that parasites or plants will be subject to any such selection, but will there not be some slots in the ecosystem that catch the cerebral upstrokes? Among the predators, if not the prey, the sequence will be "headed" for those with most "head." But this is no longer the selection merely of the fittest to reproduce, but the radically new claim that the most intelligently active are the best-adapted. Here natural selection has been transformed into, or is being surpassed by, a structural principle favoring intelligence and action.

Once again, one has to be circumspect. Does increasing intelligence yield survival power? The answers come both ways (and natural selection is majestically indifferent to the answers). *Yes* in one line of primates, but *no* in others, who survive very well with no increase in cranial capacity. *Yes* in some mammals, *no* in insects. *Yes* if one is a predator, or has predators who grow smarter and more sentient, but *no* in the absence of predation pressures. Then again, in the human case it seems that the big brains did not originate where the big brains formerly were, in the carnivores. The human brain seems rather to have come from small, tree-dwelling primates that were omnivores and needed their brains to couple with their hands and forward-turned eyes for agility in the trees, and later also for their upright life on the savannas.

Frankly, we do not have any satisfactory explanation for the arrival of the enormously sophisticated human brain; nevertheless, an evolutionist can be certain that brainpower is selected, where it is selected, because it contributes to survival. But after it is in place, it is good for so much more in the human case. We can readily envision rescuing an abandoned infant from a Borneo Stone Age people and, if she is a bright child, in due course sending her to Harvard to study astrophysics.

Evolutionary Development as a Trend Upslope

Across the main range of the theory we may now entertain the suspicion that the survival of the more fit is, at its most interesting points, (also) the selection of the advanced. This cannot be seen at nearer ranges, where it is often masked by eddies, fluctuations, and dead ends, and it is built on life-support pyramids in which higher forms depend on continuing lower ones, maintaining themselves on this biomass. But it becomes evident at broad enough a time span, just as Einstein's relativistic effects become significant on massive scales, present though not evident on a local scale, where Newtonian mechanics will suffice. Like Newton, Darwin needs now to be subsumed under a new pardigm, one that both retains the genius in him and yet rejects key axioms, those preventing explanation and prediction of ascent.

A genetic sequence is naturally selected, in part, for its capacity both to experiment and to lock onto its successes and to "make for" the upward projection of the life pulse. Perhaps we even need some new name for this macroevolutionary principle, restricting "natural selection" to the microevolutionary subroutines. So far from being reductionist, such a principle is rather compositional. So far from merely *simulating,* it *stimulates* design. It selects for complexity, function, efficiency, responsiveness, sentience, and intelligence. We can no longer say with naïve simplicity that watches imply watchmakers, and rabbits imply a Rabbitmaker. But what are we to say of a rabbitmaking system; who made the system? The question becomes ever more urgent after we learn from astronomy how easily it could have been to miss a life-producing system.

Consider a diagrammatic sketch of the evolutionary development (Figure 3.2). Every form of life does not trend upslope. The understories remain occupied, and in most of them there is no ascent, only onward movement. Life proliferates both vertically and horizontally. So most creatures are plants, fungi, bacteria, protozoans, mollusks, insects, crustaceans, and the like. In our diagram we want to avoid the look of a phylogenetic tree (these have dropped out of evolutionary biology) and to leave the connections loose, for we are not very sure how the major innovations took place.

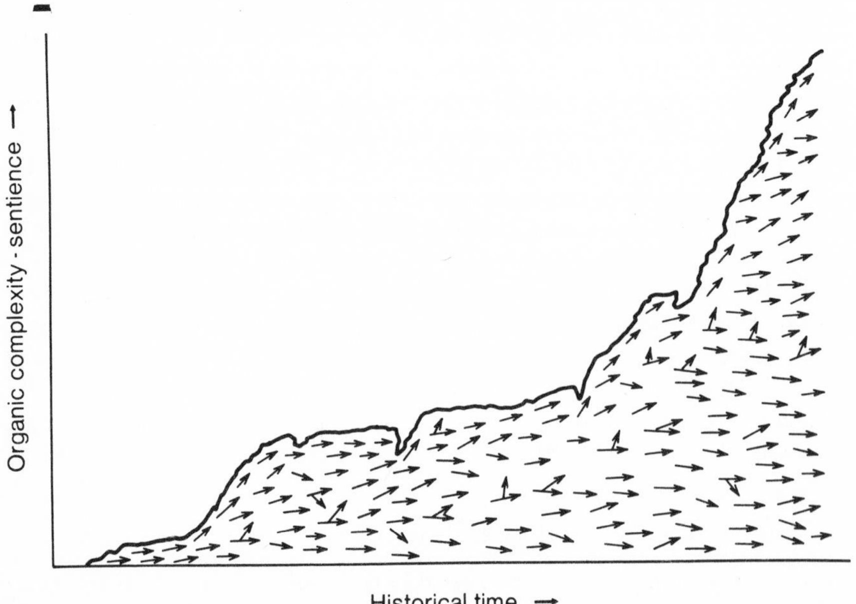

Figure 3.2 A sketch of evolutionary development

There have been steady extinctions replaced by steady innovations. Our arrows can just as well represent populations, groups, species. Some species change hardly at all in hundreds of millions of years; others change dramatically in ten years. Evolution can be very slow or very rapid; the latter is shown in the evolved resistance to antibiotics that has occurred in our lifetime.

If we follow the pathway of some single sequence of vectors in the evolving ecosystem

→ ↘ ↗ → ↗ → ↘ ↗

this may not climb at all. *Lingula,* a burrowing brachiopod, and *Nautilus,* with its exquisite shells, persist as genera from the Ordovician to the present, living fossils. This may be because they are stable in little-changing environments, but some forms (*Lycopodium*) change little even in steadily changing environments. Some forms even degenerate. Birds may lose their wings. When free-living forms become parasitic, they may lose eyes, legs, even their brains. Notice, however, that retrograde evolution in a particular organism, such as tapeworms or viruses, requires such an organism's living in an environment more complex than itself, so that it can borrow its lost skills from its hosts. In this sense, overall systemic complexity is not lost. These are minority cases that are, so to speak, conceptually (as well as biologically) parasitic on the main trend, which is systemically upward, a prolife profile, so to speak.

Some groups dead-end in extinction, but most evolve into something else. Certain present-day worms are descended from Cambrian worms, and this life line has remained horizontal. But present-day humans have also descended (ascended!) from Cambrian worms. There appears an ascent toward complexity and sentience in some of the pioneering forms, in the topmost arrows, or, if the top ranks overspecialize, in near-the-top arrows. If one focuses on those forms that, as ecologists say, are on the top trophic rungs of the pyramid, one will find upward movement. The ecosystem as a whole grows more complex, while remaining dependent on the less complex understories, where one needs the prolific reproducers and the good colonizers without regard to their "improvement." The geomorphic cycles—erosion, sedimentation, orogeny—are repeated over and over, with the result that Earth's geologically active crust maintains land above the sea and ever revises environments, building new niches, forcing life to evolve, recurrent cycles on which the spiraling advances can be superimposed.

The upward trend may be spasmodic in the short range, and unpredictable on any particular run, but it is reliable on the whole. There is a general trend, not in the sense that every life form trends this way, but in the sense that the later the system chronologically, the more it will have arrived at higher levels of sentience and complexity in its apical forms. But we do not want to draw too uniform a curve, for we have learned to be suspicious of smooth curves in the field of nature. Physics has taught us that all the laws of nature are probabilistic, and so we will not be surprised to find in biology also that statistical patterns reveal causal laws —not tight, deterministic laws, but constant, open tendencies. Nor will we be much upset if the local events occasionally have the look of random dice-throwing, seemingly casual rather than causal. Short data runs in physics can be just as scattered; physicists' curves too look jagged from certain reference frames. The global trends will be what count.

In terms of individuals and numbers of species there are even a couple of great crashes (at the ends of the Permian and the Cretaceous periods), and we can suggest this by major dips in the curve. But if we are diagramming an increase of complexity and sentience (not just species loss), it is not clear that we want much, if any, dip in the curve, because just these crashes were followed by swift resurrections, both with more diversity in species and with more complex uppermost forms.[45] Despite the closing down of life lines, something seems to open new doors. The setbacks reset life's directions in productive ways, a sort of backing up for a running jump. Sometimes in climbing a mountain we find that we can go no farther uphill unless we first go downhill temporarily. There is no such intention in the evolutionary process, but there is this result nevertheless, a sort of progress through negation. It was the wiping out of the dinosaurs at the end of the Cretaceous that triggered the age of the mammals. There seems almost at work here a principle to which we will appeal later, that the great dying yields the great renewal.

Further, it is notable how, once the life forces are under way, they to a marked extent remake the environment out of which they first came, molding it further in a prolife direction. Life is not simply subject to environmental vicissitudes, not passive before geological and meteorological forces, but is interactive with them. The soil with its humus results from what otherwise would be only mineralogical earth. The atmosphere with its oxygen, carbon dioxide, and ozone-shielding layer is a product of plant and animal life. The rivers and springs flow moderated by runoff control due to vegetative cover, plant respiration, evaporation rates, etc. Life to some extent modifies its climate. There are feedback loops set up between the organic and the nonliving world. Sometimes it almost seems that we have again what we first met in quantum biology, the phenomenon of an organism's not simply accepting a random physical nature, but selecting from its possible routes those that are more favorable to life. The physicochemical environment is rebuilt biologically.

Nor do we wish to be too anthropocentric in this sketch, even though we cherish the dramatic human story and claim persons as Earth's most advanced project. Given our understanding of ecosystems, the other life niches are indispensable in the sense that some such lower slots must be filled if the higher forms are to exist at upper ranges of the pyramid. There can be no humans without grass, no birds without insects, nothing else without the microfauna and microflora. While any particular insect, crustacean, plant, or microbe may be replaceable, unless there is a large population of such forms, the life-support system crumples. But neither should we regard this merely as instrumental to our food base, for every life form is interesting and intrinsically valuable in its own right. Nature loves otherness and richness in the system. In theological terms, God made all creatures and pronounced them good. We do not want to view everything as ancestors and fertilizer for ourselves. That would be too arrogantly humanist. The wilderness is good without us, and we may enjoy it just because it is an alien world. Nevertheless, we do sit atop a biological pyramid, climaxing natural history to date, and we need an account of this.

Complexity and sentience did happen to evolve more or less like this. But did they have to evolve so? Was this probable? Or is the directionality only chance? Cautious thinkers here are troubled by the fact that we have only one known case of evolution, here on Earth, and we are reluctant to construct a law from only one case. Still, the case we do have is a very long-continuing one. Nature steadily builds up the phyla and orders of the creatures, more loosely than (because using historical information transfer) but still somewhat reminiscent of the way nature once built up the periodic table of the elements.

There are some striking convergent evolutions. Eyes were invented at least three times—by insects, mollusks, and fish. Wings were invented at least three times—by insects, reptiles, and mammals. Learning was invented several times, and this comes to serve as the basis of intelligence. We do not know whether to say that there is only one known case, so we must not be impressed with it as being improbable. We cannot jump to conclusions from just one case. Or to see the parallel evolutions and say that we must not marvel at what is thereby proved to be entirely natural! Although one can say that the repeated cases represent multiple subaccidents, and that the long-term trend is a superaccident, it certainly seems reasonable also to search for some biotectonic principle that loads the dice. The struggle of the fittest is a struggle not merely on, but up.

The Incompleteness of Evolutionary Theory

Indeed, we are hardly going to have much of a sense of explanation, beyond mere description, until we do get some *executive or legislative principle that lures the ascent.* Something makes for life, and then makes more life, and then makes for more quality of life. But it is beginning to look as though a scientific law, mathematical, causal, or otherwise, though it may help, is hardly going to provide a robust explanation. More than with any other science (physics, astronomy, biochemistry, psychology, sociology), in evolutionary biology we are going to need a theory of history. The record does not show life courses with equally random wanderings overall, but there are great bursts of activity at certain times and places, great crashes; some routes are much more fertile than others; there are crucial turnings and crises. In short, the record takes on a storied form, over and above a much simpler pattern of random gropings or predictable scientific laws, over and above anything that upward arrows on an evolutionary time line can diagrammatically portray. *Story* is a better model than *law* (much less randomness) when one wants to derive more out of less.

Some may reply that it is too much to expect of Darwin that he should be able out of his precedents to predict the consequences (as might Newton in celestial mechanics). Evolutionary biology explains but it cannot predict. But the fact is that neither the emergents nor the upslope movements are explained even retrospectively. This theory does not allow us to say, after the events, why they happened, and why other events failed to occur. Not only do the mutants arrive at random, but there is no principle for selecting any advanced ones. There is only an explanation of why such ones as are selected (regardless of whether they are advanced, neutral, or degenerate) make for survival, nothing more. In an open system we cannot be required to specify hard causes, since an open system is not deterministic.

But even in an open system we expect the theory to enlighten about the large-scale features. If none of the significant trends are derivable from the theory, then the theory is not doing any significant explaining.

The theory does not have to settle the future, perhaps it does not even have to settle the past, but at least it could make plausible the leading features of events. The universal principle (the best-adapted survive) never entails any particulars (dinosaurs exist), not even when given initial conditions (microbes, trilobites). Or, if that is too much to expect from a scientific theory, the universal principle does not entail the outlines of natural history, and in that sense the universal does not explain this world story. Nothing has really been explained until the questioner's puzzlement vanishes.

It becomes hard to see how the theory is hypothetico-deductive in the sense of deriving outcomes out of it, at least in any larger or interesting sense that illuminates us about what did or did not happen over longer spans of time. It becomes hard to see how the theory offers any puzzle-interpreting pattern by which we make intelligible why vertebrates occupied dry land, why mammals followed dinosaurs, why learning followed instinct, why sentience followed reflex, why culture followed nature, describe these events though evolutionary science may. The theory is really vacant about how we steadily get order out of chaos, about the forces that order life up to its high levels of achievement. John Maynard Smith concludes that we need "to put an arrow on evolutionary time" but get no help from evolutionary theory. "It is in some sense true that evolution has led from the simple to the complex: procaryotes precede eucaryotes, single-celled precede many-celled organisms, taxes and kineses precede complex instinctive or learnt acts. I do not think that biology has at present anything very profound to say about this."[46]

In strong contrast to Futuyma's confidence in the completeness of evolutionary theory, cited earlier, Walter J. Bock, another evolutionary biologist, laments, "The single aspect of evolutionary biology that [has] eluded attempts for successful and convincing explanation is macroevolution—the appearance and subsequent specialization of distinctive new features and taxa. . . . One of the major failures of the synthetic theory has been to provide a detailed and coherent explanation of macroevolution based on the known principles of microevolution."[47]

"What's not in doubt is the fact of evolution, which is merely the fact of genealogical connection and descent with modification." But, admits Stephen Jay Gould, evolution could be "false in its grandiose claims. For instance, the hard version of the theory of natural selection says that almost every evolutionary event is shaped by natural selection. I don't think it could be false that natural selection operates at all. There is observational evidence for it. But it could be false that it's as strong and determinant of evolutionary events as we think," and this "would lead to a very different evolutionary theory."[48]

David M. Raup reflects, "I think it is safe to say that we know for sure that natural selection, as a process, does work. There is a mountain of experimental and observational evidence . . . which shows that natural selection as a biological process works. . . . There has been complete turnover in the biological world many times. This record of change pretty clearly demonstrates that evolution has occurred if we define evolution simply as change; but it does not tell us how this change took place and that's really the question. If we allow that natural selection works, as we almost have to do, the fossil record doesn't tell us whether it was responsible for 90 percent of the change we see, or 9 percent, or .9 percent."[49]

The main theory has not been found yet, and natural selection is only a fraction of some bigger truth. We don't want Darwinian theory cast down, but we do want it recast. Biology in its Darwinian stage stands where physics once stood in the Newtonian stage, and its Einstein has yet to appear. When she comes, just as the determinism in Newton (which had so troubled religion) vanished with the advent of contemporary physics, so the randomness and directionlessness (which now in biology troubles religion) will vanish in some biology of the future. Every big step that science has so far taken teaches that present theories are approximate and valid under limiting conditions, telling less than the whole truth, and evolutionary biology is no exception.

If an orderly innovating principle can be found, this can partially repair the incompleteness of neo-Darwinian natural selection. But it also merits notice that life appeared where there was none, and here any principle of incremental assembly has a tendency to appear to explain what it in fact does not explain. Complicated things have their precedents in simple things; but biofunction, when it appears, has no precedent at all. Once there was no smelling, swimming, hiding, defending a territory, gambling, making mistakes, or outsmarting a competitor. But these all appear gradually, also without precedent if we look farther along their developmental lines. That they appear gradually ought not to mask the fact that they appear without precedent. Mind appears incrementally but appears where, at the start, there was none whatsoever. In each quantum jump we get a little more of what was not there before, and if we integrate the differentials we get something in kind where before we had nothing of that kind.

We need a large-scale theory that predicts, or at least retrodicts, all the startling history. On the other hand, it is not too clear what sort of formative innovating principle could ever predict or retrodict all these storied events. We seem to be dealing in evolutionary nature with a lure upslope, but also with what is sometimes called, in a philosophically revealing analogy to physics, *quantum evolution.* Similar to the way in which there are indeterminacies and quantum leaps in physics, with the result that nature does not proceed continuously in full determinism from the past through the present to the future, so too in biology nature does not always or even usually evolve by increment and determinism over predictable gradients. Rather, there are quantum surprises. So we can add to quantum physics and quantum biophysics now quantum evolution. All these provide evidence of an openness in nature that may involve randomness, but that may also be a sign of potentials and forces that our scientific theories are not competent to detect.

Perhaps what we need is something analogous not only to indeterminacy, randomness, and quantum jumps, but also to the *interaction* that physics has found between the macrophenomena and the microphenomena. There can be downward causation when a participant interacts with the microevents, sometimes acting as a preference sieve catching the advantageous indeterminisms, sometimes coagulating nature to its advantage. In biology, too, not only at the level of the individual organism, but also at the level of evolutionary development, there is some interaction principle, some prolife force that acts in macroevolution to coagulate or catch any upstrokes in mutational micronature, and so to lure the ascent of life.

The assembly is of materials, complexity out of simplicity, but there comes with it autonomous life out of dead matter, biofunction out of nonfunctional antecedents, subjectivity out of objectivity. Once there was a world with only matter and energy, but later there appeared within it information centers, and later still, incarnate subjects. Molecules, trillions of them, spin around in complicated ways and generate the unified, centrally focused experience of mind. For this we cannot as yet even imagine a theory. For these appearances, natural selection, so far from providing "both a sufficient and a necessary explanation" (Futuyma), does not have any theory at all, no cause adequate to this effect, no understanding of how it can be so.

When conscious awareness arises, this will, of course, be said to have survival value. The animal can conduct trial-and-error thought experiments, and bad ideas can get eliminated without the death of the organism. But nothing in such an explanation gives insight into how subjective experience arises by the complication of mere objects, and where subjective experience has never yet arisen (as in plants), or if subjective experience had never arisen at all, natural selection theory would remain quite untroubled by its absence.

Critics have often faulted *ex nihilo* accounts of creation, it being incoherent to hold that something comes from nothing. Evolutionary theory softens that by assembly, but it continues to deliver mind and life *ex nihilo,* where absolutely none were before. Only it masks this by doing it incrementally rather than swiftly; then, by switching to explain those increments in terms of material units, the theory conceals the fact that it describes technical conditions necessary for the production of life with no account of the necessary or intelligible derivation of what emerges. Incremental qualities joined and rejoined are also reformed and transformed into novel qualities. One gets, at length, brilliance of mind by organizing armies of stupid atoms. This is something like assembling a smart computer by wiring up a million idiotic switches, none of which "knows" more than *on* from *off, flip* from *flop.* Except that such a computer is built by intelligence, which we are not allowed to posit here, never by randomness, and computers have no inwardness.

Slowing things down and putting together molecular parts does not really alleviate the lack of theory explaining how inwardness comes out of outwardness, how biofunction comes out of causality, why the life process never really runs downhill. It only spreads the inexplicable element thin, rather than asking us to swallow it in one lump. At the point of evolutionary creation, scientists, no less than theologians, in these latter reaches believe what they do not understand. We do not know how, much less why, there emerges out of the natural process this capacity we enjoy to reflect scientifically and religiously over the process, surely the most remarkable characteristic in the process itself.

Out of physical premises one derives biological conclusions, and, taking these as premises in turn, one then derives psychological conclusions. One derives meaningful action out of causal reactions. There seems still as much magic as logic in it. What is sternly preached to us in strong causal models is that the causes must be quite sufficient for the effect. What is uncompromisingly maintained in strong logical models is that the premises must precontain the conclusion. It is sometimes said that neo-Darwinism is the only credible theory and that objectors are being unscientific, their ignorance permitting them to hope that an unpleasant theory is not true. One does not use religious dispositions, however, but rather uses these theories against themselves when one complains that the theories are not "up to" the ascent of life and mind. The result (humans) is quite out of proportion to the causes (matter). The output vastly exceeds the input; one needs more explanation than randomness.

This is not, as we will later hear Freudians allege, a rationalizing, psychological demand for adequate "support" in nature; it is a rational, logical demand. If we define a miracle as a wondrous event without sufficient natural causes, so far as is known, then there remains miracle here, and we were hasty to say that under bioscience the secret of life stood explained. Man and woman arising via all the intermediate steps (trilobites, dinosaurs, primates) from the maternal Earth is not less impressive, rather more so, than Aphrodite arising from the formless seas.

We must also be careful about the demand for *like causes for like effects.* The cause of *x* need not be *x*-like, at least within limited contexts. The cause of my headache is the cold wind, but the wind is not in pain. The cause of my experience of green is an electromagnetic radiation, but the wave is not greenish. Water is not finlike, but water over evolutionary time has caused the fin to assume the shape it has. Similarly, the cause of life is not lifelike. On the other hand, no one offers a cold wind or an electromagnetic wave as a constructional cause composing sentient experience *ex nihilo,* where none was before. Such physical causes affect experience, one ingredient among others, but they do not create the capacity for *experience* in this ultimate, generative sense. Dirt and manure cause tomatoes to grow; the latter are tasty; the former are not. But we have a perfectly good theory about how nutrients are recycled, against little or none about how dirt spontaneously assembles itself across billions of years, ascending a negentropic slope to form persons who enjoy tomatoes.

We can allow, on entropic trends, that a physical cause can destroy a biological or psychological organism, as when a bullet kills. But these are short-range effects. They describe only the contribution of one factor, among many others. I can easily break something that I cannot make, and a cause can destroy something it cannot produce. In a world constantly tugged toward entropy, destructive interference is something quite different from constructive emergence. The latter is at issue on the cosmic scale —the emergence of subjective life from objective life, the mysterious prolife trend. We have a perfectly good theory of how fins are adapted to function in water, but we can only stutter about those stirrings in the ancient waters that generated fish with their fins. We need a theory of the composition of the whole, not of the contribution of one factor. Perhaps one does not need a like cause for a like effect, but one does need an executive, logically adequate cause for the creative effects, a sense of intelligible genesis. Else we must face the poverty of our explanation.

We do not wish out of any religious creed to specify for biology what these more adequately conceived causes of life must be, and we may well believe that some naturalistic pathways were and are there. But in the two preceding sections we have been trying to see that bioscience need not block religious belief. First, any causes, such as are alleged in biochemistry, are not so mechanical as to require the devaluing or desacralizing of life. Second, any noncauses, such as the randomness alleged in evolutionary theory, are neither so confirmed nor so compelling as to prohibit assigning meaningfulness to the arrival of life. From this point, we can now move on, joining evolutionary biology and molecular biology, to try a more positive account of the dynamics of life. Finally (in section 4), we will test this more positive account against some judgments of good and evil, and against some biological counterexamples.

3. THE CYBERNETICS OF LIFE

A cybernetic account of evolution and biochemistry is as promising as any now on the horizon, incomplete though this may eventually prove in the light of what we will later say about historical, passionate, and dramatic elements in nature. We do not suggest this as a final model, only as one congenial to the life sciences as we presently know them and one that opens out toward a more robust account, such as religion may wish to give, where the Divine Spirit is the inspirer of life. This account will permit the recovery of some of that teleological perspective that Darwin seemed to banish.

If earthen nature proves to be a sort of evolutionary system that gropes for information, for know-how, then we shall begin to wonder whether we should call nature a machine any more. Rather, a cognitive-organic model may seem more appropriate. At least we have a supercausal principle, one not of mere *survival,* but (to coin a term) of *supervival,* by which we denote this living upward as well as merely living onward. Such a principle is at this level only an objective one, without subjective anticipation, which comes later with mind. But this principle begins to leave behind the random walk of bare survival, and it will be less hostile to those meaning accounts with which religion desires to complement it.

In recent decades there has been between biology and cybernetics a sort of paradox. We have watched the bioscientists sternly try to read purpose out of their organisms, at the same time that the cyberneticians were reading purpose into their machines, speaking freely of memory, learning, programs, and goal-oriented behavior in their systems of electronic hardware. A partial resolution of these cross-tendencies results when we notice how organisms evolved spontaneously to do these things, and continue to do them autonomously, while computers, though they may do these things, do them only artificially, as extrasomatic products of organisms, and only in some continued interaction with their organismic operators.

Life as a Cybernetic System

There are important differences within various cybernetic models themselves. The one we need here is a full-blown model, where there is *a self-maintaining system with a control center, sustaining and reproducing itself on the basis of information about how to make a way through the world.* There is some internal representation, which is symbolically mediated, of the goal that is held forth. There is motion toward the execution of this goal, a checking against performance in the world, using some sentient, perceptive, or other responsive capacities by which to compare match and mismatch. On the basis of information received, the cybernetic system can reckon with vicissitudes, opportunities, and adversities that the world presents. Here we notice that something more than causes is operating within the organism. There is *information* superintending the causes; and without this information the organism will collapse into a sand heap.

This information is a modern equivalent of Aristotle's formal and final causes. It gives the organism "purpose," although we are now deploying that word into a broader spectrum of meaning than is visible in its everyday range of use. Organisms have ends, although not always ends-in-view. There is here an energy flow, a material flow; and physics and chemistry can trace such flows. But there is here also an information flow, not found in physics or chemistry, and in this organic form it is distinctively biological. What survives in the pulse of life is an information set inlaid, informed on matter, a cargo carried by the DNA, essentially a linguistic molecule.

Humans artificially impose an alphabet on ink and paper, but living things long before were employing a natural alphabet, imposing a "code" on four nucleotide bases strung as cross-links on a double helix. A triplet of bases stands for one of the twenty amino acids, and thus by a serial "reading" of the DNA, "translated" by messenger RNA, a long polypeptide chain is synthesized, such that its sequential structure predetermines the bioform into which it will fold. Ever-lengthening chains, logical lines, like ever-longer sentences, are organized into genes, like paragraphs and chapters, and so the story of life is told. Diverse proteins, lipids, carbohydrates, enzymes—all the life structures are thus "written into" the genetic library. The "genius" of life is recorded there, and the information content in any cell exceeds that in any book, indeed, in any set of encyclopedias.

Genetic Sets as Propositional, Logical, Normative

The genetic set is thus really a *propositional set*—to choose a deliberately provocative term—recalling how the Latin *propositum* is an assertion, a set task, a theme, a plan, a proposal, a project, as well as a cognitive statement. From this it is also a *motivational set,* unlike human written material, for these life motifs are set so as to drive the movement from genotypic potential to phenotypic expression. No book is self-actualizing. Given a chance, these molecules seek organic self-expression. They thus proclaim a life way, and with this they claim the other for self as needs may be, an assertive claim. An inert rock exists on its own, making no assertions over the environment, and not needing it (although it did not come into being on its own). But the living thing cannot exist alone. It must claim the environment as source and sink, from which to abstract energy and materials and into which to excrete them. It takes advantage of its environment. Life thus arises out of earthen sources (as do rocks), but life turns back on its sources to make resources out of them (unlike rocks), which is done because life is a propositional and motivational set.

The DNA is thus a *logical set,* not less than a biological set. We have a sort of symbolic logic, using these molecular positions and shapes as symbols of life. The novel resourcefulness lies in the epistemic content conserved, developed, and thrown forward to make biological resources out of the physicochemical sources. The presence of this executive steering core makes fitting the term "cybernetic," a word recalling a governor or helmsman. Even stronger still, the genetic set is a *normative set;* it distinguishes between what *is* and what *ought to be.* This does not mean that the organism is a moral agent; it is not. But it is an evaluative system, an axiological system. So it grows, reproduces, repairs its wounds, and resists death. Finally, we can say that the physical state that the organism seeks, idealized in its programmatic form, not only *has* a meaning but, so far as it is achieved, *is* a meaning state. The organism is inspired for and by *meanings* inlaid into symbols. *Value* is present in this achievement.

The genetic set thus forms a kind of "memory," which, though carefully to be distinguished from acquired memories resulting from individual learning, nevertheless permits life know-how from the past to be transmitted onward. These genetic sets become centers of propaganda, propagating centers. With the evolution of conjugation and sexuality, there arises the capacity for the interfusing and exchange of genetic information. Each present individual is a living fossil, an institution, surviving by its genetic power to recollect what has proved viable across previous eons of time. So far from being a random walk, life is a skilled tradition, forged before an unrelieved pressure to project and improve these bioforms that have earlier tested out successfully.

In the earlier stages of chemical evolution, prior to the formation of informational molecules, the constructional advance toward life may have been lured by selectivity intrinsic in the structures. Afterward, however, the life process becomes coded, involving an epistemic flow over time, with genetic sets tested for their skills in stable and persistent information transmission. There evolves the capacity for a sort of objective "experience" (as yet without inwardness), and for the handing on of this, although first in a Darwinian rather than a Lamarckian form.

In the embryonic development of an individual, ontogeny partially recapitulates phylogeny. The story of the individual is a telescoped retelling of its ancestral story, often done at high speed and with much editing. What earlier was the self-assembling of molecules is now the self-reassembling of information sets by means of genetic templates and their mediating enzymes. Thus, the chains of DNA form a sort of text, a scripture, a canon, not only in and for the individual, but one lengthening out over the millennia, a story that no doubt has its causal sequences, but one that also needs to be interpreted for its significance. The story of life has thus been written, in this crucial part, in the DNA. But it is a story that still needs to be interpreted for its meaning, after the decoding of its writing down has been accomplished in bioscience. This last chapter in the secret of life is still a religious and philosophical undertaking.

Randomness and Trial-and-Error Learning

But the genetic set not only "remembers," it can "learn," and here we again employ a familiar word in an unfamiliar context. Mutation, crossing over, and related permutations represent the capacity of life to experiment, to proceed by trial-and-error methods. The mutation is a trial "idea" (*idea,* form, type), oftenest neutral or detrimental, but sometimes beneficial in function because it enables the organism to handle itself better in its corresponding with the environment, or even to invade and exploit a new environment. So fins become flippers, then feet, then fingers. It is precisely this groping, blind character that strikes so many as being wasteful, dumb, and ungodly; but we have now to notice something more. Just as the capacity for stable genetic transmission has *value,* since life would be impossible without such information transfer, so the capacity for mutation has exploratory *value,* since life cannot undergo creative development without options, without opportunity.

Hidden behind that fearful word "random" is a *constructive freedom,* a resourcefulness in options tossed up for many diverse directions of movement. Mutations thus supply the imagination and revisability of which a deterministic mechanism, without possibilities in excess of actualities, would be devoid. They supply quantum jumps that permit incremental emergence; indeed, they seem in part to rest on that very openness in nature that the new physics allows. Constancy and variation are thus complementary values embedded in genetic evolution. The editing pressure is likewise of *value,* because, from many trials, the beneficial inventions are selected and the rest eliminated. Mutation forms the context of discovery; natural selection is the context of justification. Just as chance favors the prepared opportunist, chance favors the selective system that gropes for life and more life, and locks in the upstrokes. The whole operation is thus a projective, selective, and evaluative system. If we add to this how the long-range process ascends a cybernetic upslope, then we may say of nature (what was often said so paradoxically of God) that, given time, it writes straight with crooked lines. Such randomness is not noise, but news, good news. The random could veil the numinous.

It is not easy to make *judgments about the balance of this constancy and variation,* about the extent of "worthless" and "detrimental" mutations. It cannot be entirely true that nature does nothing in vain, but there is a pruning toward efficiency and resilience. There are different mutation rates at different genetic locations. Mutators and antimutators increase or trim the mutation rates as a function of population stresses. Repair mechanisms snip out certain genetic errors. Individual genetic sets are adept at pumping out their own disorder, but they do not pump out all novelty. There is a shake-up of the genes under conditions of environmental stress, so that the fastest evolution toward variant forms, often more highly organized forms, takes place almost explosively after major geologic crises.

Genes that may lie dormant are reactivated under special circumstances. Genes that in short-scope perspective seem flaws may turn out, in a deteriorating environment, to rescue a population. The genetic tendency to make a thick cuticle on leaves may work against a plant species so long as the climate is wet but save the species when the weather turns dry. Genetic information is often kept unused within all or some individuals in the population, and then released through "operons" on occasions of infrequent stimuli in crises. There are a good many conditional strategies coded into the genes, such as the production within a single species of flightless and more fecund aquatic beetles, if the water is plentiful and food is scarce, but a capacity to switch to winged forms, which are less fecund but more mobile, when something threatens to dry up the pond. There are insects that alternate generations in winter and summer; but what needs to be selected for in the summer offspring to fit them for winter is the reverse of what needs to be selected for in the same summer offspring to benefit the third generation, next summer's population.

The resulting mix of stability and mutation is not perfect, but it is impressive in view of the historical successes of evolution across billions of years. On the one hand, before deleterious or neutral mutations, misfits, monstrosities, and extinct lines, we may think that nature does trashy work. "Odd arrangements and funny solutions are the proof of evolution—paths that a sensible God would never tread." With their petals modified to insure pollination, continues Stephen Jay Gould, "Orchids were not made by an ideal engineer; they are jury-rigged from a limited set of available components."[50] But on the other hand, these trials and ancestral forms are subject to optimizing pressures and tested for their performances. What nature conserves is the best of its constructions within a particular ecological niche. Almost axiomatically within the theory, each life form has to have a comprehensive situated fitness. We presume that evolutionary pressures will tend to adjust toward a maximally favorable blending of retention and variation. The usual case is that each organism—the woodpecker, for instance—is mostly *hit* and a little *miss,* and the miss is really an experiment. The old "hit" is when the cumulative know-how of that species is successfully inherited, but also with a shot in the dark gambling to see what else there is to hit.

There is no particular reason to say that all the proto-woodpecker species suffered because of traits that they did not have, living as they did in niches that were somewhat different. As a matter of fact, they must have prospered in those earlier environments, at least enough to sire the species we now have. Experimentation is slight and conservative, and yet enough to get us the astounding development we have at hand in the world. This innovative exploration is of great value as it accumulates in the drama of natural history. The mutants and reshufflings first seem to be mistakes, but, seen again, are the key to the entire evolutionary growth.

The brown fat cell, modified from the ordinary fat cell, evolves late in evolution and is present in hibernators, seals, ground squirrels, and bats, and in the nonshivering young of many other mammals—rabbits, cats, sheep, even newborn humans, who cannot yet shiver. It provides the capacity to survive through winter cold. Such cellular tissue is in essence a heating jacket, which by a thermogenic response lets fats burn without forming ATP, thereby generating heat more efficiently.[51] But when we multiply this kind of phenomenon hundreds of times over, as can easily be done in bioscience, the cash value between design and simulated design becomes nearly nil. We do not mean that these animals were intentionally crafted and inserted into the system with their heating jackets. We only mean that their way of living is a fit form experimentally crafted for a niche within a designing system. They have systemic design. We can even hold in abeyance the question of what religious account to give of such a system, where the adversities of winter can call forth such a defense. *The present point is that in biological nature we are in the presence of a designing system.* That claim is fully as credible as the counterclaim that the presence of natural design is an illusion.

Evolution as a Prototype of Rationality

The human brain can anticipate discovery, for which there is no analogue in natural selection. Yet deliberative thought is also the launching of many trial ideas, and the selective testing of these in experience. A vast number of these innovations are abandoned; a very few of these ideas prove to add to our know-how and are worthy to be transmitted to posterity. If one thinks of the invention and engineering of the internal combustion engine or the aircraft, there lie abandoned a thousand dreams and attempts for every component that we now inherit, as there were eliminated a thousand mutations for each one now preserved in the brown fat cell. Indeed, in that sense the entire scientific enterprise and all theological endeavor, as we have sketched these in Chapter 1, move by throwing forward hypotheses on the forefront of experience, by testing these, and preserving only those few that succeed.

From this perspective, *the evolutionary process, so far from being irrational, is a prototype of the only kind of rationality that we know.* It is not babel, but there is a logic to it, not only to its information conservation, but to its random exploration and problem solving. Imagination is as necessary as is logic for rationality. Mutation scans for new "ideas," and natural selection throws out the trash and saves the gems. An "error," a mutation, is also a new idea in an information-gathering experiment. Do we then want to say that evolutionary design only *simulates* rationality? Perhaps, rather, it is a rudimentary form of cognition.[52] In terms of human imagination and logic, it is not always a waste but sometimes an index of creativity to cast forth a thousand ideas so as to sort out the single best one. Perhaps we would not want human life to operate any other way; perhaps creative logic cannot operate any other way. Perhaps we will not lament that God, if there is a God, has arranged a world in which the logic of creaturely development anticipates in this nonconscious way the logic of science and religion. God lets these creatures too (so to speak) figure things out for themselves. Perhaps even for God the only way to think straight is with crooked lines!

The earthen creatures, in their heritage from the past, are programmed cybernetic centers, under informational control. But they are also, in their prospect for the future, nonprogrammed learning centers, "programmed," if you will, to the open search for something else, to a little better fitness, or sometimes to a little advance toward complexity, mobility, sentience, or intelligence. On its cutting edge, the speciation process is (blindly?) drifting, but it is drifting through an information search, and edited for its discoveries of information. This editing is for survival, but it also scans and produces new arrivals on a climb toward complexity, sentience, and, eventually, mind. It is the production of errors that produces knowledge. The whole system is a context of instruction.

When a child learns to talk, she emits meaningless babble, but the babble is rapidly reinforced by the linguistic and cultural environment, so that the child tracks the verbal world she inhabits. The openness of the experimental babble is not a bad thing. So with the mutational babble in evolution, which is a kind of learning to speak, only there is nothing to track; rather, a creative groping for tracks never before traveled.

But this is not a stupid kind of creation. Rather, it anticipates what goes on in the most rational of the adventures that we know. The human brain and mind likewise come into being and mature under a heritage from the past, under genetic and also cultural cybernetic control. But the brain-mind too is in its essential genius a nonprogrammed learning center that openly and flexibly scans to see what else it can learn. We do not entirely wish to take the concept of mind out of precerebral nature; we want partially to deploy cognition there.

Evolutionary history is the story of the seeking, finding, and handing on of "pathways." That uses a favorite metabolic word, but the finding of "pathways" is likewise a scientific and even a religious achievement. In all problem-solving processes there is a picking of the most "viable" hypothesis from trial guesses. The wild "misfits" are rejected in favor of those that better "fit" the realities of experience. Here "to err" has its original meaning: to wander about looking for something. Life is a kind of ongoing science, a finding of and getting to know one's way around in the world through better perception and information processing. The processes of learning—gaining, storing, retrieving, and using information—are remarkable ones (as cognitive psychologists also tell us), but we notice here the even more remarkable *learning how to learn* that takes place systemically. The system produces learners.

Natural selection, by this revised account, is not so much "blind" to development as does it at crucial innovations and turnings in the upper levels of its systems "see" those mutations that are superior and select them. It tends in that direction, even though it does not intend it. Thus, the seeming absurdity of the random element can be put in a more intelligible gestalt, where it becomes a precondition of epistemic development. Afterward, it still seems to be true, when the discussion is over, that the eye was made for seeing—true in the biofunctional sense that the eye serves the organism, true in the embryonic sense that the genetic program constructs an eye for seeing, true in the evolutionary sense that the eye evolved as an adventure in seeing. Afterward, it still seems that the brain-mind, not less than the genes, evolved as an adventure in cognition.

It is instructive that life, as a pathway, requires not a bare "knowing" but one coupled with "going." The life pulse is, as it were, programmed to argue, to probe, to fight. Even when we do not yet have a living subject, the living object is a kind of object-with-will. Even the evolutionary ecosystem seems to grope for some *ought-to-be* beyond what *is.* All life has its commitments, something it values, and a kind of faith in tomorrow. When we do reach the levels of sentience and awareness, these are everywhere coupled with activity. This is never clearer than in the way the capacity to think coemerges with the capacity to act. The brain evolves together with the hand, with much of its volume devoted to the activity of the hand.

Nor is there advance only in neural design and circuitry; the skeletal muscle cell is one of the supreme achievements of evolution, as is the fat cell, both of which represent motion and energy. This coupling of knowing and acting, theory and practice, which fits well the cybernetic notion of a helmsman, is initially biological, but it deploys well into those two great human wisdoms that we are here featuring: science and religion. In science we cannot know the world just by thinking, devoid of experiment, action, and participation in it; and in religion any pathway that recommends contemplation devoid of action ought to be suspect.

Evolutionary Theism

The *creative action,* once reserved to God's special creation of fixed species, must now be reallocated to include a vast self-creativity within the creatures, lured upslope over this long evolutionary process. The tracks (fossils) left on a sandy playground by a child whose father was teaching her to walk would seem erratic and meaningless to an outside observer, uninformed about what was going on. But they would in fact be the tracks of a significant and lovely process. The evolutionary advance, with its paleontological record, requires a still more sophisticated keying in to providential pathways.

What theologians once termed an established order of creation is rather a natural order that dynamically creates, an order for creating. The older and newer accounts both concur that living creatures now exist where once they did not. But the manner of their coming into being has to be reassessed. The notion of a Newtonian Architect who from the outside designs his machines, borrowed by Paley for his Watchmaker God, has to be replaced (at least in biology, if not also in physics) by a continuous creation, a developmental struggle in self-education, where the creatures through "experience" become increasingly "expert" at life. This increased autonomy, though it might first be thought uncaring, is not wholly unlike that self-finding that parents allow their children. It is a richer organic model of creation just because it is not architectural-mechanical. It accounts for the "hit and miss" aspects of evolution. Like a psychotherapist, God sets the context for self-actualizing. God allows persons to be imperfect in their struggle toward fuller lives (as we will see in Chapter 4), and there seems to be a biological analogue of this. It is a part of, not a flaw in, the creative process.

God is not molding the material, craftsmanlike. But God "from below" microscopically creates the energetic, prolife materials that bubble up trials. "From above," systematically and environmentally, God via selectivity intrinsic in the processes coaxes forth living organisms and via natural selection selects the best-adapted. At an intermediate level, the organism has its freedom, selecting "from above" the quantum states below that maintain its biological program. But transcending the selective capacities of the individual organism, the environment gradually selects for improvement of the organic form from the mutants that also "bubble up" from below. One can say, restricted to science and with maximum economy, that only the environment acts to shape the organism. But one can say, religiously and with maximum insight into the meaning of what is taking place, that this formative force, which elicits so much from so little, mind from life, life from matter, is a divine force in the world.

Now the problem of nonpredictability in evolution looks rather different. Randomness seemed before a barrier, but it is now seen as confrontation with, even a carrier of, inventive creativity. One can seldom predict the next big invention, for that would be to make it in advance. Real discoveries typically take us by surprise in human society, and likewise in the evolutionary process. The so-called randomness is hiding some innovative forces, which our theory is incompetent to detect. This shows up as nonpredictability. The randomness, which once blocked explanation, is more comprehensively embraced in a generative process. One has to ask, "Why is this mutation there?" and to answer, "By random chance." But one has to ask, "Why is randomness there?" and to answer, "It serves as an idea generator in the evolutionary process."

We are prone first to think that the random cannot be reasonable, and to worry that the statistical is something imperfect in rationality. But not so. The randomness is God's giving freedom to individuals at the same time that God shapes the destinations of the whole, for what are random processes individually can often in the aggregate yield predictable order. The individual paths are left open; the course of history as a whole is hewn by God. But that is the divine love of freedom, the divine love in freedom. God is veiled in the probabilities and possibilities.

Randomness is there with (and for) creative results. Randomness plays its necessary part. Small-scale disorder (mutation, genetic drift) allows the emergence of more complex large-scale order (dinosaurs, primates, persons). In some veiled way, the *meaning* of randomness is creativity, and thus the *very absence of causality* serves the production of meaning. At every level, the selection from a subroutine of random proposals, like the eliciting of quantum states to accord with an organic program, need not be random. It can be the work of genius. God throws dice after all, but the dice-throwing is really idea-searching. Even in God's logic, ideas may appear at random with respect to their significance, and be preserved for their significance. A. R. Peacocke, a biologist and theologian, concludes, "Chance is the search radar of God, sweeping through all the possible targets available to its probing."[53] Events are often perceived as being random if their complexity exceeds the sophistication of our detection apparatus and interpretive theories.

We do not forget (what was learned in Chapter 2) that God is ever the Divine Ground in which matter with all its movements is sustained. In the materializing of the quantum states, in the compositions of prebiotic molecules, in the genetic mutations, there is the throwing of dice, but the dice of God are loaded. There are selective principles at work, as well as stabilities and regularities, which order the story and perpetuate a swelling wave over the transient particles. This portrays in some respects a loose teleology, a soft concept of creation, and yet one that permits genuine, though not ultimate, integrity and autonomy in the creatures. What comes to pass wells up from below, congealing out of the quantum states. But we gain with organism a further truth that the higher levels can also come to superintend the lower, responding to potentials presented there. But what is true of the individual organism can likewise be believed of the life process overall. We have in the life adventure an interaction phenomenon, where a prolife principle is overseeing the affairs of matter.

Recalling how the life secret is coded in linguistic molecules, what is an appropriate response to this "writing" that once appeared without precedent in the spiral threads of the DNA, and has continued to spell out the agelong dance of life? It seems entirely rational, past any naturalistic causes, and in view of the experienced meaningfulness of life, to say that this writing is the reception of a message, an inspiration from a *Logos.* The molecular self-assembling is a sort of self-actualizing, but it is also a response to the brooding winds of the Spirit moving over the face of these earthen waters. Even those who find this theistic account too strong are likely to find attractive some other religious metaphor. Mother Nature—or Brahman, *śūnyatā,* the Tao—is a womb generative of diverse life forms. Ernst Mayr, an evolutionary biologist, says, "Virtually all biologists are religious, in the deeper sense of this word, even though it may be a religion without revelation. . . . The unknown and maybe unknowable instills in us a sense of humility and awe."[54] The question here is but the biological form of one we will address (in the next chapter) as a psychoanalytic question, what to make of the generative, extended "parenting" in nature.

This matrix nurtures some of these forms into ever more sentience, awareness, and intellectual power. The appearance of mind can then be seen not as an epiphenomenon but rather as a climax in this process, though we should not obscure the fact that anticipatory purpose, which blossoms in the mind, has no known counterpart in the evolutionary dynamics. Nevertheless, what an epistemic adventure this is that assembles atoms so that they may think about themselves! The evolution out of the process of the capacity to apprehend the process is stunning indeed, and this conclusion to the story is the chapter most in need of explanation.

We need an account of matter equal to the account that matter has given of itself. One may well suspect that, if ever an adequate causal theory is found for this, it will not conflict with the deepest of those meanings that religion may wish to discover in the advent of life and mind. "I would say," concluded Loren Eiseley, an anthropologist, "that if 'dead' matter has reared up this curious landscape of fiddling crickets, song sparrows, and wondering men, it must be plain even to the most devoted materialist that the matter of which he speaks contains amazing, if not dreadful powers, and may not impossibly be, as Hardy has suggested, 'but one mask of many worn by the Great Face behind.' "[55]

4. THE LIFE STRUGGLE

Before accepting such judgments, however, we need to appraise the role of struggle in this genesis of life. In bioscience we meet not only a functional capacity unprecedented in physical science, but something still more novel. Matter can be meaningless, as when so much cosmic material seems tossed forth in waste; but it cannot suffer. Living things can suffer, most obviously as sentient forms in their subjectivity; but also, in some weakened sense, even nonsentient forms struggle bodily, objectively to avoid death. They have needs and endure stress.

So much of Earth's life seems tossed forth in waste, only now the process seems cruel, at least at its advancing levels. This torments the possibility of divine design, reducing natural history to a desolate, evil scene. Emptiness and vastness in an oversized universe is the challenge to interpret from modern physics. The time span of ceaseless struggle is the challenge to interpret in biology. Something stirs in the cold, mathematical beauty of physics, in the heated energies supplied by matter, and there is first an assembling of living information centers, and still later suffering subjects. Energy turns into pain. Is this now ugliness emergent from the first time? Or is it a more sophisticated form of beauty?

We are here on nonscientific ground, for bioscience as such can only amorally and nonaesthetically describe what has happened, and to assess whether this is good or bad requires valuational judgment. But "struggle" is integral to Darwin's description of this history. Biology in the last half-century has not been particularly comfortable with the word "struggle," which has largely disappeared from biology texts, being replaced by the notions of "adaptedness" and "fittedness." Reproductive efficiency is all that natural selection requires. Competitors may never even meet, much less kill each other; they are tested against their niche, only rarely in combat. Still, plenty of "struggle" remains in biology (although the switch in emphasis is revealing), and when philosophical participants find that they themselves have ascended via this struggle, they are confronted with the question whether such a struggle can be meaningful.

Much can be argued in explanation here, although we may suspect that the question of the worth of the struggle for life cannot altogether be rationally answered. It must be existentially answered in oneself; here too there is a "going" coupled with "knowing," and any answer has to be projected with both risk and faith across the entire adventure. At this point, the cybernetic model of life, promising though it is, begins to seem doubtfully competent. For what account is a living cybernetic system to give of itself and its world when it is called to suffer?

Suffering in the Prolife Struggle

Whatever one may think of the detail of its operation, *the evolutionary dynamics is,* on the whole, *a prolife principle* in that it stubbornly demands survival before all other costs. The absolutely nonnegotiable criterion is *procreation,* and in that sense it is *pro creation,* just because it edits all forms for their tenacious capacity to pass their creaturehood on, and up. This may seem a context of blind judgment, but it does allow through only those forms that are "right-for-life," and perhaps that in some sense have a "right-to-life," that is, are most lively in their combination of aggressive pursuit of life and inclusive environmental fitness. Perhaps we cannot say that the evolutionary principle "values" the life that it passes through so stern a testing. Nevertheless, the outcome is that life is critically tested and thereby kept vigorous.

Adversities make life go and grow. The pressure, before extremity, for doing better is steadily a blessing in disguise. The cougar's fang has carved the limbs of the fleet-footed deer, and vice versa. The brown fat cell evolves because of the need to survive the fierce wintry cold. Such a milieu of struggle might seem to suppress life, but there is a paradoxical reverse result—that life is disciplined, strengthened, and improved. These evolutionary dynamics are, further, *prolife* in the sense that they are *prolific,* fecund in producing a diversity of forms.

All those marvelous life features once ascribed to God's careful design—from the bill, toes, and tail of the woodpecker to the human hand and mind—do in fact gradually result from this ambitious evolutionary process. If the process is to be adequate to the produced result, it must somehow lure forth those effects. Those values ensuing must be an implication of possibilities of discovery and powers of justification laid into the process. Else we pass to an analogue for the logic of meaning of the *non sequitur* that troubled us in the logic of causes. If the process is entirely evil, then once more we have an illogical result: that out of evil premises we derive good in the conclusion, that out of worthlessness evolves worth.

Among world religions, Western monotheism has made the heaviest investment in endorsing the phenomenal life process as a divine creation and scene of providence. The whole natural history is somehow contained in God, God's doing, and that includes even suffering, which, if it is difficult to say simply that it is immediately from God, is not ultimately outside of God's plan and redemptive control. God absorbs suffering and transforms it into goodness. There is ample preparation for this conviction in Judaism, but it reaches its apex in the crucifixion and resurrection of a suffering Messiah, who produces life out of death in his followers. For we must be careful here. *It is not simply the experience of divine design, of architectural perfection, that has generated the Christian hypothesis of God. Experiences of the power of survival, of new life rising out of the old, of the transformative character of suffering, of good resurrected out of evil, are even more forcefully those for which the theory of God has come to provide the most plausible hypothesis.* We shall return to these themes in our concluding chapters.

In the cybernetic evolutionary advance, we do not find any anticipatory intention located in the individual creatures, but we nevertheless find logical achievements when life forms become more cognitive in their environments. So here again we do not find—what only comes later in the penetrating religious mind—an anticipatory faith in the positive result of struggle, but we nevertheless find a beneficial result. Earlier, we met a *causal* puzzle, one of *creatio ex nihilo.* How could life appear where absolutely none was before? We add to that now a *meaning* puzzle, one of *creatio per passionem.* Life arises in passionate endurance. Struggle is the dark side of creation.

This existential fact, discovered by sensitive souls, can now be seen as a truth written into life's creation, but obscured by the facile Newtonian notion of a Divine Designer fabricating his world machine. Organic life requires an entirely different model, one of suffering through toward something higher. Only later on, in humans, can this goal be consciously entertained. Prior to that, there is only an instinctive biological drive to survive at the cost of ordeal, present at every biostructural level. If irritability seems at first an unwelcome, adventitious intruder into the life project, by this switch of gestalts it becomes part of the biologic and logic of meanings.

It is difficult to keep all components in weighted balance here. Life is a good, as evidenced by the struggle to retain it. Every being that is pressed by need in face of death has also been handed life, so that the adversities are a posteriori to, and come in defense of, an a priori natural heritage. *Life is the first mystery that comes out of evolution, and death a secondary one.* But the same life that is given, and enjoyed, must be earned to be kept. We are given life out of nature, out of God, and yet only by sweat do we eat bread. So each creature lives by labor, in mixed success and failure. In sentient forms, this brings pleasure and pain.

Life preys on life; all advanced life requires food pyramids, eating and being eaten. Humans are degenerate in the sense that we cannot synthesize all that we need, while our remote ancestors were autotrophs. But in such degeneracy lies the possibility of advancement. If the higher forms had to synthesize all the life materials from abiotic materials (also degrading their wastes), they could never have advanced very far, not even as organisms, much less as humans in culture. The upper levels are freed for more advanced synthesis because they depend on syntheses (and decompositions) carried out by lesser organisms below. Heterotrophs must be built on autotrophs, and no autotrophs are sentient or cerebral.

In the end, any individual must die, by accident or by internal collapse, and here the death of earlier creatures makes room for later ones, room to live and, in time, to evolve. If nothing much had ever died, nothing much could have ever lived. Just as the individual overtakes, assimilates to itself, and discards its resource materials, so the evolutionary wave is propagated onward, using and sacrificing particulate individuals, which are employed in, but readily abandoned to, the larger currents of life. Thus, the prolife evolution both overleaps death and seems impossible without it.

The vast number of creatures sprouted, hatched, or born are, of necessity, more or less well endowed genetically and emplaced in a more or less congenial environment, despite or including the fact that in their environment they are spurred to earn their way. Even though most will not live to maturity, that task is a reasonable natural ideal, a *telos* or form for which they are fully programmed. Plant and lower animal forms, seeds, gemmae, and spores may be dispersed to impossible locations, and but briefly germinate, if at all. Sentient and mobile forms, those that can suffer, have more control over their circumstances. Indeed, the capacity to suffer is generally accompanied by possibilities of avoiding suffering, and some freedom and self-assertion. Animal forms have more or less, but to some degree without exception, a motile period in their life cycle during which they can partially choose their environments. They select the environments that will select them.

There are lethal mutants and monstrosities, but these bad ideas, as it were, are aborted immediately without further experiment. If not, they survive in about that proportion in which they are viable, so that life is sustained in any individual in some relative proportion to its fitness for it. Thus, in their evolutionary development, there never were any really ill-fitted woodpecker species. All the preceding forms that enjoyed life did so in that extent to which they were impressively well formed, 99.999 percent hit, and .001 percent miss or mutational gamble, "blessed," we might say, with the cumulative tradition of a billion years, "cursed," if you will, by fractional mutations, random variations, sometimes detrimental, usually harmless.

But it is just the "curse" they bear in which lies the possibility of there being woodpeckers better yet, or even continuing on, in the future. The mutational element is very minor in any viable individual; the major thrust of life is remarkably stable. But flawless repetition in reproduction would not only prohibit development; it would mean certain extinction in a changing environment (as all environments eventually are). Variability *is* stability in a changing world.

The long-term growth rate of any population has to be nearly zero, except in very unusual circumstances. A population that has a growth rate of .07 percent per year will increase a million times in twenty thousand years. Given this, if there is to be any selection at all over mutants, there must be a surplus of young, most of which are cut back by premature death, although even these shortened lives may have flourishing stretches between generation and demise. But what is premature death from one individual's point of view, and thus an evil, can be the source of life, and thus a good, from its eater's point of view. Even cannibalism can be an efficient use of resorbed material. From a still more systemic point of view, we see the conversion of a resource from one life stream to another—the anastomosing of life threads that characterizes an ecosystem. Plants become insects, which become chicks, which become foxes, which die to fertilize plants. In the ecosystem as a whole, for all the borrowing and spending, little is wasted in biomass and energy. The surplus of young is efficiently used as resource material in alternative life courses.

Thus, the surplus is doubly beneficial. It permits mutational advance and it permits the interdependent syntheses of biotic materials with higher forms at the top of the ecological pyramid. Here again waves overleap the particles, and overlaid on these interconversions of energy and nutrients is the natural selection by which life is edited for evolutionary advance. Like the particle annihilations and re-creations in the submicroscopic world, here in the macroscopic world, too, the living materials make their zigzag trips and do their identity flips, destroyed to be re-created. But now all this is with a conservation at a higher level than any known in physics, a conservation of the life process, simultaneous with its historical development. The ecosystem (again, as was said of God) writes straight with crooked lines.

The massive cutback in offspring is reduced in rough proportion as one goes up the phylogenetic scale. So this "waste" (which is really the systemic interconversion of life materials) is trimmed so that the premature dying is reduced about as the capacity to suffer elevates. An oak produces a million acorns to regenerate one oak, but none of the acorns suffer. An earthworm produces hundreds of worms to regenerate one worm, and all of them suffer slightly. A robin lays thirty eggs to replace one pair of parents. But the human mother, on the average through history, has borne four or five children to see two survive to maturity, and here, without denying the tragedy of infant and childhood mortality, it is hard to see how the rate could be cut any lower and still let natural selection operate over mutants.[56]

In general, the element of suffering and tragedy is always there, most evidently as seen from the perspective of the local self, but it is muted and transmuted in the systemic whole. Something is always dying, and something is always living on. For all the struggle, violence, and transition, there is abiding value. The Darwinian account of the life process has, on the one hand, increased our sensitivity to the struggles in nature. Life is a contest that most must lose early and all lose eventually. But on the other hand, when we appraise the suffering that attaches to struggle, Darwinism needs also to suppose a natural selection for the maximally beneficial pain, at least within certain rough limits. Pain in dysfunctional proportions (too little of it to register alarm, or too much of it disorienting the organism) will be selected against. The pressures will be for enough of a good thing, or, seen another way, for the minimum of a necessary evil.

Every life is an unceasing adventure in endowment and risk, and all organic being is constituted—to employ a scientific metaphor—in a mixture of environmental conductance and resistance, where the world is both resource and threat. To adapt the Psalmist's religious metaphor, life is lived in green pastures and in the valley of the shadow of death, nourished by eating at a table prepared in the midst of its enemies. Struggle is a driving motif, but then again, its product is life forms selected for maxim adaptation to their environmental niches, and the harmony that comes out of the struggle is quite as impressive as the struggle.

The Backup Pelican Chick

But, in keeping with good methodology for both science and religion, we ought to seek out anomalies, even if we distrust those that fly in the face of prevailing paradigms. We next choose *two discordant cases,* one from the macrolevel, one from the molecular level. These are the kinds of observations that accommodate poorly to a governing model, like ER-1470 on the anthropologist's curve of human evolution, or Jesus cursing the fig tree despite his being portrayed as a normative figure,[57] or like the offending points on the physicist's smooth curve.[58] Neither of these two cases involves the large-scale, long-term trends about which evolutionary theory is so unhelpfully silent, but they both involve the local subroutines of evolution. If the parts are too ugly, it may be difficult to fit them into a godly whole. Let us see what we can make of them.

White pelicans, large and majestic in flight, nest in skittish reclusion on remote islands found at large inland lakes.[59] Of interest to us here is, first, their seemingly rather inefficient nesting habits and, second, the treatment of the hapless chicks, illustrating the uncaring indifference and the suffering that so often attend the evolutionary process. The anatomy that makes the birds graceful when airborne and gives them great fishing skills renders them awkward on the ground, driving them to nest in isolation. But on islands they have no large predators, and seem counterproductively skittish. If the birds are disturbed, eggs and chicks are readily deserted, becoming food for gulls. What is waste in the pelican life stream is nutrient within the gull life stream. Even undisturbed, about one-fourth of nests are abandoned, often for no obvious cause. The parental instinct is easily disrupted.

The pelicans form large colonies composed of pairs in adjoining nesting sites. Females lay two eggs on open ground, the second about two days after the first, and parents by turns incubate them by wrapping a webbed foot over each egg. Few parents can raise two young; the earlier-hatched chick, more aggressive in grabbing food from its parent's pouch, becomes progressively larger, attacking the smaller sibling, and the resulting abuse and starvation are the major cause of chick loss. So the second is reduced to a backup chick, surviving if the first is lost, or if the second is lucky. It has only one chance in ten of fledging. The second chick is often driven to the edge of the nest by its sibling, only to fall or wander out, whereupon it will not be allowed to return, seemingly unrecognized by its parents, a refusal that protects against adopting alien chicks and wasting parental care on unrelated genes. The pelican mother, like the Bible's ostrich, "deals cruelly with her young, as if they were not hers."[60] Later, when the chicks wander freely over the open ground, parents recognize and feed only their own offspring.

These pelicans, being especially ancient birds, have outsurvived many now-extinct bird species, for some of which we can presume more efficient nesting habits. We have to say that long-lived birds such as these, which can breed for up to twenty-five years, do not have to be very good nesters to replace themselves. On the other hand, they must be subject to steady selection for efficiency. If many nests are too readily abandoned, if much time, energy, and food is spent on the chicks, including the second chick, only to have them routinely lost to gulls, or to have the older invariably destroy the younger, the overall efficiency of the pelican life stream will go down.

As they nest in large colonies, there is severe competition for fish. Pelicans must transport food up to sixty miles, and at the time of fledging a chick must weigh 20 percent more than an adult, the extra weight enabling it to survive the period when it learns to secure its own food. Feeding two chicks is a challenge, especially for younger parents. Older, more experienced parents learn to reduce sibling rivalry by keeping the chicks well fed, thereby sedating aggressiveness. To some extent, at least in good years, parents will need to learn to discern and feed the hungrier, more needy chick regardless of its size; and genes that enable this discretion will be selected for.

Genes for fratricide in the older sibling, if such there are, might help it to survive the juvenile months, when it pushes its brother out of the nest. But later, when it is a breeding adult over a quarter of a century, those same genes when passed along to offspring would curtail the number of offspring such an adult is likely to leave in the next generation, and would be selected against. So too with genes for being overly aggressive. They might benefit the older chick as a nestling. It grabs all the food. But, passed along to offspring, they would reduce its reproductive success, for such a parent will raise only one very aggressive child, while the second one starves. Natural selection will select against inordinate aggressiveness. Even the older chick has half its genes duplicated in the younger chick, and so it is in the older chick's genetic interest for the younger also to survive, quite as equally as it is in the interest of the parents to put two chicks into the gene pool.

The backup chick, normally lost, is insurance and must be treated sufficiently well for the fail-safe population to be there in a crisis when the first chicks are lost. Alternatively, in an especially good year, even the "runt" chicks will make it, to the genetic advantage of both parents and sibling. From a populational viewpoint, every present-day first chick is an inheritor of some genes from previous-day second chicks, and strengthened by them. Parenting cannot be too careless, else reproductive success lowers. We may not want to say that the process here is well designed; rather, at some points it is crude.

Nevertheless, there are powerful biological pressures out of the governing laws that prevent the nesting from being really dysfunctional. Pelicans have remained relatively stable for thirty million years. To some extent, both evolutionary theory and any religious interpretations that complement it are alike troubled unless they can find tendencies that keep the chick rearing logical and successful. Often the same features that contest theistic theory by making a biological process ill designed likewise strain evolutionary theory by making it maladaptive. Efficiency is a biological criterion, comparable to the way design is a theistic one.

It is a mistake to view the struggles or sufferings of pelican chicks too anthropopathically. Animals and birds typically have fewer nerve endings per surface area unit of skin, for instance; and the level of consciousness, self-awareness, or experience, or whatever is the proper name for their experiential state, is very different from, more subdued than, less intense and coherent than our own. Further, we do not want to judge pelican behavior by human moral codes, faulting the parents or the older sibling, since moral criteria are irrelevant to nonmoral forms of life. The male spider, after having accomplished his paternity, submits to being eaten by his mate, and so feeds the womb that nourishes his young. What is so bad about that? In their own alternative ways, human fathers do as much themselves. It would be a mistake to suppose that spider mates ought to behave like human lovers. Some birds of prey routinely hatch one more chick than they can probably raise, just in case there is a good year, and also they use the extra chicks as a form of food storage, feeding the runts to the older chicks in times of crisis. This would be thought evil in humans, but is it evil in hawks? Or in the God who is supposed to have made hawks?

Nevertheless, we cannot help judging the process as being more or less makeshift, and in particular the elements of struggle and suffering as being worthwhile or futile. For these are, after all, subroutines in the larger evolutionary process, although no moral agents are included, and they must be functional and good, even godly (if we can make them out so) in the executive programs into which they are incorporated. Here we find, in the pelican case, what must be viably adaptive mechanisms—skittish birds, abandoned nests, parental instincts disrupted, rivalry between chicks, a backup chick as insurance—that seem brutal and callous. Can one make religious sense of this element in nature, after one has managed to make some biological sense of it? The luckless backup chick suffers and dies, a minor pelican tragedy, but this sort of thing, amplified over and over, makes nature seem cruel and ungodly. If God watches the sparrows fall, God must do so from a great distance.

Sickle Cell Anemia and the Hemoglobin Molecule

The *hemoglobin molecule,* which we earlier admired, is subject to a disease of Central African origins.[61] By the genetic alteration of a single amino acid in the sequence of two of the four chains that fold to compose it, a defective hemoglobin "sickles" or precipitates into fibers under low-oxygen conditions, preventing the hemoglobin from transporting oxygen. Sickle cell anemia is present in homozygous individuals, who usually die before becoming adults, but in heterozygous individuals symptoms are not present (under typical conditions). To the contrary, the gene is present with high frequency, often in over one-fifth of the population in parts of Africa, a frequency somewhat correlated with areas where malaria is a common cause of death. Fewer sicklers than nonsicklers die of malaria. The malaria parasite seems unable to live inside red blood cells that contain the abnormal hemoglobin. Thus, it is hypothesized that sickle cell trait in the heterozygote conveys protection against malaria at the same time that it causes death in the homozygote.

Protection against malaria for many with one copy of the gene is purchased at the price of anemia and death by the few who get two copies of it, resulting in a balanced polymorphism, natural selective pressures at odds with each other, editing out a course between malaria and anemia. The mutation is simultaneously beneficial and harmful, but in different individuals. The sickle cell mutation, it should be noted, is an extreme. Most of the known mutant hemoglobins are harmless, although there are other genetic diseases of hemoglobin. Some mutants are beneficial; the Duffy blood group negative is an evolutionary adaptation that conveys resistance to malaria.[62]

Like most of the infectious and parasitical diseases, malaria is a disease of agriculture and civilization. More than 99 percent of the human years in Africa have been spent in hunter-gatherer tribes, and such diseases, which are among the worst scourges of civilized humankind, are little problem where populations are small, mobile, and scattered. Chances of reinfection are low, and there are natural selection pressures on disease organisms not to kill their hosts. Malaria became a problem in Africa only in comparatively recent times.[63] All agricultures do not cause increased malaria equally; widespread malaria seems to have come only with the Malaysian agriculture introduced about two thousand years ago, though it has since been perhaps the worst killer disease. Thus, the hemoglobin deformity, preserved because it chances to offer protection against malaria, is a short-range feature on an evolutionary time scale. It is a molecular disease, but it is one resulting from recent civilization too. We therefore do not here see evolution in progress in a natural form, working slowly and incrementally, but rather upset by the rapid cultural advances of an imported agriculture causing increased human populations and increased breeding of the mosquito. Cultural innovations, we can notice, have continued to upset stable biological regimens, often with human detriment, as seen currently in the ecological crisis.

Both the anemia and the malaria attack children. Some malaria-stricken children die; others survive, and most adults have become relatively resistant to further attack. When a population moves out of a malarious area, as when Negroes moved to North America, the hemoglobin deformity is rapidly edited out. Carriers of the gene have declined from perhaps 22 percent in the early slaves to 9 percent in present blacks, ten to twelve generations later. In a typical population in a malarious area, some children will die from malaria, some from anemia, some of other causes, and fewer than half will live to reproduce.[64] Nevertheless, the juvenile death rate in such populations is not substantially different from that in other premodern human populations.

It is also important to caution here that the whole phenomenon is not well understood, and we may get a revised estimate in subsequent years. Some recent studies point out that many homozygotes lead normal lives, many are only mildly affected, and the severity of the disease is influenced by many other factors.[65] Nor should we forget the many misgivings about natural selection theory earlier expressed. The evolutionary account here seems convincing, but then we have available no other paradigm with which to interpret what is going on. If something else, a more kindly-looking process, had occurred instead, the theory would have explained that too, equally well. The paradigm we have has a way of demanding an interpretation to fit it, and absorbing all the evidence. The ugliness that here looks as though it is in nature might be the artifact of a bad theory. Meanwhile, few will want to say that the process, as nearly as we now understand it, is well designed. It is not only clumsy, it is cruel.

Even within the evolutionary paradigm, the hemoglobin case is atypical. Given the interlocking of protection against malaria and liability to anemia, natural selection will, for the population, have edited out a course that favors a maximum human population, although this requires, under present genetic parameters confronting the novel cultural conditions, the persisting birth and untimely death of a percentage of the population, the homozygotes. Natural selection has fallen into a strategy in which to produce a majority who are fit it must continue to produce a minority who are quite unfit; this is a queer sort of optimizing of fitness over the population. But then again, malaria as a major scourge in recent cultures is an unusual problem for natural selection to work on. A double dose of the gene is lethal, but a single dose is beneficial under the constraints of a particular environment, a nasty one with malaria in it.

What can seem (in the usual mutational cases) groping exploration seems here a tragic genetic misfortune. In view of this wastage, if any genetic mechanism had appeared to unlock the balanced polymorphism, it would at once have been selected for, simultaneously conveying protection against both malaria and anemia, while preserving more of the young. Hemoglobin is a sophisticated but delicately balanced molecule, difficult to "improve." There are a number of other mutant hemoglobins (exploratory trials, as we are viewing them) that function normally except in high fever, when they degenerate. These hemoglobins are present worldwide in nonmalarious regions, but have been selected out of malarious areas, where they are especially disadvantageous during the fevers produced by malaria. Selection can work beneficially (as we have just noticed) over a few generations in favorable circumstances, but here the new cultural situation seems to have presented the already fine-tuned biological processes with a problem that could not be any better solved in eighty generations or more. But, uncompounded by culture, a merely biological problem might have been evolved past more smoothly.

Meanwhile, what is awkward but understandable under the scientific paradigm is still more difficult to handle in religious accounts of meaning. In the anemic African child, a certain blessing is linked with a curse, so that the larger life process moves by the sacrifice of unfortunate children. Though largely successful and conservative, the evolutionary procedure at its cutting edge is a trial-and-error process in which every mutational trial risks dysfunction and disease, usually less, sometimes more, an experiment in which a little fitness for a present few, carried on to the many in the future, is purchased at the cost of misfitting in the deleterious mutants. Emile Zuckerkandl and Linus Pauling wrote of this process: "The appearance of the concept of good and evil, interpreted by man as his painful expulsion from Paradise, was probably a molecular disease that turned out to be evolution. Subjectively, to evolve must often have amounted to suffering from a disease. And these diseases were of course molecular."[66]

The truth is rather more complicated than this. Evolution is a molecular miracle too, creating all the life and health we so greatly value. Good and evil do emerge out of the molecular evolution. There is a painful emergence from the primitive, nonfeeling states of presentient life and inorganic matter. It does indeed seem that subjectively to evolve is invariably to suffer. Yet the suffering is both corollary and cause in the larger currents of life. The basic question here changes its scope and form. We want to ask not whether Earth is a well-designed paradise for all its inhabitants, nor whether it was a former paradise from which humans were anciently expelled. The question is not whether the world is, or ever was, a happy place. Rather, the question is whether it is a place of significant suffering through to something higher. Malaria and sickle cell anemia are coupled with the evolution into agriculture and culture.

All of us carry some genetic load (genes that reduce health, fertility, effectiveness in some degree), though proportionately as these genes are serious they tend to die out in the population. They are, however, steadily being replaced by new mutations, and so we are always carrying some load of half-bad ideas, as it were, workable but less than perfect genes. Reproduction is, overall, marvelously successful. The possibilities of going wrong in human reproduction are enormous. Yet serious inborn errors persist in infrequent individuals. Still, there are mistakes, often tragic ones, that challenge any notion of divine design. But we can view such mistakes as evils that are the necessary product of a good system, where the load is the same as the exploratory process, the trial-and-error, learning dimension.

Adaptation is imperfect, but if it were perfect evolution would cease, nor could we have evolved to where we are. It is the imperfection that drives the world toward perfection, and in that sense it is a necessary evil. Nor, for most of us, does it follow that these (usually slight) handicaps impair our ability to lead productive, normal lives. To the contrary, learning to live with one's weaknesses, as well as strengths, is a growing experience. Natural selection requires evil (bad ideas) in its exploratory processes, but selects against them, to leave the good ideas in place, so far as this is possible under local genetic and ecological constraints.

Likewise, natural selection requires pain in its construction of concern and caring, but any population whose members are constantly in counterproductive pain will cease to exist or develop some capacities to minimize it. In this sense, natural selection, so far from needlessly increasing malformation and pain, rather edits them out of the system, so far as the remaining system can remain developmental. Evil is self-eliminating, except as it is productive of a subsequent good. We cannot show this in the detail of every case, and there are troublesome anomalies. Nevertheless, this is a general tendency in the evolutionary system. But this position is not inconsistent with a theistic belief about God's providence; rather, it is in many respects remarkably like it. There is grace sufficient to cope with thorns in the flesh.[67]

After sentience comes, with suffering on its shadow side, there also emerges with further sentience the capacity for learned behavior, and that reduces the suffering by employing the skills of achievement. With still more sentience there emerges subjective awareness, even animal thought, and this at once steps up the capacity for suffering but simultaneously allows the animal to conduct more of its trial-and-error learning in thought experiments, discarding the misfit ideas, rather than itself dying as a misfit. The suffering seems to be driving some of the progress upslope.

Even the genetically well-endowed, as with the luckless pelican chick, live in prospect of tragedy. Most are born to an early death; the few survive to prosper. But an aggregate host of life flourishes in outcome. We must not indiscriminately read either pain or suffering, as humans experience these, into nonhuman life contexts. These are present only in partial forms, diminishing as one moves downward and rearward on the scale, disappearing in nonneural forms, although there is everywhere effort and duress. Nevertheless, nascent analogues and precedents are there, and so far as struggle can be accommodated to a religious creed, we can recognize here a principle both of redemptive and of vicarious suffering, one whereby success is achieved by sacrifice. We cannot deal with this principle in its pronounced and existential forms until we have advanced to the level of mind, reaching there layers of meaning untouched in nonhuman nature. But we can already notice that the biological process anticipates what later becomes paramount, and this forces us already to ask about the meaning of suffering, although that question is one which biological science is incompetent to answer.

Seen historically over evolutionary time, the woman in childbirth was caught in a bind. Selective pressures increased the adult cranial capacity from about 500 to

1,500 cubic centimeters.[68] Big-brained babies were selected for, but difficult to deliver. If slightly oversized, the infant's skull could not pass through the pelvic opening, and both mother and child perished. Nor can the pelvic opening evolve steadily larger without maladaptively reducing the woman's powers of movement. Is this the curse of childbirth and nothing more? Perhaps. But one result was steadily to deliver the infant earlier. About a year is taken away from the uterine period in comparison with similar-sized mammals. The child is born less preformed, more helpless, both with greater neural potential and less neurally specialized. (Overspecialization, especially in large animals, has often doomed a species to extinction.) This year is shifted from a uterine environment, poor in stimuli, to one in the mother's arms, rich in stimuli. Further, the juvenile period is extended across a decade and more.

But this is exactly the context of our becoming human, of the remarkable growth of the human mind, of family life, caring, culture, language, and education. The tragic evolutionary situation, without ceasing to be that, becomes a story of the suffering to achieve something higher, the emergence of the human person.

The Way of Nature as the Way of the Cross

Here Christianity seeks to draw the harshness of nature into the concept of God, as it seeks by a doctrine of providence to draw all affliction into the divine will. (Judaism and Islam do so as well, though perhaps not so centrally.) But it requires penetrating backward from a climaxing cross and resurrection to see how this is so. Nature is intelligible, as we have earlier argued. Life forms are logical systems. But nature is also *cruciform.* The world is not a paradise of hedonistic ease, but a theater where life is learned and earned by labor, a drama where even the evils drive us to make sense of things. Life is advanced not only by thought and action, but by suffering, not only by logic but by pathos.

This pathetic element in nature is seen in faith to be at the deepest logical level the pathos in God. God is not in a simple way the Benevolent Architect, but is rather the Suffering Redeemer. The whole of the earthen metabolism needs to be understood as having this character. The God earlier met as the divine wellspring from which matter-energy bubbles up, as the upslope epistemic force, is here the suffering and resurrecting power that redeems life out of chaos. The point is not to paint the world as better or worse than it actually is in the interests of a religious doctrine, but to see into the depths of what is taking place, what is inspiring the course of natural history, and to demand for this an adequate explanation.

The secret of life is seen now to lie not so much in the heredity molecules, not so much in natural selection and the survival of the fittest, not so much in life's informational, cybernetic learning. The secret of life is that it is a *passion play.* Things perish in tragedy. The religions knew that full well, before biology arose to reconfirm it. But things perish with a passing over in which the sacrificed individual is also the carrier of a wave. The anemic African child, like the pelican chick, is delivered over as an innocent sacrificed to preserve a line, a blood sacrifice perishing that others may live. With both the chick and the child we have a kind of "slaughter of the innocents," a nonmoral, naturalistic harbinger of the slaughter of the innocents at the birth of the Christ, all perhaps vignettes hinting of the innocent lamb slain from the foundation of the world.[69] They share the labor of the divinity. In their lives, beautiful, tragic, and perpetually incomplete, they speak for God; they prophesy as they participate in the divine pathos. All have "borne our griefs and carried our sorrows."[70]

To return in closing to the opening question of this chapter, the abundant life that Jesus exemplifies and offers to his disciples is that of a sacrificial suffering through to something higher. When one is caught in the anguish of this struggle, as in a fight against disease, insanity, infertility, or moral evil, one may still cry out for strength and meaning to the strong son of God, beyond all recourse to biochemistry, beyond all description of the evolutionary struggle. We may say on the lower level that the whole evolutionary and biomolecular struggle is for better living through better biochemistry. But we may say on a higher level that in this struggle there is something divine about the power to suffer through to something higher. The Spirit of God is the genius that makes alive, that redeems life from its evils. The cruciform creation is, in the end, deiform, godly, just because of this element of struggle, not in spite of it. There is a great divine "yes" hidden behind and within every "no" of crushing nature.

But this kind of evaluation makes heavy demands on the spirituality of the judge. It is not the kind of conclusion that can be reached just by recording natural history, by reading biochemical instrumentation, or by deciphering hemoglobin structures. These are not judgments of mathematical elegance or symmetry such as may be found in physics. They are not judgments from the hardworld of science, but from the hard world of tragedy. These are judgments about the worth of life, about the worth of sacrifice, about ugliness transformed into beauty, about travail and splendor. They are judgments in which one prays not to be blinded by one's paradigms, not to accept evil too lightly. But one also prays for enough sophistication to see beauty and elegance in ardor and passion, in fear and trembling, in patience and in suffering, to see God in the seemingly God-forsaken places, to see how God writes straight with crooked lines. In evolutionary biology, even more than in physics, one needs to share the faith Einstein once expressed, when puzzled and looking for a new paradigm. "Subtle is the Lord, but malicious He is not."[71]

God is in, with, and under the plasma that grounds all, supplying all the material and energetic possibilities. But out of the particle play there emerge the living creatures of the biological story. These creatures are permitted their own integrity and freedom; the contingency in creation guarantees this. God is not the creator of a passive, inert nature, but of a creative nature that is given an efficacy of its own, in which the creatures are given efficacy of their own. But genuine as these self-creativities in a world of chance are, they are yet bound under the sign of struggle. God, who is the lure toward rationality and sentience in the upcurrents of the biological pyramid, is also the compassionate lure in, with, and under all purchasing of life at the cost of sacrifice. God rescues from suffering, but the Judeo-Christian faith never teaches that God eschews suffering in the achievement of the divine purposes. To the contrary, seen in the paradigm of the cross, God too suffers, not less than his creatures, in order to gain for his creatures a more abundant life.

To translate from evolutionary science to theology, just as the suffering at Calvary was human, creaturely suffering, out of which new life on Earth was redemptively to come, and yet, seen more deeply by Christian conviction, was the very suffering of God for the creation, so in the natural course there is creaturely suffering, autonomously owned, necessitated by the natural drives, though unselected by those caught in the drama. Yet this drive too may be construed, in the panentheistic whole, as the suffering God with and for the creation, diffused divine omnipresence, since each creature both subsists in the divine ground and is lured on by it. The Son of God is an innocent led to slaughter, and his production of new life for the many climaxes a via dolorosa in which not only the luckless pelican chick and the anemic child but also even the struggling survivors stand under the divine watching over.

"The creation," affirms Paul, "was subjected to futility not of its own will but by the will of him who subjected it in hope. . . . The whole creation has been groaning in travail together until now." But this is not the absurd, unending labor of Sisyphus. Rather, it is a constructive, redemptive drama of birth, survival, and maturation. Indeed, in just this suffering does the hope of redemption lie; and the Apostle concludes, in words that we here deploy into the biological sector, that "in everything God works for good with those . . . who are called according to his purpose."[72] In some way that we mixedly believe and dimly understand, the biology of the world, not less than the physics of the universe, is a necessary and sufficient habitat for the production of caring sentience and, at length, of suffering love in its freedom. Life is a paradox of suffering and glory, and this "secret of life" remains hidden in God, unresolved by biochemistry or evolutionary theory. The way of nature is the way of the cross; *via naturae est via crucis.*

NOTES

1. John 10:10.
2. Francis Crick, *Of Molecules and Men* (Seattle: University of Washington Press, 1966), pp. 10, 14.
3. James D. Watson, *Molecular Biology of the Gene,* 3rd ed. (Menlo Park, Calif.: W. A. Benjamin, 1976), p. 54.
4. Estimated from data in J. M. Orten and O. W. Neuhaus, *Human Biochemistry,* 10th ed. (St. Louis: C. V. Mosby Co., 1982), pp. 8, 154.
5. Lubert Stryer, *Biochemistry* (San Francisco: W. H. Freeman and Co., 1975), p. 90.
6. Partial exception has to be made for identical twins, and medical contrivance can suppress these mechanisms and make transplanting possible. In lower invertebrates and plants grafting is easier, but even here considerable distinctive individuality remains. See Jacques L. Theodor, "Distinction between 'Self' and 'Not-Self' in Lower Invertebrates," *Nature* 227 (1970): 690–92.
7. Eugene W. Nester et al., *Microbiology,* 2nd ed. (New York: Holt, Rinehart and Winston, 1978), p. 142.
8. W. Maxwell Cowan, "The Development of the Brain," *Scientific American* 241, no. 3 (September 1979): 112–33.
9. Michael Ruse, *Darwinism Defended* (Reading, Mass.: Addison-Wesley, 1982), p. 124.

10. Darwin, in a letter to Joseph Dalton Hooker, quoted in Gavin de Beer, *Reflections of a Darwinian* (London: Thomas Nelson and Sons, 1962), p. 43. In other moods, Darwin can find the process impressive and beautiful.
11. True birds seem already to have evolved at the time from which the presently known *Archaeopteryx* fossils are dated, but this nevertheless serves as a missing link, even if already a living relic in Jurassic times.
12. David M. Raup, "Conflicts between Darwin and Paleontology," *Field Museum of Natural History Bulletin* 50, no. 1 (January 1979): 22–29, citations on pp. 22, 25.
13. David B. Kitts, "Paleontology and Evolutionary Theory," *Evolution* 28 (1974): 458–472, citation on p. 467.
14. Stephen Jay Gould and Niles Eldredge, "Punctuated Equilibria: The Tempo and Mode of Evolution Reconsidered," *Paleobiology* 3 (1977): 115–151, citation on p. 115.
15. Gould and Eldredge, "Punctuated Equilibria," pp. 115–51.
16. C. H. Waddington, "Evolutionary Adaptation," in Sol Tax, ed., *Evolution after Darwin*, vol. 1, *The Evolution of Life* (Chicago: University of Chicago Press, 1960), pp. 381–402, citation on p. 385.
17. Steven M. Stanley, *Macroevolution: Pattern and Process* (San Francisco: W. H. Freeman and Co., 1979), pp. 192–93.
18. Karl R. Popper, *Objective Knowledge: An Evolutionary Approach* (Oxford: Clarendon Press, 1972), p. 241. Popper has had some misgivings about his misgivings. Natural selection might be, and has been, falsified by genetic drift. Nevertheless, many biologists hold it as a structural principle in a way that is nearly tautological, holding to it as long as they possibly can. It forms the presumption of their research program. See Karl R. Popper, "Natural Selection and the Emergence of Mind," *Dialectica* 32 (1978): 339–55.
19. R. H. Peters, "Tautology in Evolution and Ecology," *American Naturalist* 110 (1976): 1–12, and "Predictable Problems with Tautology in Evolution and Ecology," *American Naturalist* 112 (1978): 759–761, citations on pp. 3, 760. See also the vigorous discussion by others in *American Naturalist* 111 (1977): 386–94.
20. John Maynard Smith, *On Evolution* (Edinburgh: University Press, 1972), p. 85.
21. Charles Darwin, *On the Origin of Species* (Cambridge, Mass.: Harvard University Press, 1964), p. 171.
22. Theodosius Dobzhansky, *Genetics and the Origin of Species*, 3rd ed. (New York: Columbia University Press, 1951), p. 21.
23. C. H. Waddington, in a discussion of "Inadequacies of Neo-Darwinian Evolution as a Scientific Theory," in Paul S. Moorhead and Martin M. Kaplan, eds., *Mathematical Challenges to the Neo-Darwinian Interpretation of Evolution* (Philadelphia: Wistar Institute Press, 1967), pp. 5–19, citation on p. 14.
24. Motoo Kimura, "The Neutral Theory of Molecular Evolution," *Scientific American* 241, no. 5 (November 1979): 98–126.
25. Kenneth J. Hsü et al., "Mass Mortality and Its Environmental and Evolutionary Consequences," *Science* 216 (1982): 249–256. Asteroid or comet collisions triggered cycles of extinctions over fifty thousand years, beginning with reduced photosynthesis.
26. D. M. Raup, "Size of the Permo-Triassic Bottleneck and Its Evolutionary Implications," *Science* 206 (1979): 217–218.
27. Ernst Mayr, "Cause and Effect in Biology," in C. H. Waddington, ed., *Towards a Theoretical Biology*, vol. 1 (Chicago: Aldine Publishing Co., 1968), pp. 42–54, citation on p. 50.
28. Maynard Smith, *On Evolution*, p. 89.
29. Jacques Monod, *Chance and Necessity* (New York: Random House, 1972), pp. 112–13, 138, 180.

30. George Gaylord Simpson, *The Meaning of Evolution* (New Haven: Yale University Press, 1949), p. 292.
31. S. E. Luria, *Life: The Unfinished Experiment* (New York: Charles Scribner's Sons, 1973), p. 148.
32. Stephen Jay Gould, "Extemporaneous Comments on Evolutionary Hope and Realities," in *Darwin's Legacy, Nobel Conference XVIII,* ed. Charles L. Hamrum (San Francisco: Harper and Row, 1983), pp. 95–103, citation on pp. 101–102.
33. Gavin de Beer, "Evolution," in *The New Encyclopedia Britannica,* 15th ed., *Macropedia,* vol. 7, pp. 7–23, (London: Encyclopedia Britannica, 1973–74), citation on p. 23.
34. John C. Kendrew, *The Thread of Life* (Cambridge, Mass.: Harvard University Press, 1966), p. 107.
35. Eccles. 9:11.
36. Stephen Jay Gould, *Ever Since Darwin* (New York: W. W. Norton and Co., 1977), p. 45.
37. Douglas J. Futuyma, *Evolutionary Biology* (Sunderland, Mass.: Sinauer Associates, 1979), p. 449.
38. Chapter 2, p. 60.
39. Michael Polanyi, *Personal Knowledge* (New York: Harper and Row, 1964), pp. 382–87.
40. Melvin Calvin, "Chemical Evolution," *American Scientist* 63 (1975): 169–177, citation on p. 176.
41. Ibid., p. 169.
42. George Wald, "Fitness in the Universe: Choices and Necessities," in J. Oró et al., eds., *Cosmochemical Evolution and the Origins of Life* (Dordrecht, Netherlands: D. Reidel Publishing Co., 1974), p. 7–27, citation on p. 9.
43. Manfred Eigen, "Selforganization of Matter and the Evolution of Biological Macromolecules," *Die Naturwissenschaften* 58 (1971): 465–523, citation on p. 519.
44. W. Ross Ashby, *Design for a Brain,* 2nd ed. rev. (London: Chapman and Hall, Ltd., 1960), p. 233.
45. The loss of numbers of species, especially marine species, has been large at four or five anomalous mass extinctions. At the close of the Permian period this is quite startling (some say 75 percent or even 95 percent of all species). These disasters are so contrary to normal evolutionary trends that many paleontologists look for extraterrestrial causes (collision with an asteroid, a nearby supernova explosion, oscillations above and below the plane of the galaxy). Others think that the extinctions were slower and the causes more terrestrial (changes in climate, continental drift). In any case, the biological processes on Earth are still remarkable for their powers of recovery, and the recovery produces advanced ranges of sentience and complexity. See D. M. Raup and J. John Sepkoski, Jr., "Mass Extinctions in the Marine Fossil Record," *Science* 215 (1982): 1501–1503; Dale A. Russell, "The Mass Extinctions of the Late Mesozoic," *Scientific American* 246, no. 1 (January 1982): 58–65; J. D. Archibald and W. A. Clemens, "Late Cretaceous Extinctions," *American Scientist* 70 (1982): 377–85.
46. Maynard Smith, *On Evolution,* p. 98.
47. Walter J. Bock, "The Synthetic Explanation of Macroevolutionary Change—A Reductionistic Approach," *Bulletin of the Carnegie Museum of Natural History,* no. 13 (1979): pp. 20–69, citation on p. 20. Bock nevertheless believes that macroevolution can be reduced to microevolution.
48. Stephen Jay Gould, interviewed in "A Geology Professor Answers Questions on Creationism," *Unitarian Universalist World* 13, no. 2 (February 15, 1982): 1, 8, citation on p. 8.

49. Raup, "Conflicts between Darwin and Paleontology," pp. 25–26.
50. Stephen Jay Gould, *The Panda's Thumb* (New York: W. W. Norton, 1980), pp. 20–21.
51. Olov Lindberg, ed., *Brown Adipose Tissue* (New York: American Elsevier Publishing Co., 1970).
52. Ashby, *Design for a Brain.* That natural selection is a kind of learning, with resemblances to the development of mind, goes back at least to the last century. See James Mark Baldwin, *Mental Development in the Child and the Race (1894),* 3rd edition revised 1906 (reprinted New York: Augustus M. Kelley, 1968), pp. 161–171.
53. A. R. Peacocke, *Creation and the World of Science* (Oxford: Clarendon Press, 1979), p. 95. Peacocke's account is among the best of the dialogues between biology and theology.
54. Ernst Mayr, *The Growth of Biological Thought* (Cambridge, Mass.: Harvard University Press, Belknap Press, 1982), p. 81.
55. Loren Eiseley, *The Immense Journey* (New York: Vintage Books, 1957), p. 210.
56. Biologically a woman is adapted to bear perhaps twelve children, breast-feeding, without contraception, and if well fed on a fat diet (the case in some early American farm communities). But more typically in history a woman has had adequate protein but little fat, and the birth interval rises to four years, with an average family of about five. When agriculture appears, the birth rate may rise, and disease takes a greater toll. Despite the benefits of agriculture, agricultural peoples are, on the whole, hardly healthier than hunter-gatherers. Settled life is more conducive to infectious diseases, which cull the extra children. Adam's curse of tilling the soil is not incidentally connected with Eve's increased ordeal in childbirth.
57. Chapter 1, p. 15.
58. Figure 2.2; 2.3, pp. 37–38.
59. Fritz L. Knopf, "A Pelican Synchrony," *Natural History* 85, no. 10 (December 1976): 48–57; "Spatial and Temporal Aspects of Colonial Nesting of White Pelicans," *Condor* 81 (1979): 353–363; "On the Hatching Interval of White Pelican Eggs," *Proceedings of the Oklahoma Academy of Science* 60 (1980): 26–28.
60. Job 39:16.
61. A. C. Allison, "Sickle Cells and Evolution," *Scientific American* 195, no. 2 (August 1956): 87–94.
62. H. C. Spencer et al., "The Duffy Blood Group and Resistance to *Plasmodium vivax* in Honduras," *American Journal of Tropical Medicine and Hygiene* 27 (1978): 664–70.
63. Davydd J. Greenwood and William A. Stini, *Nature, Culture, and Human History* (New York: Harper and Row, 1977), pp. 300ff., 428.
64. S. L. Wiesenfeld, "Sickle-Cell Trait in Human Biological and Cultural Evolution," *Science* 157 (1967): 1134–40; F. B. Livingstone, "Malaria and Human Polymorphisms," *Annual Review of Genetics* 5 (1971): 33–64.
65. Darleen F. Powars and Walter A. Schroeder, "Progress in the Natural History Studies of the Clinical Severity of Sickle Cell Disease: Epidemiologic Aspects," in Winslow S. Caughey, ed., *Biochemical and Clinical Aspects of Hemoglobin Abnormalities* (New York: Academic Press, 1978), pp. 151–64.
66. Emile Zukerkandl and Linus Pauling, "Molecular Disease, Evolution, and Genic Heterogeneity," in M. Kasha and B. Pullman, eds., *Horizons in Biochemistry* (New York: Academic Press, 1962), pp. 189–225, citation on p. 190.
67. 2 Cor. 12:7–9.

68. Figure 1.4, p. 15.
69. Matt. 2:16–18; Rev. 13:8.
70. Isa. 53:4.
71. Albert Einstein, quoted in Abraham Pais, "*Subtle is the Lord . . .*": *The Science and the Life of Albert Einstein* (New York: Oxford University Press, 1982), p. vi.
72. Rom. 8:20–29.

Chapter 4

Mind: Religion and the Psychological Sciences

As soon as the phenomenon of biofunction emerges, in the early genesis of life, we are halfway into the realm of reasons, beyond causes; we ask *why* in addition to *how*. Biological organisms superimpose physiological functions on physical chemistries; they are motivated, learn, and have their active logic beyond the passive intelligibility of inorganic systems. But we do not enter fully into the domain of reasons until, after three billion years and passing from genetic through neural learning, we reach in human beings the flowering of mind. With the intentional self-assembling of a subjective ego, reasons become so pronounced a principle that we no longer know just how to retain those objective causes that still seem to be underpinning human affairs. An agent conducting herself in and corresponding with her world does so out of some meaning structure. But life-orienting meanings are traditionally a realm of religion, while science from its Newtonian inception has been more at home with material effects and causes.

If a science of persons merely extends the paradigms so successful in natural science, will it be competent to its advanced subject of study? Can there be a science of meanings? Is this an experimental or an experiential science, a *natural* science as well as a *human* one? What are its novel paradigms? If the biologic of life was already relevant to religious concerns, how much more will any logic of the human "spirit" (Greek: *psychē*) or of conduct (behavior) be of import for theological anthropology. "Know thyself," advised Socrates. With a science of personality? Or have religion and philosophy their contributions still? We turn first to the sciences of the human person, and then, in the following chapter, to the sciences of society, the cultural wave of which the particulate individual is a transient part.

1. THE POSSIBILITY OF A HUMAN SCIENCE

A Science of Persons?

By comparison with physics and biology, it can seem strange to devote a whole science to only one species, *Homo sapiens,* a nonphysical and nonbiological science focusing on *humanitas.* Perhaps it is not the species but the phenomenon of psyche, mind, that requires special attention, so startling and different is it from its antecedents in matter, energy, and metabolism. Physics established a science of matter, from the seventeenth and eighteenth centuries onward. Biology established a science of life, from the nineteenth century onward. But how much science of persons can we have? This twentieth-century question is as yet unanswered.

The body and brain are certainly natural products; perception and consciousness are shared with animals. Neurophysiology is to all outward effects a natural science. We can learn about various levels of nervous integration—spinal, autonomic, volitional—or that sensory nerves are dorsal, motor nerves ventral in the spinal column. We can learn the localizations of certain mental traits in the brain, or investigate its bilateral symmetry. We can map its tripartite structures—the neomammalian, paleomammalian, and reptilian brains—reflecting (as is claimed) a progressive evolutionary development.[1] So far as there is nature in human nature, this must be amenable to descriptions at the level of natural science.

But the tangled web we call the human brain contains some 10^{11} neurons, each of which may have hundreds or thousands of synapses, and the possibilities of diverse circuits on 10^{14} synapses, with hookups not merely linear and sequential but cross-wired and holographic, become almost limitless. This most organized structure in the universe, so far as is known, generates the self-aware human mind in a way for which we can give as yet little or no account. A multiple net of billions of neurons objectively supports one unified mental subject, *a singular center of experience.* How far can that mind understand the brain that supports it?

We can start with animal brains, beyond which humans have much advanced, and come to understand their simpler neurophysiologies. But we have difficult, veiled access to the subjective life of animals, and are unable to cross-check with their psychic experience. So far as our ethics permits, we can compare our own conscious experience, to which we have direct access, to the neurophysiology with which we are experimenting. We can experiment with perception and behavior. We can discover that the ratio of the difference in stimulus that we can just barely notice to the total stimulus is a constant in the normal ranges of taste, pitch, brightness, loudness, or judgments of weight (Weber's law). Or we can plot how persons forget the most the soonest, and more gradually afterward, the Ebbinghaus forgetting curve. Behaviorists plot their extinction curves, graphing how partially reinforced behavior is more persistent than continuously reinforced behavior. Psychology has a few, but relatively few, such laws, and they operate at rather trivial levels so far as significant insights into the higher mental operations are concerned. The formulas apply only to the subroutines of perception, memory, and behaving.

Indeed, if we press toward the fundamentals of perception and conception, another problem looms. Scientists are using their brains to understand their brains, and while we can well suppose that the brain might understand itself in part and in outline, can any logical system transcend itself exhaustively to critique its own structures? We run afoul of a new kind of indeterminacy, a new limit to our resolving power, namely, that a system of great complexity can perhaps not be wholly understood, predicted, and controlled either by itself or by some observer of the same type and complexity. There is a kind of bootstraps problem, a problem of the science of equals, where the subject of study is as rich in powers as are the resolving powers of the scientist who would define him.

No one will deny limited powers of self-understanding or of the understanding of equals. But a consummate human science would seem to need an outside epistemic point, some more advanced logic in which to subsume ourselves as we analyze and evaluate our own reasoning equipment. Here too there is a problem of access indeterminacy, owing to the privacy of experiential states and even of brain states, for the latter cannot be observed without surgical tampering. There may well be limits to our capacities to observe without distorting the brain, so far as it is an open, self-determining system that overlays macroscopic neurophysiological states on microscopic quantum states, with this giving rise to inner experiential states. There are, of course, limits to our capacity to observe without distorting electron states of any kind.

Perhaps brain processes can be computer-modeled in part. But the brain evolved so that first animals and later persons could better eat and reproduce, activities in which computers have little interest, and artificial logic might not fully simulate the natural logic. Indeed, in our evolutionary origins lies part of the puzzle. The brain evolved as a pragmatic instrument of survival, but in serendipity it has become so much more, lately a theoretical instrument probing quasars, time dilation, acetylcholine as a neurotransmitter, or building computers on which to model its logic. Eugene P. Wigner, a mathematician, calls the mathematical facility humans have achieved a "miracle in itself," and comments, "Certainly it is hard to believe that our reasoning power was brought, by Darwin's process of natural selection, to the perfection which it seems to possess."[2]

The natural sciences are so impressive that one wonders whether we can further construct a superscience of the marvelous intellect generating such sciences. If we can, that would be the final marvel of self-transcendence. On the one hand, since we can reason at ranges over which we surely were not naturally selected to reason (about quarks and the big bang), perhaps we can successfully reason about persons and their brain bases in dramatically self-transcending ways. On the other hand, all other sciences study a simpler other, while in psychological science mind tackles itself. That may imply limits to the possibility of a human science.

Causal Laws and Personal Agency

Nevertheless, causal sequences sometimes obviously affect persons, at both molecular and molar levels, alike objectively and subjectively. If I take a lithium drug, then I will not have episodes of excitability and depression. If I have three chromosome copies at the twenty-first chromosome, I will be mentally retarded. If a microelectrode stimulates a certain brain region, I will have flashbacks of forgotten experiences. If my corpus callosum and other commissure structures connecting the two halves of my brain are surgically severed, I will not be able to name or describe a spoon seen with my left eye only, although I can with my hands pick a spoon like it from a varied group of objects.[3] If I undergo certain childhood traumas, then in adult life I will probably have certain neuroses. Sociobiologists may or may not be right in their claim that the degrees of altruism that we are free to embrace are genetically influenced, and thus, say, that my views on private property or on Christian ethics have genetic parameters. Repeatedly, the human sciences describe causes of our behavior. How far should we extrapolate to suppose causes for all human experience?

Here we meet the possibility of self-refuting and self-fulfilling predictions, unprecedented in previous science. When a zoologist predicts that grizzly bears will become extinct, the bears cannot become involved and consciously change to increase their chance of survival. But when an ecologist predicts, owing to environmental degradation, that humans will become extinct, humans, who can imagine their own extinction as bears cannot, can intervene to avoid that outcome. Predictions of holiday traffic deaths are partly descriptions of what will happen, partly broadcast in the hope that this alarm will arouse behavior to refute the predictions. Predictions, which are difficult enough to make in the human sciences in any case, seem to take on a different form from those in the natural sciences. The logic is not merely one that connects cause (C) and effect (E), but one that involves decision (D). *If* C, *then* E—*unless* D is decided. Jesus predicted and believed that he would launch the Kingdom of God, and his kingdom has partially come in Christendom as a result of what he decided and what those who followed decided about him. Marx predicted revolution, and invited it, and his writings, widely read, prove to be causes of their own fulfillment. The predicted eclipse occurs whether we want it or predict it; but the predicted revolution may quite depend on whether we like it, work for it, even on whether we predict it.

Nonhuman organic life already involves a self-fulfilling drive, but not one that involves the conscious making of predictions and making them come true or defaulting them. We are dealing not so much with *predictions* as with *projections* of where we are headed, *if* we do not revise our heading. That last *if* is hardly a causal *if*, certainly not a straightforward causal one; it means something more like "but will we decide to revise our heading?" It registers uncertainty about the future decision, as much as certainty about the causal chains over which we can exert our decisions. Unbeknown to a person studied, we might discover some predictions that held with high probability, but what exactly would be the causal basis of this anomalous sort of prediction if it were effective only when kept secret from the agent? Predictions of individual human courses of action are sometimes altered when the subject becomes aware of the content of the prediction.

I may perversely refute what you predict of me. Such a prediction has to take itself into account; it can change things, but then again it may not, depending on the decisions of the actor. One does not know whether to make a still further, revised prediction after the first prediction has been made, using that as an additional cause not previously available in the calculations. If unconstrained by abnormal biochemistries or psychological neuroses, intelligent humans can often challenge or consummate predictions made about them, alertly entering on the basis of new information and self-reflection into the causal sequences operating within themselves. Even if I can predict your actions in isolation from mine, unless I can also predict my own responses in interaction with your actions, and your responses in turn to my responses, I cannot predict the outcome of the exchange, and so your determinism gets interlocked with my autonomy; we are both or neither predictable.

Unconscious forces lose much of their force once we become conscious of them. Women can discover that, unawares, they prefer security over adventure in decision-making contexts, while men are prone to prefer adventure over security, either from social conditioning or from genetic tendencies. But once this fact is known and we understand the causes behind it, women and men can take this disposition into account and decide to overleap it or compensate for it. I learn that psychologists have found that human aggression rises with increasing ambient temperatures. On the next hot day when I feel angry, I remember this, modify my behavior, and thus alter the facts. So humans do not seem to permit the maximal specification that physics and bioscience more often achieve, although we have noticed with the uncertainty principle and evolutionary mutation that even the natural sciences must settle for less than complete resolution of their objects of inquiry.

Every form of life is already linguistic in the coding of information into genetic molecules. Some animals also have complex communication systems. But human language is dramatically developed, rich in complex conceptual symbols, as is quite evident in science and in religion. The latter are activities impossible even to the most advanced of primates or porpoises. One may suspect that such a sophisticated being, when participating in the sacrament of communion or abstracting the philosophy of psychology, will not be reducible to causal explanation, owing to the emergence of these properties.

The most baffling symbolic logic that we confront is not that of the mathematicians who write equations and metricize things, not that of the logicians who abstract into symbols portions of our thought processes. The ultimate symbolic logic is language itself, ordinary language, the languages of science and of religion, where words and texts become such powerful symbols of the world, of the world-logos, and of our place in the world. Humans have a double-level orienting system: one in the genes, shared with animals in considerable part; another in the mental world of ideas, as this flowers forth from mind, for which there is really no illuminating biological analogue.

If humans with their language are to be free to think, then they must have thought options and select the best ideas for good reasons. Perhaps this is an extension of natural selection, only here it seems crucial that there be room to select the fittest idea, and to discard the misfits, on *internal* grounds in a deliberative process unlike any previously present in the *external* environmental editing, an emergent form of cybernetic control. In one sense, we do not have any options as to how we shall think, either in science or in religion. We must think as we are compelled to think by the external world. True thought is much constrained by the reality it studies; it corresponds with it in some degree, if only approximately, selectively, or symbolically, only functionally or pragmatically. In Luther's conclusion, "Here I stand; I can do no other." But, given the possibility of error, of better or worse modeling, there is a second sense, required for the first, in which we cannot be compelled by reasons in evaluating world views if we are blocked by psychological or other causal compulsions, if we are programmed by "laws of learning" to think in the way we do without genuine consideration of options.

If I can predict on causal grounds what you will say or not say, as I can predict what you will digest or not digest, then I must wonder whether to trust what you say when you attach reasons to your beliefs. Perhaps the ideas that come into our heads are caused; perhaps, given a decision to act on an idea, we operate causally to execute a change in the world. A reason in this latter way becomes a species of cause. But unless we somewhere choose between our ideas by logical deliberation, it hardly seems that we could be justified (or unjustified) in holding them, since "justification" would no longer have any meaning. On the other hand, if I can evaluate your reasons, predict what you rationally ought to say and sooner or later logically must say, then I can rid us both of error. I am operating in a rational mode, not a causal mode.

What sort of human science can handle the transcausal intellect? Rather paradoxically, science—supposedly the most rational of all enterprises—desperately needs a scientific model of the human person that takes rationality seriously, as the very basis of science itself, to say nothing of rationality's role in religion. If we are to believe some psychologies, *mind* is a *soft,* subjective concept that needs to be set aside before any hard science can begin. But *mind* is just that reality which makes good *hard* science possible. One must have a keenly analytic and synthetic mind, and the better the *mind* the more objective the science: the concept of "mind" does not get in the way of clear, scientific thinking; "mind" rather provides this.

Physics describes at the electronic level various superposition states that present possible outcomes, and finds that sometimes one sequence of states is arranged for by the decision of an interactor. Biological organisms likewise decide for some quantum states and not others by overtaking physical chemistry with their genetic information. In mind we have a further advance, as some among the possible neurophysiological superposition states are decided for and others not taken. By mechanisms as yet unknown, thought is imposed on the microchemical processes. Rather than being predetermined by biochemistries, it is sometimes crucially the other way around: logic determines which superposition states will materialize, what the electronic course of the neurophysiology shall be. The mind has its role in forming the microevents of its nature. Only if there is some logical superintending of these electronic circuits does the sort of rationality we demand in science and religion become possible. Here the evolutionary upshot, which during its development along the way is so often said merely to *simulate* rationality, does at the end *stimulate* rationality, since humans are enabled and provoked both to reflect over their world and to decide what ought to be beyond what is.

Thinking is matched with activity, and we are dealing not only with cognitive but with agentive capacities. These agentive capacities seem to leave something "up to" processes that are inner-directed, up to the "person." Even in animals we already wonder if this is not so, but, given access to our own inner life, in any human science the question becomes one of the first magnitude. No doubt we are repositories of natural drives—hunger, fear, libido, anxiety, wants, and so on—but is the person by "laws of motivation" only a resultant vector of these subforces and counterforces, conscious and unconscious? Is the neural dynamics not different in kind from, though more complex than, hydraulic or current systems such as those we meet in fluid mechanics or electronics? What more is the agent than stimuli, genes, and drives? Do voters in booths choose candidates more or less as rats in cages choose food?

Personal agency is often noticed as a worry to religion, and indeed so, for the moral responsibility that religion (or ethics) supposes seems to imply that a particular author could have done otherwise, *if* she had owned other desires, *and* that she could have had other desires, but did not choose them. But the agency question is also worrisome to science, for scientists also hope (being agents themselves) to be the masters of the sciences that they originate and use. Indeed, this becomes an urgent issue just as we do bring human conduct under causal prediction and control. Who are the ultimate agents that will employ this science? We cannot reach a trustworthy human science unless we have moral room to act valuationally in consequence of it.

Objective Science and the Subjective Life

Beyond this, humans are culturally defined. The question "Who am I?" can be given an initial biological answer, and we may worry about descent from the apes or desires preset in our genes. But humans are not defined entirely by nature. "Who am I?" has further to be given a cultural answer, arising from my tradition, my nation, my family, my roles, my career, my story. Any of the myriad cultures by which humans define themselves, like any of the languages they speak, can typically be emplaced in any newborn. The human is born at once more complex and less finished than any other animal species. This complexity lies in how humans must finish themselves through nurture. What lies innate in our heads is critical here, but the cultural input is more critical still.

Neither "the American experience" nor "the black experience," nor much in even "the feminine experience," is transmitted in genes. One does not need Greek genes to have a self determined by Plato, nor Jewish genes to be a disciple of Christ or Marx. Both science and religion, which so much define us, belong to our social experience, not to our natural roots. But how much science can there be of affairs of meaning that may require an ongoing self-definition? Even a cultural answer, illuminate who I am though it may, does not reach that unique specification of myself that lies in my proper name.

Science takes an outside view; it looks on its object of study, notices correlations between variables, finds covering laws, presumes causation, predicts, and succeeds marvelously in physics and in biology. Psychologists have often supposed that their model is physical science. Donald J. Lewis, writing a scientific psychology, claims: "Psychologists study their subject matter in very much the same fashion that other scientists study their subject matter. They make precise observations and they conduct experiments. Even though the object of study may be another human, the psychologist must treat this human objectively, in the same fashion that physicists, chemists and biologists treat their subject matter. As far as the science of psychology is concerned, the fact that the subject matter is frequently the human being makes no difference. The science-wide rules of objectivity and precise measurement still apply."[4]

But perhaps it is not so simple as this. When science is brought around to myself must I not also take an inside view? Not only psychologically but logically, I must demand that it be a science of subjects, in that sense a subjective science, an anthropomorphic science, science with a human face.[5] Still, what this subject claims to know of itself must be negotiable, for error may lie within the misperceiving self. In keeping with good methodology in both science and religion, I want objective correction of blind spots in my views of myself and my world. Here there is no sharper clash than when I see myself an agent acting for reasons, while science contends that I rather respond to causes, whether external or internal, and thus prevents *my* doing a thing as a deliberating agent. If my every behavior is propelled, all is compelled. There is no "insider" to do any self-propelling. Any fully objective account seems to preclude an adequate subjective account.

One might at the start have pursued the unity of the sciences, both in their covering of all events and in their being of a common kind, hoping by interconnecting laws under some unified theory smoothly to transpose from physics to biology to psychology to sociology. But when we actually come to the human sciences that unity seems entirely out of sight, and even within psychology itself there is more denominationalism than unity. The phenomena are too complex, perhaps in part beyond the limits of science, and any major theory is going to be crude.

We must rest content with a piecemeal science, a multiple-paradigm science, one full of minitheories. It will cover events statistically while glossing over indeterminacy. Have not even physics and biology learned to live with this in a different form? But electrons and maple trees are all pretty much alike, easy to treat as a class, with few interesting divergencies. Humans are not so, if we are dealing with their humanity beyond their biochemistry. There are enough similarities still to write a psychology, but there are great divergences within the class, owing to cultural and to personal self-definition. So the human sciences will inevitably be softer than the hard natural sciences.

Still, we are entitled to insist on this much of a unified science: that any psychological covering model be one from which the advocating scientist does not have inexplicably to exempt himself. A logical requirement, independent of any subjective biases, is that these theories can be self-appropriated, else such a crucial negative counterinstance will be anomaly enough to refute the theory. Science has to exist within some minds, and if a scientist cannot apply his paradigm to himself, or show some clear reason why he is exempt, at least after his discoveries if not before, that will be a self-contradiction of a logical as well as psychological kind. In a parallel way, we would be skeptical of any theology from which a theologian were to exempt herself.

We must be wary of split-theorizing—holding one theory for others, another for oneself. This will prove an especially useful criterion in judging causal explanations of human behavior. The proposer of a theory presumes that he himself and all colleagues whom he invites to criticize his work, ourselves included, will rationally examine it. But if his theory prevents him or us from doing that, then we shall be right to reject it, finding that we cannot even begin rationally to accept it. This rather modest requirement will prove a serious issue both in Freudian psychoanalysis and in behaviorist psychology.

Even though psychology discovers few laws that graph or metricize well, it might gain some broad principles that nevertheless give fertile insight into what is going on, like natural selection in biology or uniformitarianism in geology. But then again, since psychology deals with native ranges of experience, and (predominantly) with only one species, with which we are already familiar, it may be difficult to gain insights into logic, personality, or character that are impressively past common sense. Sampling the discipline, we next turn to three central streams within psychology, each of which does propose some paradigmatic principle about how humans work, and each of which has a bearing on religion. These schools are Freudian psychoanalysis, behaviorist psychology, and humanistic psychology. Of a fourth school, cognitive psychology, we will have less to say.

2. RELIGION AND FREUDIAN PSYCHOANALYSIS

The advancing of science opens up realms of which prescientific humans are entirely unaware. Astronomy with its telescopes revealed a previously unseen universe. Biology with its microscopes revealed invisible bacteria, viruses, chromosomes. Microphysics with its spectroscopes detected atoms and electrons. Paleontology and geology disclosed vast sweeps of unknown time and life. But there remains a massive inner ignorance, one that psychoanalysis proposes to remedy. One might have thought, in nonscientific naïveté, that if we know anything at all, we know our own minds. But we can no longer be so sure about what was once thought the first fact of experience.

Indeed, by Sigmund Freud's account, the conscious mind is but the tip of an iceberg. "You feel sure that you are informed of all that goes on in your mind if it is of any importance at all, because in that case, you believe, your consciousness gives you news of it. . . . Come, let yourself be taught something on this one point! What is in your mind does not coincide with what you are conscious of; whether something is going on in your mind and whether you hear of it, are two different things. . . . Turn your eyes inward, look into your own depths, learn first to know yourself!"[6] But this is more than an affair of knowledge—it is a challenge to human dignity, or, as Freud puts it, to our self-love. Copernicus dealt a cosmological blow: humans do not live at the center of the universe; Darwin struck a biological blow: humans are not divine but animals; and now, in Freud, we suffer a psychological blow, the most humiliating of all: we persons are not masters of our own minds.

The Unconscious Mind

Some mental events are subconscious. We drive an automobile thinking little about it. We have subliminal skills and half-forgotten memories. Ideas and dreams appear from unknown depths. If, under hypnosis, I am instructed to wash my hands at a certain hour on the following day, then later when I am awake, though unaware of the hypnotic suggestion and giving some alternative reason for my conduct, I will wash my hands. Humans rationalize, deceiving themselves with a pretended set of reasons, while motivated by desires hidden to themselves. Adult character may yet be ruled by voices from childhood, by juvenile trauma or parental dependencies supposedly outgrown, repressed, and presumed forgotten.

From such phenomena as these, building on clinical experience, Freud proposed a complex unobserved mind, which he modeled by mixing hydraulics, metaphors, and even mythical ideas. In this great reservoir many drives mix and flow, a labyrinth of impulses striving independently of one another. Like vapors rising from the fluid in a flask with a narrow neck, a few drives migrate out of the deep penumbra into the open focus of the mind. Broadly, a tripartite tension results from interactions between the *id,* the reservoir of desires, mostly prohibited and repressed, and the *superego,* the prohibiter or nay-sayer to the *id,* with conflicts adjudicated by the *ego.* Freud put all this together to explain neuroses, psychoses, dreams, slips of the tongue and other parapraxes, moods, depressions, complexes—and religion.

By way of a cautionary note, we may observe that Freud's science is both nonmetric and nonstatistical. It is not based on representative surveys studying normal human subjects. It rather arises from clinical experience with the ill. Freud's models are developed there and projected to normal life. As a result, his theories may fit best in abnormal psychology and work progressively less well as the subject is mentally healthy. Freudianism may be a theory of diseased minds, and not of normal-type specimens, although he claimed for it universal deployability. On the other hand, it may be that extremes furnish a useful context of discovery (a not uncommon case in other sciences, too) and that the study of personality disorders suggests hypotheses to be tested in everyday life.

Freud was a fertile proposer of hypotheses: the Oedipus complex for males, the Electra complex for females, the libido, penis envy, repression, oral, anal, and genital psychosexual stages, the neuroses of civilization, the neurotic basis of totemism, and psychological cannibalism (by which the early Hebrews slew Moses and later came to worship him). No critic accepts all of Freud's speculations, and it is common in the early stages of a science to find bizarre fantasy mixed with fertile hypotheses, as with the young Kepler's belief in the special divine nature of the sun, or with the elaborate astrology in Newton. Mystical beliefs guided Dmitri Mendeleev in his discovery of the atomic periodic table, and Auguste Comte launched sociology as a science with himself as the Pope of Humanity. With Freud, as generally in science, we must separate out a context of discovery from subsequent justification, and test his proposals by critical scrutiny, if indeed he will permit to us that possibility. We do this here in the case that interests us most, the origin of religion.

Operating from the Darwinian paradigm, Freud held that we humans evolve out of nature and find ourselves set nearly helpless in struggle before a hostile nature, a nature that we fear and incline to personify, since we know how to appease mindlike beings. Humans form culture as a defense against nature, even though in culture each person is opposed to the other, which generates further psychological frustrations. But cooperate we must, or else die alone. The family is the primary social unit. In it the child depends first on her mother, but longest on her father, a figure both benevolent and authoritarian. So there arise feelings of dependency, vastly unlike anything known in nonhuman species. The earthly father proves limited, sooner or later fails, and cannot protect his children. But we repress this thought, unable to face the stark reality of our cosmic isolation.

Religion as Psychological Projection and Illusion

Our unconscious minds project the illusion of a heavenly Father (historically first gods and then God) who will deal with the terrors of nature, save us from death, and repair the privations of culture. "When the growing individual finds that he is destined to remain a child for ever, that he can never do without protection against strange superior powers, he lends those powers the features belonging to the figure of his father; he creates for himself the gods whom he dreads, whom he seeks to propitiate, and whom he nevertheless entrusts with his own protection. . . . The defence against childish helplessness is what lends its characteristic features to the adult's reaction to the helplessness which *he* has to acknowledge—a reaction which is precisely the formation of religion."[7] In this projection lies both the origin of religious belief and our subsequent tendency to accept it when it is culturally transmitted to us.

Thus Freud discovers "the psychical origin of religious ideas. These, which are given out as teachings, are not precipitates of experience or end-results of thinking: they are illusions, fulfilments of the oldest, strongest and most urgent wishes of mankind. The secret of their strength lies in the strength of those wishes."[8] Man makes God, and not the reverse. Freud concedes that he finds only the origin of the God idea, and that strictly speaking the idea might be true. But he also holds that, after we have been shown the belief's unconscious origin, we have displaced all good reasons for thinking that God actually exists. With his psychiatric science he strives to destroy this illusion, believing that his science is the truth, and that in science lies all the protection we have against nature. Religion *"is nothing but psychology projected into the external world."* "Psychical factors and relations in the unconscious" are "mirrored . . . in the construction of a *supernatural reality,* which is destined to be changed back once more by science into the *psychology of the unconscious.*" We "transform *metaphysics* into *metapsychology.*"[9] Here we no longer have any pretense of a value-free science. Freudian psychoanalysis is not nontheological; it is antitheological.

Rationality, Religion, and Unconscious Motives

The whole notion of an unconscious mind is difficult to deal with because we have neither empirical nor introspective access to it, and this even though psychoanalysis thinks to bring to the conscious surface enough of the buried material to see it for what it is, and rationally to deal with it. Depth psychology shares with other scientific theories, and with theology, the theoretical postulation of unobservable entities, from which certain observations may be deduced. In this respect "the unconscious mind," "God," and "neutrinos" are alike. Nevertheless, this particular subterranean mental entity is an especially tricky potential reservoir, where (among the circuitry of 10^{14} synapses) one can place innumerable ad hoc drives designed to explain this or that behavior. Freudian theory proves here quite resilient, but ought to be suspected just because it resolves to accommodate so much observational material by repairing its account of what goes on in our unconscious minds. "We must make a dogma of it, an unshakeable bulwark," Freud insisted to Carl Jung, pleading that he not abandon the theory of the unconscious psyche and the role of the libido there.[10]

Sometimes Freud can be modest about the limited scope of psychoanalysis. "Psycho-analysis has never claimed to provide a complete theory of human mentality in general, but only expected that what it offered should be applied to supplement and correct the knowledge acquired by other means."[11] But no theory is very modest that rests on the supposition that all conscious phenomena are not what they seem but are a facade hiding deeper, unconscious determinisms and purposes of the self.

For those wedded to the Freudian paradigm, we have heard Karen Horney (a once-Freudian analyst) say, "it is difficult to make observations unbiased by his way of thinking."[12] As the theory of natural selection supposes that there must be some benefit for any surviving biological phenomenon, so psychoanalysis supposes that there must be unconscious drives, sublimations, associations formed between related or even quite unrelated experiences, something there responsible for every belief or behavior.

We begin to worry (as we worried with natural selection in evolutionary theory, and we will shortly worry again with radical behaviorism) that this has become a hard-core paradigm, exempt from falsification. With Freudian theory too, one has to ask what evidence could refute it, to try to keep the paradigm negotiable, and, especially since the theory tampers with our powers of reasoning, to hit it hardest where it has a weak underbelly.

If we posit this theory of the unconscious mind (T_u), *then* there should follow by implication certain observed human conduct (O_c). If T_u, then O_c. That theory will need to be tested against abnormal conduct, which is outside the scope of our inquiry here, but it will also need to be tested in its implications for and its congruence with normal human conduct, including the sort involved in doing science and doing religion. The theory of repression, for instance, might be tested, but then repression is not distinctively Freudian. Carl Rogers's humanistic theory, which is quite opposed to Freud's, contains repression in the form of denial, which has to be brought to the surface and faced in order for therapy to take place. Or perhaps the theory can be tested by an analysis of dreams. In our case, we will wish to test it against religious experience.

The psychological origin of religious belief can well be true or partly true in some nonhostile, weak version, but the theory would then prove too little, that is, it would not solve any of the ultimate truth questions that Freud did think he was settling. It might be the case that the idea of God comes into some theists' minds as an unconscious projection of, by analogy and extension from, their experiences of an earthly father. The father is, after all, a classical model for God. It will be necessary to "clean up" the detail of Freud's theory at this point, but it is not necessarily prejudicial to an incarnational theism (to take the Christian kind) to find that religious insight develops in a person's total emotional, motivational, even unconscious life, in all his psychosomatic experience, not solely in his intellectual life.

It does not violate any traditional property in God to allow that the parenting experience is the natural source of the concept of God. Religious belief has functions that run deeper than those of which we may be consciously aware, and some beliefs form at tacit levels. God might use the unconscious mind as a secondary cause to suggest the idea of himself. It might be (as Carl Jung held) that God gains entrance through the unconscious mind. The unconscious is in some sense an idea generator from which inspirations sometimes come. The voice of God may "bubble up" from the unconscious, mediated by mythic symbols (amidst other "noises" and random effects), both in children and adults. But we commit the genetic fallacy if we confuse an account of the childhood origin of a belief with an adult assessment of its validity. Fatherhood might be the context of discovery of the idea of God, but we have not yet reached the context of justification. We have yet to inquire whether and in what form such an idea can be demanded in a logically adequate explanation of experience.

But Freud's theory cannot be true in a strong sense without jettisoning all belief, and here the theory proves too much and becomes self-refuting. "A deeply rooted faith in undetermined psychical events and in free will . . . is quite unscientific and must yield to the demand of a determinism whose rule extends over mental life."[13] Psychoanalysis insists "that even the apparently most obscure and arbitrary mental phenomena invariably have . . . a causation."[14] Or, if the vocabulary of causes seems too physical and not psychological enough, the person has unconscious motives, not reasons but motivations (psychological forces). Such unconscious motives trump reasons. But if all belief is determined by an unconscious projecting, and all supposed freedom to reason consciously is really an inconsequential rationalization disguising this, there remains no possibility of rationally evaluating beliefs at all. The holder of one theory might, of course, be giving correct reasons, while her competitor might not. The holders of some theories might have conflicts between their conscious and unconscious drives, while the holders of other theories might not.

But in each case, hidden behind the show of reasons, a necessary and sufficient causal set operates rigidly, and we cannot rightly attach the reasons displayed to either theory, to ours or our opponent's. But Freud is unwilling to suspend his determinism anywhere. Psychological events occur with the same unexceptional causality as do all other natural events. "If anyone makes a breach of this kind in the determinism of natural events at a single point, it means that he has thrown overboard the whole *Weltanschauung* of science." "The intention is to furnish a psychology that shall be a natural science: that is, to represent psychical processes as quantitatively determinate states."[15]

Freud thus puts down religious belief as caused by an unconscious projection. But by the same *ad hominem* stroke, why cannot the theist put down Freud's rival belief? Antireligious "scientific" belief is really the same in kind, a governing *Weltanschauung,* and we can easily postulate for it some unconscious rebellion against one's parents, some desire to be free from guilt and moral commands, drives about which Freud had much to say. Perhaps Freud was moved by a longing for ordered security or for self-sufficiency, expressed in a wish-fulfilling scientism. Perhaps Freud was unconsciously affected by some anti-Semitic insults from the community of his childhood, reinforcing a vacuum of religious education at home, or driven by some vision of science as savior. Some mixing of drives counteracted the usual theistic outcome and produced atheism instead, now being rationalized in his psychoanalysis. By an inverted sort of *tu quoque* argument, "you too" hold your belief as you say I do. Supposing that you are right, you commit the same offense as I do, and can in fact no more be right than I am. We assess the scientific believer, not his beliefs, with the same skeptical assumptions he applies to the religious believer. Neither person can hold beliefs any more competently than the other.

Thus, this first science to study human intelligence finds it vastly overrated, an extraneous veneer over subterranean processes that make occasional half-sense, sometimes defending meanings that are in our interests, but for the most part irrational, yielding erroneous associations and ungrounded beliefs, all in countertendency to Freud's hope by psychoanalytic science to make life fully rational for the first time. The psychoanalyst will say, of course, that she has been released from the unconscious determinism of her ideas, while the believer remains enslaved. Freud claims, "From that bondage I am . . . free."[16] But what arguments provide the analyst with a credible exemption? How do we know which, if any, theories "given out as teachings" are really "precipitates of experience or end-results of thinking"? Explanations by the psychiatrist can no more be regarded as true than the systems of religion under scrutiny, not unless those few, and only those few, who have undergone his esoteric therapy can come correctly to reason.

Freud seems to have believed that a person is free when, after psychoanalytic treatment, the unconscious forces are not in conflict with the conscious forces. When we have made the unconscious conscious, we can do what we consciously will to do, although not, or not simply, because we consciously will to do it.[17] The ego is dominant over the id and superego. Nevertheless, unconscious as well as conscious forces remain operative in us all, and all behavior is fully causally determined, as are all thoughts. We cannot will to think or do otherwise. Humans have no genuine options. But such Freudian freedom does not really solve the logical puzzles here. Let us posit two persons, each free in the psychoanalytic sense but making conflicting judgments about the truth of a theory, each contender's beliefs being fully determined, and then ask how they could settle their dispute on rational grounds. Neither is free to change his beliefs for good reasons, although both have Freudian freedom.

The difficulty in the path of psychoanalysis is not self-love. It is the need for logical room to select the fittest hypothesis, a capacity we absolutely must have to evaluate religious belief and Freudian belief. But here the strong Freudian theory is a juggernaut that rolls over its holders. All our ordinary education in either science or religion is but the playing out in later life of repercussions, repressions, conflicts sustained in childhood. All our reasoning, whether before or after psychoanalysis, is so much fluff over processes that are determinate and ultimately nonlogical. No one can think otherwise than she does. Reasons have little to do with which view comes to be adopted. Or, if this is not so, how do we draw the lines between those who can (really) think and those who cannot, or when we are (really) thinking and when we are not? Here the psychoanalytic model breaks down. Reasons count, consciously, cognitively. If you did not believe that, you probably would not be reading this book.

Morality and the Unconscious

Freud's account of theism can be usefully compared with his related account of morality. The child's morality develops as the internalized commands of his father.

Later, adult morality, so called, seems to take place in a world where love, mutual trust, and concern for others are both possible and prominent. But in fact humans are driven by unconscious drives, rising from the id and from instincts. Adjudicate these though the ego and superego may, whatever we do sooner or later seeks our self-interests. We may seek the interests of others (the account we often consciously give of ourselves), but this will coincide with our unconscious seeking of our own interests. When enlightened, egoism may take considerable pleasure in the well-being of others. So our conduct is multidetermined; it has both conscious motives and different unconscious causes. But interests may not always be mutual; and where there is conflict the unconscious determinations will override seemingly conscious "decisions." We are psychologically unable to stretch love very far from our circle of intimates, those who can supply us with (more or less enlightened) gratifications, reciprocity, and self-esteem.

Freud accordingly objected to the golden rule as a psychological impossibility. Humans are not so constituted that they can, much less ought to, love their neighbors as themselves (to say nothing of enemies), not at least in any radical sense.[18] In fact, all genuine altruism is a sham, although a pseudoaltruism (conscious altruism coinciding with unconscious egoism) is not only possible but widespread. Humans accept only such "moral" checks as are (consciously or unconsciously) in their mutual, enlightened self-interest. All conduct (as well as belief and thinking) is really driven by mainsprings beneath the level of consciousness, and it is rather incidental that this sometimes coincides with a conscious wish to act in the interests of another.

That the sun is setting is a great sham, only an appearance, though it fooled the world for centuries. Galileo and his astronomy exposed that. That color, sound, taste, smell exist out there in the world is only an appearance, and Newton showed their mind-dependence, leaving only matter-in-motion. That the world is designed is a great sham, only an appearance, though it fooled the world for centuries. Darwin and his evolutionary biology exposed that. That humans are, or can be, genuine altruists is a great sham, though all the classical religions have advocated their versions of the golden rule. Freud with his psychoanalysis has exposed that.

But a philosophical ethicist will reply that this account of morality is too juvenile and incomplete. We would not regard as a moral adult someone who mimicked her father's commands. That is only the first stage of childish morality. Many persons mature into critical, reflective stages. This rational reflection, coupled with other-regarding capacities, constitutes moral agency. Better psychologies of moral development must take this into account. We do not regard as a moral adult someone who is unconsciously driven by self-interest while he supposes that he is acting altruistically. In a mature ethic, self-love need not be forbidden; to the contrary, it can be encouraged (as the golden rule permits). But it cannot be the covert determinant of all actions.

Meanwhile, we encounter some persons who impress us as moral adults. In the presence of an autonomous, rationally sensitive, altruistic conscience, Freud's account of its origin and operation is unconvincing because he does not have a complex enough account of what full-blooded morality is. Indeed, in most persons, whether children or the morally mature, there is a deep desire, unconscious long before it is conscious, for meaningful relationships with others. Our behavior is driven by these other-regarding desires, which contain significant dimensions of altruism, involve our fulfillment though they may (as neo-Freudians have often recognized). Other persons are more than instruments of our gratification and self-esteem; we relate to them as the persons they are. My conscious or even unconscious desires often override more primitive satisfactions. Altruism is no sham, but one of the deepest expressions of human nature. (None of this denies that self-interest is also a problematic good in human nature.)

But just as Freud's account of morality is juvenile and incomplete, so too is his account of religion.

An Incomplete Account of Religion

Freud's theory of religion needs to be clipped of pretensions to completeness, pruned back by empirical evidence. Theravada Buddhism posits neither God nor gods, but rather *nirvāṇa.* Mahayana Buddhism finds an ultimate emptiness, *śūnyatā.* Advaitan Hinduism posits a nonpersonal Brahman, altogether without fatherly characteristics. Taoism interprets all as the interplay of the *yang* and the *yin.* Mystics typically seek an oceanic consciousness, abandoning the self for absorption into the divine, and Freud's hypothesis becomes awkward in any advanced religion outside of the Judeo-Christian West and Islam. Even confined to mature theism, the search for God is not a retreat to juvenile securities; it is an adventure in search of self-transcending Reality.

As will the behaviorists after him in a different way, Freud makes of the religious impulse a reactive and defensive device, not a positive and creative developmental urge. But others, even in the Freudian tradition and those who regard themselves as atheists, can find that Freud's account is very incomplete and one-sided, largely because it does not account for the constructive, healthful, and positively orienting meaning forces within religion.[19] Psychiatric patients can be expected to foster the illusion that God will fulfill their egocentric desires. But others find that God corrects and redirects their wishes, establishes a concern for moral integrity, demands self-sacrifice, agape love, crucifixion of the juvenile self and resurrection of the adult person. Religion is a powerful source of altruism, of the sense of justice, of prophetic voices within society, of the forgiveness of sins, of peacemaking, of humility, of the redemptive use of suffering and tragic defeat. But none of these dimensions in religion is explained very well if religion is only juvenile projection of an illusory Father on which to lean.

There are, at the same time, other places where the religious disciplines can join Freud in noticing a darker side to human life, perhaps our secret sins, our fallenness and capture by bestial drives, our defensive conduct, a biasing self-love that leads us to irrational conduct or to do the right thing for the wrong reasons, while proposing supposedly purer motives. Theism much reflects over the experience of guilt, and even partly shares the Freudian belief that evils infest the unconscious, with self-love as a secret and omnipresent sin. But it does not find all guilt illusory, all freedom missing, altruism a sham, or forgiveness unnecessary.

That there are persons who are divided against themselves, or persons whose unconscious minds drive them to self-defeating behavior, or persons who cannot help being cruel or weak-willed, or jealous or afraid, provides no contradiction to theism. It is rather what Christian theism has often taught—that persons become prisoners of sin and helpless without divine grace. On the matter of illusions, Christianity holds fewer illusions than most liberal-minded psychiatrists about the extent of the evils that infest the heart, and whether reformation is possible on the basis of an innate human goodness that only needs to be released in a suitable social environment for human nature to flourish.

Both Freud and Christianity agree that humans cannot be rational until we have undergone a cathartic experience that releases our ideas from their embedding in our self-interest. Only the one proposes rescue by psychoanalysis, the other by forgiveness and conversion. Possibly too the one is right in abnormal circumstances, and the other more broadly applicable to the ordinary, if fallen, human condition. Perhaps some biblical categories—those of alienation, pride, fallenness, self-deceit, or self-salvation—are a key to understanding Freud, quite as much as Freud is a key to understanding the Bible.

At this point, we see on the one hand how a true believer in Freudianism might by ad hoc hypotheses adjust for all the nontheisms of the Orient and elsewhere, for the polytheisms and animisms of preliterate peoples, and in our own Western culture adjust for all the varying strengths and varieties of both mature and immature theisms, for all the mixtures of egoism and altruism. But we see on the other hand how such a theory, originating in the therapist's clinic and deployed across all religious life, is not really any longer a hypothesis derived from experience. It has become a prejudgment brought to experience, in the light of which all phenomena are being reinterpreted. It is what we called a "blik," quite as much as are alleged to be the religious beliefs it sets out to criticize. It is not science; it is a pyramid of conjectures.

The Search for a Parenting Cause

When we come to deliberate over whether the parenting experience supports the God idea that it may have formed, we do not find this question so illogical as Freud supposed. While the immediate father is finite, we stand created and maintained in an impressive historical and evolutionary succession. The surrounding culture, with which Freud found the individual to be in frustrated tension, is the crucial tradition without which our self-definition is impossible. We are children of our forefathers and may well respect the splendid multigenerational "parenting" power in several millennia of Judaism and Christianity. In this traditional sense, though our fathers die, we remain inescapably under their heritable influence and provision.

The surrounding nature, which Freud found a hostile terror to be kept at bay, is the further context of this parenting. Our morphology, biochemistry, and genetic information reach back across billions of years. So far from being a local affair between son and father, this extended "parenting" is that global event that most of all stands in need of explanation. Finding ourselves with life handed to us, surviving through ills and all, both by our own wits and by wisdoms genetically programmed and culturally transmitted, we are borne along on a wave of life greater than any particulate self. Nature moves through us, pushes us on. This generating force is no illusory projection, but rather is a movement anchored in experience, and this "secret of life," which finds its nearest symbol in the immediate father, is quite logically deployed into the whole ecosystem in which life is projected and provided. The familial father is a parent, culture our grandparent, and mother nature a grander parent still.

And behind that? "The common man," laments Freud, "cannot imagine this Providence otherwise than in the figure of an enormously exalted father. . . . The whole thing is so patently infantile, so foreign to reality."[20] But confronted with this deepest mystery of life-giving, is there not a quite sane insight that sees beyond and within nature and nurture "an enormously exalted father" whose presence has created and sustains life in the midst of its hostilities, and before whom the "juvenile" submission is quite appropriate? With this alternate account, the bearing of religion's origin on the validity of our belief assumes a different color. The larger "parenting experience" is the gist of what is to be appraised, and psychoanalytic science, so far from redeeming us from superstition and leaving us rational in our world view, gives us no help whatever in solving this crucial question of the meaning of cultural and natural history.

Religion may first begin, in any one individual's career, in the babe's wish for the mother's breast, in the child's wish for the father's strong arm, an infantile wish projection toward a God who stands over the world. Yet this ontogeny of belief is not so causally separable from the allied demands for a logic to the universe, for a Parenting Cause ontologically and phylogenetically adequate in ages past to generate and even yet to sustain the life process. Nor is religion so easily dismissible, after the childish stages are known. Everything collapses into wishful thinking by the Freudian account; but by a richer account the childish stages elevate toward some sterner demands of logic. A child's primeval hypothesis—that there is a ground of support transcending one's own being—is confirmed in the juvenile years in the environment of the home, but afterward confirmed in the harder, colder world of the adult years, through which in faith she later suffers and struggles.

Faith begins in instinct, prelogical but not illogical. The core hypothesis triggered by these childish thoughts has, of course, to be critically tested in the grown-up world. But when that comes to include judgments about the world that has negentropically grown up over the evolutionary course, about the adult world as a historical and dramatic moral theater, one very reasonable account is to hold that inquiry about this transcending ground of support does not come to rest until it rests in God. Some gratitude toward primordial power and goodness is appropriate. At least the question is logical and real, not merely psychological and illusory.

Consider, for instance, the searching questions of two prominent astrophysicists, confronting the scientific evidence about the interconnections between humans and the "anthropic constants and coincidences" in the universe. Bernard Lovell asks whether this is not "indicative of a far greater degree of man's total involvement with the universe" than we have ever before realized, whether "human existence is itself entwined with the primeval state of the universe."[21] John Wheeler asks "whether the particles and their properties are not somehow related to making man possible. Man, the start of the analysis, man the end of the analysis—because the physical world is in some deep sense tied to the human being. . . . We are beginning to suspect that man is not a tiny cog that doesn't really make much difference to the running of the huge machine but rather that there is a much more intimate tie between man and the universe than we heretofore suspected."[22] Before such acute questions, the issue of our "parenting" in and by the universe hardly seems "patently juvenile," a matter to be explained away as the product of our psychological needs, which are unconsciously forcing on us (even on these astrophysicists) the illusion of a connection between ourselves and the powers that have fathered us.

The religious result in the Freudian view stems only from primitive motivations that stumble into illusion. But the end result in the saint's view is primal insight into the most fundamental ecology of all—the everlasting arms underneath. Life is first a gift, and only afterward a struggle, and in this sense the father hypothesis is justified.

Nature is nevertheless an arena of tragedy. In suffering through, and coming to believe that one is enduring by the divine grace, do we fall into illusion? Or do such experiences and the convictions they generate lead to an understanding of the meaning of the world and the divine will in it? Like the Newtonian model of an Architect fabricating his world machine, the Freudian parody of a Father providing easy security is too simple. The cruciform character of God, and of that life in God in which the Christian participates, the experience of redemptive suffering pivotal in Judaism and in Christianity—these are almost entirely missing in Freud.

The believer feels that by powers which he but dimly comprehends all his forebears must have suffered through and prevailed, so far as they have thrust life forward to him. He stands as the beneficiary of a vast providence of struggle, with life led to green pastures through valley after valley of deepest shadow, finding a table prepared in the midst of its enemies. He looks to such procreative forces to carry him on, so long as his life shall last, and to carry the wave of life on beyond his particular existence. The continuing mystery is not death but life, and what the Freudian doctrine of nature can only regard as an absurd, epiphenomenal miracle, the theist by the parental account can meaningfully explain in God. Only an amazing grace could have brought us this far; only grace leads us on.

What if, among the psychological conflicts that nature and culture decree, there are some that we can solve only religiously? Convictions about meaningful orientation in the world would not then be the irrational, neurotic outcomes of evolutionary and cultural pressures, but rather their logical resolution. Nature would drive us toward meaning, expressed in religious culture, using our psychological structures to do so. The spiritually restless soul can be sometimes given a psychopathological explanation. Indeed, it will have to be so explained by those who presume to live in a merely secular world, for they have no other available categories of explanation. But to those who live in a richer environment, where the sacred suffuses all secular things, this restlessness provides further perceptive power, working rationally but also working in an underworld of the mind to sense an underworld of spiritual support.

Freud's unconscious forces are not so much causal as they are drives in the quest for meaning, drives often greater than our conscious powers of appraising them. If we read Freud between the lines, he can seem almost to know this.[23] Here many of the things we do, including seemingly senseless slips, dreams, fantasies, and fears, can be efforts toward a pattern of meaning, however distorted or illogical, obsessed or compulsive. The religions have brought this meaning drive to reflective awareness and to creedal expression in better and worse ways. So far from being blind illusions to be displaced by science, they have to be appraised by a complementary joining of scientific and religious logic. But the therapists will have to move beyond the competence of psychoanalytic science if they wish to finish this inquiry.

"In psychoanalysis," concludes Philip Rieff, "Freud found a way of being the philosopher he desired to be, and of applying his philosophy to himself, humanity, the cosmos—to everything, visible and invisible, which as a scientist and physician he observed."[24] And, though Freud insisted that his opinions were scientific and empirically based, in contrast with those of the theologians and philosophers he opposed, good theologians and philosophers are quick to realize metaphysics beyond science when they see it, even in the guise of psychoanalysis.

3. RELIGION AND BEHAVIORAL SCIENCE

Psychology is the only major science to become embarrassed by its name. While Freud was discussing the tyranny of the unconscious mind, John B. Watson was proposing to do human science without the mind altogether: "Psychology is the science of behavior." "I believe we can write a psychology," he predicted, ". . . and never go back upon our definition: never use the terms consciousness, mental states, mind, content, introspectively verifiable, imagery, and the like. I believe we can do it in a few years. . . . It can be done in terms of stimulus and response, in terms of habit formation, habit integrations and the like . . . The plans which I most favor for psychology lead practically to the ignoring of consciousness."[25] In order to make psychology a natural science, behaviorism makes a clean break with the whole concept of consciousness. "The behaviorist cannot find consciousness in the test tube of his science. He finds no evidence anywhere for a stream of consciousness."[26] This revolutionary manifesto has been endorsed by a generation of psychologists, leading to the wry quip: "Psychology first lost its soul, then its mind, then consciousness; but strangely enough, it still behaves!"[27]

But that complete psychology has never been written; indeed, after two-thirds of a century, we begin to wonder if, under these parsimonious assumptions, it can be written. The most determined effort to do so has been made by B. F. Skinner. We will here be examining behaviorism in its most aggressive and conceptually pure form, the radical behaviorism represented in its most celebrated spokesman, and widespread in contemporary psychology. Some notice of cognitive modifications of behaviorism, and of rebellion against it, will conclude this section.

Stimulus, Response, and the Conscious Life

Things "behave" in the sense used by psychology only if their responses can be conditioned. Having drive and motivation, they can learn to modify behavior. Evolutionary learning is in the genes and is intergenerational, but when evolution reaches sufficient neural complexity the individual can learn from its own experience.

This capacity is present in rudimentary form in *Planaria* and matures in the impressive human capacity to adapt. One almost inescapably judges that there is "somebody home" inside pigeons and rats, intuits something of what it is like experientially to be such an animal, and finds no such existential experience in a vegetable.

But to this inwardness, behaviorists warn, we have no scientific access. We cannot really do psychology, only behavioral science. In the *if-then* mode, we observe first the *stimulus* (S), then the *response* (R), the environmental input to the organism and the organism's output to its environment. By painstaking observation and experiment, simplified for systematic observation in the laboratory, we can formulate S-R laws, beyond those unlearned reflexes studied by biological science, "psychological" laws of learning and motivation. Molecular neurological changes underlie such learning, and about these the psychologist may consult with the physiologist, but molar behavior itself is reducible to scientific law.

One might have thought that the central task of psychology is to construct a science of immediate experience. But not so. Mentalistic concepts must be translated, made operational in terms of their physical counterparts. We do not say that the pigeon or person "expected" or "hoped" or "felt" that this act would yield that effect; we simply describe empirical behavior. The pigeon or person acted in this way with that effect. Especially to be banished are end-state strivings in the organism; we substitute their behavioral equivalents.

We are not to suppose that some mentalistic process or entity, such as interest, instinct, intelligence, or paranoia, *causes* the behavior as its *effect.* Everyday mentalistic concepts simply are prescientific *names* for a certain class of behavior, never *causes* of it. "The judgment of contemporary psychology seems to be that mind is not a particularly useful concept for the empirical explanation of behavior."[28] Events may pass through consciousness, and we know that this is often so in human behavior. Or they may bypass it. But consciousness is not a determinant of behavior.

The causes lie further back. We discover the real determinants of behavior in environmental stimuli, and here, after eliminating intervening mental variables, we can find hard causes. Psychologists wary of disputes surrounding causality may prefer to speak of "functions"—the word has both a mathematical and a biological twist—between independent variables (causes) and dependent variables (effects). Skinner writes, "The external variables of which behavior is a function provide for what may be called a causal or functional analysis. We undertake to predict and control the behavior of the individual organism. This is our 'dependent variable'—the effect for which we are to find the cause. Our 'independent variables'—the causes of behavior—are the external conditions of which behavior is a function. Relations between the two—the 'cause-and-effect relationships' in behavior—are the laws of a science. A synthesis of these laws expressed in quantitative terms yields a comprehensive picture of the organism as a behaving system."[29]

As behaviorism becomes more sophisticated, the organism is found to be more than a mere hyphen between the S and the R (S-R), and we can write

$$S \rightarrow \boxed{O} \rightarrow R,$$

so as to recognize that an organism (O) receives the stimulus and emits the response. In operant behaviorism, following some initial stimulus, there is an initial response (an operant), a behavior, which operates back on the environment to produce consequences, and these consequences become a subsequent stimulus on the organism to produce further behavioral response. In learned behavior, this enters a feedback loop, so that various responses produce various consequences, these becoming in turn new stimuli, the responses being modified in directions that are found to be reinforcing. Thus, behavior gradually tracks in the direction of reinforcement, somewhat as an automatic target-seeking missile by feedback tracks its target, only in more complex fashion. Behavioral increments (operants) are being blindly and contingently emitted, and selected for by the environment. There is operant conditioning, or behavior shaping.

An outside controller can enter this process to adjust the stimuli and arrange for the consequences, thus reinforcing certain behavior and extinguishing other behaviors. The experimenter draws or trains the behavior in certain directions, but it is not necessary to posit here any consciousness or goal seeking in the organism. It is "purposive" only in the sense in which a heat-seeking missile tracks its target. The behavior is being caused by the training stimuli. It is perhaps too simple to conceive of this in terms of a tight, Newtonian determinism, for the organism does have to venture the more or less random mutant behaviors. We make some allowance for trials "volunteered" by the organism, and here the process bears some interesting analogy to genetic mutation. Indeed, unless there is some contingency we cannot get the novel operants on which the possibility of adaptive learning depends. Nevertheless, any looseness in the S-R principle is averaged out, and the organism is to be understood causally. Necessarily this and not that behavior will be reinforced, and with high probability the organism can be both predicted and controlled. "In a system of psychology completely worked out, given the response the stimuli can be predicted; given the stimuli the response can be predicted."[30]

In the genes nothing can be altered within an individual's lifetime, although natural selection alters genes over generations. The information is stored in "read-only memory." In the learning organism, where behavior can be shaped, information is stored in "write-read memory." Experience can be written in, and afterward read out to modify future behavior within the life span of one individual. The individual has in memory an access to what it has previously done in experience. Along the lines of such a causal and somewhat cybernetic model we are to understand learning. In drive-reduction learning, the organism responds so as to reduce its primary drives, coming to learn what makes it less hungry and thirsty.

The Logic of Behaviorism

Let us next try to follow the logic of this behavioristic "thinking," if indeed that is permitted to us! The behavioral model certainly has its truths. Organisms respond to stimuli, learning to repeat successful responses. By common sense, which we can assume has been selected for over evolutionary time, an intelligent organism will shape its behavior to fit its environment. More learning capacity makes it more fit. Learning by reinforcement, or success, seems to reach back perhaps a third of the way across the evolutionary process. It represents, in fact, the second great stage of that process, when neural or psychological learning by individuals is added on to merely genetic, intergenerational learning. Indeed, given an organic system with incremental learning capacities, it is hard to see, in the broadest outlines, how learning could take place any other way than by environmental feedback. We have to learn by increasingly adopting what works. It seems absurd to think that we could learn by continuing with unsuccessful behavior, which fails to be reinforced, just as it seems inconceivable that, over time, the best-adapted should not survive.

But this does not yet address the question of what to make of the inner states that emerge with enough neural complexity, for these can well add new directions, so that the process is not as blind, objective, or causal as it was before. There is not merely the cumulation of conditioned habits, but there is an inner formation of a consciousness that accompanies the learned information. It may not always be, at the levels of interaction where persons are involved, that the environment coerces; sometimes it can win our acceptance, even our love. Sometimes it can present us with options and decisions. This may force a revised account.

When we reach humans, there is one sense in which persons furnish no exception to the broad principle that an intelligent organism will shape its behavior to fit its environment. Humans may be different in the extent to which they learn, vastly more than animals, to do the reverse; to shape their environment to fit their behavior. Their cultural and built environments are products of such knowledge. Nevertheless, the person acts under environmental incentive and must adapt or perish. Enormous stretches of behavior can be treated under a causal if-then formula, partially by statistical generalization, overlooking what is irrelevant for those interested only in producing effects. *If* an employer wishes to increase productivity, *then* she offers her employees additional pay by piecework (a stimulus), and, within limits, productivity will rise (a response). A simplified analysis, such as an economist may want, can skip all mentation in the workers, ignore how they wish to use their added income, their decisions about whether it is worth the extra effort, concerns about fairness in pay, quality in their work, etc., and abstract out a causal relation. We may find this satisfactory in explaining segmented conduct and subroutines, and use it for a piecemeal technology of behavior modification.

Further, it is a welcome fact that the environment has its strong impact on the organism, which becomes what it becomes under environmental pressures. These demands, like opportunities and threats, are the milieu of life. The organism and its environment, dichotomized from an organic perspective, are also to be integrated in an ecological perspective. Life is "in" the organism, but then again life is a relationship "between" the organism and its environment. When biological evolution gives way to culture, this principle continues in a new ambience. Unless the person can respond to the stimulus of his culture, the transmission of a heritage—including science and religion—is not possible. To be cultured is to be shaped by educational forces that do not originate within ourselves.

Here the behaviorist claim that aversive stimulus (punishment) is less effective than positive reinforcement in shaping behavior is congenial with the longstanding Judeo-Christian maxim that loving one's enemies is a faster and more ethically desirable way of changing their behavior than is hating and taking vengeance on them. Along these lines, we may understand the discipline of parents over children, and of society over individuals. In these senses we are all tied together, as surely as by electromagnetism and gravity, by laws of reinforcement that hold us in mutual interaction.

What rational persons want, both in their beliefs and in their behavior, is to have their ideas and their conduct "controlled by," that is, pulled into suitable accord with, the real world. To learn truth is functionally to have one's information correspond with the way the world is, and the doing of science and of religion alike are a testing of our theories against the world, a revising of them in directions of positive reinforcement and an extinguishing of them with enough negative reinforcement. Only those who are blinded by belief will foolishly persist in conduct that is not in this sense continually reshaped by environmental feedback. It is the insistence of the real world that helps us to correct those errors into which we are led by our unthinking beliefs or (with Freud) unconscious instinctual drives. But the question remains whether the nonmental causal model of this learning is adequate and whether the environment, natural or social, is an exhaustive determinant of behavior. For behaviorists want to preserve a causal logic, to omit consciousness and see the resulting behavior, without loss, as response resulting from stimulus, nothing more. There is no rational, self-conscious agency.

The behaviorist paradigm is deployed across animals and humans alike in the conviction that basic principles, learned in animal experimentation, can be applied to human conduct, fortified with auxiliary modifications but retaining the basic theory. On the one hand, experiments can be tightened up until repeatable results are attained, with tamed animals preferred over those that are "wild." On the other hand, positive reinforcement, negative reinforcement, aversive stimuli, deprivations, extinction, and the like can be expanded—in terms of social stimuli, token stimuli, verbal stimuli—loosely to include an unlimited range of secondary reinforcers and learned drives. By such "et cetera" extensions there is no conduct that cannot so be interpreted, at least in programmatic outline.

As external behavior becomes more complex in cosmopolitan society, any conduct that does not first fit the dogma can be handled under the supposition of unknowns in the history of reinforcement. For habitual motor skills, we posit internal, fractionalized S-R chaining, "proprioceptive stimuli in tendons and muscles." For more contemplative behavior—the writing and reading of this book—involving the use of nonspoken language, we can extrapolate "covert behavior,"[31] a self-reinforcement different from overt behavior and now contrary to the earliest hope of having everything observable. But now, at no cost, we can always put in some subdermal "epicycles" (recalling how the old astronomers faced with embarrassing planetary motions put in epicycles to retain Ptolemaic theory), conceive of them causally, shuffle them around, and save the theory.

Logical Difficulties Assessing Behaviorism

But is not the behaviorist regauging all evidence to fit a creed now become impregnable, one that, like badly conceived theological bliks, seems now to include everything but that is in fact empty? It invents and supposes stimuli wherever it finds responses, and vice versa. Despite the operants that are emitted more or less randomly by the animal or person, only those that are reinforced are preserved, so that the persisting conduct is essentially a kind of recoil from the environment. The central dogma—that behavior is causally shaped by environmental stimuli, with no need to consult consciousness—is not really negotiable. Auxiliary hypotheses, such as those about what goes on in unknown reinforcements or inside the skin, are going to be adjusted so as to save the big theory.

But if we look more critically at the attempt to handle all human activity under this model, we find that it is really as much a promissory note as an empirical science, and that it operates as much by an inflated ideology as by a set of verified principles. There is a great deal of praising by anticipation successes not yet achieved. There are lots of et ceteras, ellipses, and details to be filled in later—if ever.

Let us in a thought experiment try to assume the complete truth of behaviorism, and we will see how problems of self-implication become insurmountable. If all learning is really S-R learning, causally shaped, then what of our learning of the S-R theory itself? Whether it is a thoughtless or thoughtful response does not matter, for mind is epiphenomenal in the causal process. We can no longer say that we hold or reject the idea deliberately, in free thought having fully considered the theoretical options, and selected the fittest. Now, as with Freud before, whatever reasons are offered to go with the theory are burdened by those precedent causes that pre-fix the belief and its reasons, and neither the behaviorists nor we can tell whether these are in fact good reasons that could independently support the belief.

Once again, we must press an *ad hominem* objection to throw into relief a logical difficulty. Does Skinner hold his beliefs, and conduct himself in a scholarly way, driven by environmental determinants? Behaviorists praise the rational, scientific status of their beliefs, often with the accompanying complaint that religious explanations of behavior are crude. But what privileged exemption can be allowed for behaviorists to hold their beliefs credibly, which is not also granted to theologians and scientists being invited to criticize the theory? What is the point of behaviorists *urging* us to share their theory, if the *urging* is only to be subsumed under a causal reinforcement schedule? Where is any place for the urging of reason? It is remarkable to see how Skinner can go to such trouble to persuade us to choose his point of view, which is that choice is illusory. Among the most complex of human behaviors are those of our learning in science and in religion. But, having evicted "the traditional fiction of a mental life,"[32] behaviorists have lost all authority convincingly to assert the adequacy of their own theories, much less to criticize inadequacies in competing scientific or religious ones.

It seems almost as though in scholarly debate we have only noises coming out of those persons involved, symbolic noises perhaps, but noises that, if we go far enough backward or outward, are caused by their environments. The speakers do not and cannot mean what they say, not deliberately. The same stroke that takes away the purposes that others may have, replacing intentions with responses that causally track stimuli, takes away the grounds that behaviorists may have for their theory. The behaving subject and the behaviorist scientist alike lose their reasons. They only track causes. What the subjects "think" about all that happens to them and about the responses they emit, or about those responses that are selected for, is irrelevant. It does not matter what persons think, as it does not matter what pigeons think. All that matters is the administered reinforcements. But what the scientists "think" is just as irrelevant. The theory here defended can find no addressee, no mind, to hold or critique it. There is no meaning left in "justified theory." Behaviorists too are passive recipients of their theory, caught in their own net. The whole behaviorist scheme is a massive explanatory net through which mind and agency slip unnoticed. But a genius caught up in a systematic error is difficult to correct. He will be ingenious at finding devices to ignore unpleasant evidence.

Behaviorists can only say, in consummate deploying of their paradigm, that their scientific behavior must be "right" because it has been reinforced, found more "successful" in competition with those who emit nonbehaviorist behavior.[33] So the scientific enterprise is just the most successful form of operant conditioning! But that such behaviorist type of scientific behavior will prevail is an unfulfilled prediction, not an observation. While we grant that organisms that can learn are fit, and that science has its impressive successes in fitting us for life, as does religion, the further claim that rigorous behaviorists are more successful in science or in life than are nonbehaviorists is something that no one has begun to attempt to demonstrate.

Moral Freedom and Responsible Decision

We have no one to hold morally responsible, and again (as earlier with Freud) a valuational problem follows a logical one. Humans are complex piano keys who cannot do other than they do, given the pressures under which they play. "Science ultimately explains behavior in terms of 'causes' or conditions which lie beyond the individual himself. . . . As such explanations become more and more comprehensive, the contribution which may be claimed by the individual himself appears to approach zero. Man's vaunted creative powers, . . . his capacity to choose and our right to hold him responsible for the consequences of his choice—none of these is conspicuous in this new [behavioral] self-portrait. Man, we once believed, was free. . . . He could initiate action and make spontaneous changes of course. . . . But science insists that action is initiated by forces impinging on the individual, and that caprice is only another name for behavior for which we have not yet found a cause."[34]

"Inner entities or events do not 'cause' behavior, nor does behavior 'express' them. . . . In an acceptable explanatory scheme the ultimate causes of behavior must be found *outside* the organism."[35] Skinner concedes that a person may be "aware of his purpose," but the person does not choose it. "The basic fact is that when a person is 'aware of his purpose,' he is feeling or observing introspectively a condition produced by reinforcement."[36] Again, events may pass through consciousness, but consciousness is not a determinant of behavior.

The agent is, in short, only a reagent. "A scientific analysis of human behavior dispossesses autonomous man and turns the control he has been said to exert over to the environment." Science moves us "beyond freedom and dignity."[37] These are myths used to allow society to hold persons responsible, to praise and blame them, but they are illusions covering up the fact that all human behavior is shaped deterministically from the cradle to the grave by reinforcements from the environment. Thus, Skinner joins Freud in thinking that the responsible person in the classical sense is a myth to be displaced by science.

But perhaps behaviorism has not so much moved us beyond freedom and dignity as it has found freedom and moral dignity inconspicuous, because these are systematically overlooked by the paradigm under which it operates. Are there not some anomalies that can alert us to troubles at the core of the model? If the control of persons has really been turned over to their social and natural environments, what is the point of Skinner's fervent urging of his readers to apply behaviorism and to exercise the controls now available from the new science? Is he only changing my reinforcers? Am I falling victim to some scientific special pleading involving irresistible new stimuli? Or am I being treated as a responsible person?

If I believe the theory to be true, there seems to be no self left capable of taking advantage of the theory. Granting that we know how to eliminate unwanted behaviors in others, and even in ourselves, by identifying the old reinforcers, eliminating them, and introducing new reinforcers for competing and wanted behavior, how do we tell which behaviors in others and in ourselves are unwanted? Indeed, the very word "control," of which Skinner is fond,[38] seems a hopelessly teleological word, since anyone who sets out to control another, or even herself, must have an end state in mind. Yet we are not told by the theory how to gain those end states of mind, but rather prevented from taking them seriously. It is hard to conceive of an engineered environment without end-state strivings in some engineers.

In physics, the more sharply one focuses on the momentum of a particle, the more fuzzy the particle's position becomes; the more clearly one measures the position, the less clearly one can measure the momentum. In psychology, it seems that behaviorism has brought the reactions of persons into so sharp a focus that their agency has become invisible. Yet for critics alert to anomalies and contradictions within the theory, the concept of agency is not to be argued away, but rather reveals the weak underbelly of an otherwise aggressive theory. "A scientific analysis shifts the credit as well as the blame to the environment, and traditional practices [of holding persons responsible] can then no longer be justified." Behaviorism wishes to "strip away the functions previously assigned to autonomous man and transfer them one by one to the controlling environment."[39]

But why are we being urged to abolish our old ideas—myths of responsibility, freedom, dignity—and step forward into a behavioristically engineered world, unless it is in fact the case that ideas do make a difference, and that we can decide to act in consequence of our ideas? When behaviorist psychology supposes that the stimuli shape all the responses, science gets only reactions without agency, and seems to know nothing of what even physics can call interaction phenomena, in which a participant-observer calls forth a world event.[40] Rather, it finds only a world environment that produces an observer. This behaviorist psychology seems to know nothing of what biology knows about information coding and learning in an agentive organic center, nothing about the programs and goals an organism defends.

So far as humans, moved beyond or falling short of freedom and dignity, can be brought "under control," the question of the uses to which this science will be put becomes urgent, a moral question to which this radical science neither provides an answer itself nor permits a place for other disciplines to supply an answer. Let us suppose that we desire the benevolent use of behavioral modification toward a good life and a planned society. If behavioral law covers all the members of that society, we have no leaders, no reformers, no therapists, no engineers, no educators *free* to decide what the character of that reformed life shall be, what to set as values for the perfected society. Nor can we allow any prescribers in our theory without compromising its paradigmatic character. The question is not merely who will be the controlled and who the controllers, but whether there can be any decision makers, since every player is bound by the stainless steel causal chains that make marionettes of us all. There is no place made for, much less power that makes for, righteousness. Or love. There are only managers, who in the end are not really managers at all, but managed themselves.

More pluralistic and decentralized accounts of the behaviorist society do not envision a hierarchy of controllers and controlled, but rather a sort of reciprocal controlling and countercontrolling of each on the other. We do not picture an unidirectional controller and his controlled, but a multidirectional and reciprocal determinism in which each is both controller and controlled in a great web of stimuli and responses. But this does not escape the problem. For if the experimenter controls the rat but the rat also somewhat controls the experimenter, if the teacher controls the students but the students also the teacher, if the employer controls the employees but also the employees the employer, if we control others as others control us, there still remains both the constant assigning of responsibility external to the agent and no account of any guidance of these controls we exert over others.

Everything is rebutting everything else; nothing is instigating anything. All the activating stimuli come from without, and, once again, what anyone is "thinking" is not a relevant question. "A person does not act upon the world, the world acts upon him."[41] There is no doubt which half is uppermost in the two-way process: it is the environment that shapes a person's behavior. There is never any answer to the question "What sort of shaping shall I do?" Indeed, the question is not even permitted by the theory. Each of us oscillates between shaping and being shaped, but no one "thinks" about it. Everyone responds, but no one is responsible.

In practice, behaviorism runs counter to our experiences of rational deliberation and moral decision. In theory, the model becomes internally incoherent. We have at the crucial points no one to reason and no one to hold responsible, no one to write the policy, whether in government, industry, science, or religion. So the master theory will not do the really heavy explanatory work we need for what we observe, what we experience, and what we must think.

Behaviorism and Religion

A further human gift, beyond rationality, morality, and valuing, is the capacity to love. These virtues have often been identified by theology as marks of the divine presence in humans. But if the loves we have for each other are reduced to output responses to causal stimuli, they hardly seem human, much less divine. It is true, of course, that love requires stimulus and response, give-and-take, and this results in behavior shaping as love develops between persons. But can we envision this process in disregard of the intentional consciousness that enjoys and drives these experiences of love? It may not matter whether pigeons think as they peck levers gaining food in result, but it matters greatly what persons think as they seek goods from and share these goods with one another. Behavior described without analysis of the intentions that produce it has lost its meaning. Worse, if such behavior is causally produced by environmental stimuli in such way as to make these intentions both irrelevant and inevitable, then love seems vacuous.

Further, our loves are of many kinds. Perhaps behaviorists can partially succeed in giving a causal account of what moralists have called erotic love. This involves a self-fulfilling wherein the loved object stimulates a lover's drives and afterward a sharing possession of the object reduces them. But what are we to say of what theologians call agape love, a sacrificial love freely chosen and conferred on the other? Such love operates, one might respond, upon the call of the needy other, a stimulus, and the person loves in response. We do not operate on our own resources but by taking in the divine grace. But if this occurs without decision and awareness, if it is not in any sense "up to me," but rather predetermined by my reinforcement history, then to call this sacrificial love seems a fiction. There is no grace here, no nobility, only scientific causation. God is the divine One who loves in freedom, according to a definition from Karl Barth that we will later examine.[42] Humans are children of God, imaging God as they are loving and free. But these are the dignities that behaviorism thinks it has shown to be incoherent and that it wants to move us beyond.

A passive, mechanistic view of human conduct secularizes life, ignoring or denying the dimension of the sacred. Religion is seen, on the causal side, as a historic reinforcer. This function is served by its institutional forms, where religion is a behavior shaper. Seen as an effect, religion in the individual is a response to socioreligious stimuli. Religious beliefs, such as the conviction that life is sacred, that divine forgiveness is available in Christ, or that the world is God's creation, cannot be rationally trusted, for they are products of reinforcements that form necessary and sufficient causes of these beliefs. So far as the classical religions have served as behavior shapers, molders of normative conduct, they are typically judged by behaviorists to have functioned rather naïvely, ineffectively, and nonscientifically, particularly because they were prone to use mentalistic and superstitious explanations and because they suffered from the illusionary ideal of human agency and responsibility. Skinner claims, "A god is the archetypal pattern of an explanatory fiction."[43]

The classical religions did contain certain moral codes that had survival value for the group, and to some extent these ought yet to be kept in place. But they also contain taboos and dogmas that have become accidentally connected with these codes, or that have served to sanction them in the absence of more rational reinforcers. Such "religious behavior" can now be eliminated in favor of "scientific behavior." Positive ethical response is disconnectible from its religious backing and can be achieved by scientifically engineered human conduct, better living by better behaving through behavioral science. Here behavioral science is quite melioristic. Through it we can more speedily evolve a more humane society. Behavioral science becomes a substitute savior, and proposes its utopias and models of the good life (*Walden Two*) in the place of the Kingdom of God. "We have," says Skinner, "the physical, biological, and behavioral technologies needed 'to save ourselves.' "[44]

When behaviorists propose the reformation of life and society, theologians find them often naïve about sin and evil, since these dimensions of life are not available in the behaviorist paradigm. They not only forget the heights to which humans can rise in rational, moral, loving freedom; they forget the depths to which humans can descend in selfishness, malice, and iniquity, especially when life is emptied both of the sense of the sacred and of moral reponsibility. Thus, a theologian fears all the more the uses to which behavioral technology may someday be put, whether by dictators or well-meaning bureaucrats, advertisers or educators, or by any governors who gain effective control over the lives of others.

There is ample evidence throughout history, continuing in contemporary affairs, of the existence of sets of persons at higher and lower levels of government (or other social relations) who will, if they can, exercise their power over others in their own self-interest. Ecclesiastics have sometimes been prominent among them, using heaven and hell as powerful reinforcers and punishers. But now the scientific discovery of advanced powers for human manipulation, in the absence of any attention to their probable misuse—indeed, by a theory that belittles the concept of responsible use, and strips responsibility from misusers—bodes ill for the future. This is quite contrary to the behaviorist hope for a better world through behavioral science.

A radical behaviorism points up the valuational incompetence of science. We have human conduct proposed to be brought under *causal* law so rigorously as to allow no quarter in which to judge life's *meaning,* an area that has been the classical province of religion. Meaning is closely allied with mind, but excising all mentalistic phenomena leaves no agent to own meaning, no judge to appraise values. If offered as a complete account of human life, both descriptively and normatively, behaviorism not only runs into incoherence and contradiction, it robs life of its humanness and sacredness. It leaves life void. In evolutionary theory, one has to be wary of accounts that keep warning that there really isn't any direction to evolution, only chance and the illusion of design. So too here in psychology, on logical not less than experiential grounds, one needs to be wary of theories that keep assuring us that there really is no mind directing behavior, no person maintaining his own meanings, no individual with her valuable contributions.

One needs to resist being reduced to a flesh-and-blood robot. The simple theories in physics, in which all was rigidly predestined (Newton), have been replaced by more sophisticated ones in which there is openness, incomplete specification, and limited access (Heisenberg and in part Einstein). Behaviorism, analogously to early physics, has simplified its investigations in order to make them operational. In so doing, it has driven out much freedom and subjectivity. But this strategy must be supplanted by more inclusive theories that do less violence to firsthand experience, recognize the complexity of the person, and permit some openness and self-specification, some room creatively to reason and to act.

Cognitive Psychology: Persons as Cognitive Processors

The evident oversimplicity of radical behaviorism has led in recent years to a cognitive turn in psychology. It is perhaps premature to assess the significance of this trend, in terms of its adequacy as a psychological paradigm and also in its impact on religious belief. Nevertheless, this impressively advances over both psychoanalysis and behaviorism, and it serves as a sort of halfway house to the humanistic psychology that we next examine. The cognitive emphasis is marked by a return to a type of vocabulary that earlier psychologists wished to expunge from science, one to which orthodox behaviorists such as Skinner remain resolutely opposed.[45]

In animals and in humans we are dealing with a *cognitive processor,* not merely a stimulus-response mechanism. Psychological models need to be appropriately cybernetic. But (say the more rigorous "scientific" cognitivists) we can still keep these models firmly tied to empirical findings. There is no need to consult consciousness or introspective experience. We can still show our results, on statistical average, repeatable in experiments. Indeed, we may still be dealing with rats running mazes and pigeons pecking for food, although using sophisticated techniques designed to reveal cognitive maps or learning that is not routinely expressed in behaviors.[46]

As we earlier saw biology do, psychology now takes a cybernetic turn. In genes, the DNA involves more than the causal forces of physics and chemistry; it forms an informational system. At a still higher level, conditioned learning is not just a mechanistic, causal affair, but is a problem-solving, cognitive process. The advent of computers, especially the prospect of artificial intelligence, provides useful models. In fact, rather curiously, cognition returns to psychology largely by way of machines, computing machines! (Once again, psychologists seem willing to embrace a model only if some harder science licenses them to do so. They believe in cognition only if they can see it work on magnetic tapes and transistor feedback loops.)

Psychology can omit consciousness, but it cannot omit cognitive processes. Any competent theory must include the "cognitive maps" in pigeons, rats, or persons, although nothing is required about including conscious awareness—just as in understanding a computer one needs to describe the programs by which it sorts and orders data, though without assuming any consciousness in the computer. C. Alan Boneau, a psychologist, claims, "Despite some of the connotations of cognitive theory, there is no need to deal with concepts such as consciousness in order to get one in operation. The cognitive processor need only be a mechanism that categorizes inputs, computes probabilities, assesses values, and makes decisions, a feat that can be accomplished in a simplistic way by computers."[47] John Haugeland, a philosopher of psychology, writes that this "is the fundamental idea of cognitive psychology: intelligent behavior is to be explained by appeal to internal 'cognitive processes.' . . . Cognitivism, then, can be summed up in a slogan: the mind is to be understood as an IPS [information processing system]."[48] Some call this "cognitive behaviorism," but others are prepared to abandon the behaviorist label and even to think of themselves as rebelling against behaviorism.

For complex animal behavior, the most satisfactory explanation requires supposing various cognitive structures formed, though we know little about what sort of representational system the animals employ to do this. Animals can recall from memory, insert new information into an existing cognitive system, and make good guesses in the face of ambiguity. They even seem to have something like concepts, as when pigeons discriminate Charlie Brown from the other "Peanuts" cartoon characters—Linus, Marcia, Snoopy, and so on—regardless of whether Charlie Brown is wearing a sweater, jacket, cap, snowsuit, or swimsuit, whether he is lying down, crouching, climbing, somersaulting, or partially hidden behind a tree. They can demonstrate this by pecking for food appropriately.[49]

Such phenomena are not causal in physicochemical or even biological senses, but on the other hand we do not need to refer, in any subjective sense, to the pigeon's psychological experience. A stimulus produces a response, to be sure, but the intervening variable is a cybernetic psychology in the organism. The stimulus does not merely provoke a response, it *informs*. The comprehensive picture includes some "comprehending" in the animal, which somehow "figures out" the contingencies. The information state of the organism makes a crucial difference in the response. Meanwhile, no reference to mentation or experience needs to be made. If we do permit this, it is redundant to the real explanation in terms of objective cognitive systems, not existential subjects.

It should not be overlooked at this point that, although our computing machines impress us with how cognitive processing can be a mechanical and objective process, there is a vital difference between the living organisms that are the actual subject matter of psychology and those human-made computers that serve as a model. The latter are always the products of the former; that is, the machines are artifacts designed by intelligent humans; the organisms are natural products of evolution. On the cosmological scale, outside the province of psychology, we do have to puzzle over the evolutionary origin of these organic computing machines. We still need a theory of nature competent to their generation.

In human cognition, we are likewise storing, retrieving, and using knowledge. But animals are very specialized information processors, humans very unspecialized ones. Here cognitivists, while hoping to remain thoroughly scientific, become ambiguous about the need for consulting inwardness in the cognitive processor. Michael G. Wessells, a psychologist, notices, "Most cognitive theories are mechanistic, and, in keeping with the working assumption that human cognition is a species of information processing, . . . many theories are stated in the form of computer programs."[50] But the human is evidently a *conscious* cognitive processor, and the conscious dimension seems in some measure to be crucial to the advanced power for cognition displayed by humans. If so, this is no longer behaviorism, for the inwardness that behaviorism required that we eliminate has been reinstated.

Some will say that, even admitting consciousness, we can still do scientific psychology. By a mixture of experiment and reflection over experience, cognitive psychology can illuminate such processes as short-term and long-term memory, pattern recognition, selective attention (an operationally manageable equivalent of "consciousness"?), problem solving, concept formation, and the use of language.[51] All this is impressively more adequate to human experience than is radical behaviorism with its referring of everything back to the environment, more adequate than is psychoanalysis with its referring of everything to the unconscious mind. We do get an active agent. We get reasons, past causes. We seem to move beyond a deterministic system to a rational one, although the question of freedom remains rather unsettled in cognitive systems.

But have we reached a model competent to the whole human person? At this point limitations in the cybernetic model begin to appear. Storing, retrieving, and using information are certainly important. But are these the only, or the central, features of personality? Even in terms of a biological model, cognitive processors as such do not suffer, and here we could wish for more awareness within this psychology of that dimension in experience of which evolutionary biology has left the organic world almost too full, of the cruciform nature of life. In terms of a human model, cognitive processors do not feel ashamed or proud; they do not have angst, self-respect, fear, or hope. They do not get excited about a job well done, pass the buck for failures, have identity crises, or deceive themselves to avoid self-censure. They do not resolve to dissent before an immoral social practice and pay the price of civil disobedience in the hope of reforming their society. They do not weep or say grace at meals.

Cognitive processors do not have emotions or feelings—to use categories that other psychologists have thought important. R. B. Zajonc, a psychologist, laments, "Contemporary cognitive psychology simply ignores affect. The words *affect, attitude, emotion, feeling,* and *sentiment* do not appear in the indexes of any of the major works on cognition."[52] Cognitive processors do not act in love, faith, or freedom, driven by guilt or seeking forgiveness—to use categories that theologians have thought fundamental. Can one imagine a computer being overcome with anomie, swept along in mob psychology, or making a confession of faith? Can one imagine one information-processing system insulting or praising another? Telling jokes? In the crucial humanistic category to which we next turn, cognitive processors do not have egos. In a pivotal sociological category, they do not form cultures. In an even richer historical category, one toward which we are progressively moving through this work, cognitive processors do not have unique careers that interweave to form storied narratives. They do not have heroes or saviors. They do not die for the sins of the world, launch the Kingdom of God, or fall into other passionate ideologies about the meanings of life and history. The model of the cognitive processor, while necessary, is yet insufficient for the human personality.

4. RELIGION AND HUMANISTIC PSYCHOLOGIES

An effort to take seriously the notion of a self with its conscious inwardness characterizes a family of theories collectively called *humanistic psychologies.* These are often disparate, but join in thinking the behaviorist omission of consciousness and the Freudian undermining of it inadequate. They want something beyond cognitive processing. "Personality" is a level of being we readily ascribe to humans, but never (in an unequivocal sense) assign to rats or pigeons. The phenomenon of personality displays an emergent richness of rational, moral, emotional, cultural, and valuational experience. Any principles gained in animal experimentation will be at best preliminary, merely parameters within which the laws of personality will distinctively emerge. These laws can be discovered in scientific studies of personality that are neutrally but empathically conducted.

Nevertheless, much of the concern to develop this psychology has come neither from academic psychology nor from experimental laboratories, but out of clinical experience. A therapist may be content merely to treat the patient with stimuli for behavior modifications, but he may also, at a deeper level of concern, prefer to see the patient as an actor, not a reactor, at least when well, if not when ill. The therapist may observe that causal treatments are sometimes unsatisfactory. In practice this is because they leave the patient at the mercy of his environment. Though the therapist can temporarily introduce variant reinforcements, permanent therapy becomes impossible in an irreformable environment. In theory the causal model fails because patients do sometimes reform in unaltered environments, and here personal initiative, perhaps facilitated by the therapist, can be very striking. Persons can sometimes pull themselves up by their own bootstraps. They become self-generating.

Finally we reach psychologies that give us each, and want for us each, a mind of our own as a determinant of behavior. Behaviorism and Freudian psychoanalysis have both given us persons as determinate systems, persons with closed minds so far as they have minds of their own at all, products of their environments and their unconscious minds—at least before if not after their reformation by psychology. But humanistic psychology hopes to give us persons as open systems, persons with open minds in quest of logic, novelty, values, meanings, self-affirmation. In this sense, the humanists want what the behaviorists think does not exist—self-controlled, responsible persons.

One need not deny significant truth in the preceding psychologies to find them overblown and to clip them back under more sophisticated theories of mind and conduct. Every science should draw its models from its subject matter, and reform these following more adequate encounter with the subject matter. So physics has moved from clockwork to quantum mechanics and relativity. Biology has posited natural selection and the cybernetic organism. Psychology has yet to draw its full-blown model out of its subject matter, but first tried to borrow mechanism from physics or organicism from biology. More recently some cognitivists have begun to look for an irreducibly human model, and there have steadily been in psychology a minority who seek to draw a model for personality more immediately from our experience of personal life.[53]

Here it is not the necessities that impress us, though we want to learn these and find how far deterministic psychoanalysis or behaviorism may be true. It is the contingencies and the way we elect a route through them for which we need explanation. Life is an open story. Counterposed to what must be is what we make of life's opportunities. We each write our own history. Superimposed on the physics is a biology, and on the biology a psychology, and on the psychology is superposed a narrative career.

The Self in Personality Theory

Humanistic psychologies posit a *self* concept, some autonomous centeredness holding together the human individual. Here the word "self" has a specific existential use, though biologists sometimes speak more broadly of organic selves. What is designated is an *ego,* a psychological self, that sort of subjective self which constitutes a person, as persons may be distinct from phenomena not present (or only nascent) in dogs and chimpanzees. Gordon Allport, writing a psychology of personality, notices, "Many psychologists have commenced to embrace what two decades ago would have been considered a heresy. They have reintroduced self and ego unashamedly and, as if to make up for lost time, have employed ancillary concepts such as *self-image, self-actualization, self-affirmation, phenomenal ego, ego-involvement, ego-striving,* and many other hyphenated elaborations which to experimental positivism still have a slight flavor of scientific obscenity."[54] The notion of a self is puzzling to elucidate, perhaps just because it is so close to experience (as also are such key notions as cause, meaning, truth). The self is what is at the center of our coordinate systems of perception and conception, the owner and presupposition of phenomenal experience, the binder of our experiential set, the subject that knows objects. In the mature person the self becomes a most indubitable fact of ongoing experience. It is our dynamic "being there." The inward turn here leads some to call these *existential psychologies.*

Personality is a *dynamic value system,* issuing in meaningful behavior, not so much pushed by causal stimuli as self-impelled and drawn in pursuit of values. Skeptics who are perhaps reluctant to say that the self is a first fact of experience can at least easily admit that goal-seeking behavior is a first fact of experience. Even the skeptic has to explain this *trying* to see whether the self is a meaningful concept and if so how to value it. That search seems teleological enough to serve as introductory evidence of the larger, continuing quest to make sense of things.

We not only think about values, we act on these thoughts, defending the ego, its experiences, relations, and possessions. Unlike the behavioral translation of values into covert intervening variables produced by external reinforcements, values are here importantly owned by the person. The mentalistic dimensions in human values are irreducible to behavior, though expressed in it. Andras Angyal, outlining a science of personality, finds value axiomatic in life: "Any sample of behavior may be regarded as the manifestation of an attitude. Attitudes may be traced back successively to more and more . . . general attitudes which are unquestionable, axiomatic for a given person. . . . When such axioms are intellectually elaborated we may speak of *maxims of behavior.* The axioms of behavior form a *system of personal axioms.* The system of maxims may be called a *philosophy of life.*"[55] Persons, says Gardner Murphy, in a text on personality, "set up complex personal systems of wants which are relatively enduring and maintained even in the absence of external reminders; they carry around within themselves personal systems of values. Personality is in large measure this personal value system."[56] But now this personal science has become unashamedly teleological. The "determination" at work is internal and prospective; normative, not merely environmental. Allport adds, "People, it seems, are busy leading their lives into the future, whereas [empirical] psychology, for the most part, is busy tracing them into the past."[57] One looks within and ahead, not merely without and backward, to understand this subject-in-the-world.

Though this human science is as yet merely descriptive of how persons function, notice how it has come to recognize the *place of prescriptive norms.* The self can distinguish between where it *is* and where it *ought to be,* under the rules it maintains. We can bring an ideal self to confront a real self, as well as to deliberate what this ideal self shall be. The person is an *ambiguous being,* with a gap between real and ideal achievements. We are not satisfied so much by things, sustenance, and drives reduced (as animals largely seem to be); but we are satisfied by a self-image, an ideal approximated in the real, satisfied by becoming, not by getting.

The centrifugal pull of the environment contains destructive threats that will destroy the person, unless she steadily recomposes herself. The gap between what is and what might be, portraying a future of more or less value, is both psychological fact and source of the main motivating power in human behavior. The matching and mismatching of the world with our value sets is what prompts us to move through it. This is an environmental affair, but it is equally a valuational affair, one of *oughts* beyond what *is*. We are thus kept under stress from without, but this tension, felt within, follows from those values for which a self-will fights.

Disvalues also originate within, for persons are not so unified as they wish, but are beset by guilt, anxiety, frustration, and estrangement. Torn by competing drives, both conscious and (as Freud recognized) unconscious, persons are indecisive; they succeed but also fail. They actualize their potential only partially. Of many trials, a few work, and competence here increases with learning. But only exceptional individuals (1 percent by the estimate of Abraham Maslow) achieve really high levels of integration.[58] The most remain half themselves in mixed integration and brokenness, partly oriented, partly lost, only relatively finding or achieving their goals. Others break down and collapse into various abnormal psychopathologies.

Further, as some goals are achieved others are newly set up, so that even with success this tension is renewed from within, stepped up and kept up, in addition to whatever other tensions may be thrust upon us from without, tensions that we seek to reduce. Personal learning proceeds by goal production, if also sometimes by drive reduction. Complete drive reduction would be quiescence, like the Buddhist *nirvāṇa;* but persons enjoy their drives, even if they are also frustrated by failures. They have capacities that clamor to be used. There is a clamor among too many wishes and a thrust toward integration. Every equilibrium is upset by opportunities projected from within or by the onset from without of more opportunities than we can use.

This view of the person can be connected with, though it is not reducible to, evolutionary theory, where on life's cutting edges there is a trial-and-error openness, with many failures and a few successes superimposed on a basic life process that is steadily successful and innovative. But the equivalents here of what biologists call mutants, or behaviorists call operants, though they appear sometimes by chance, are preserved as intended self-developments. We recognize subjects with inwardness, a phenomenon seldom before admitted and never made central by any science. In these subjects, the context of discovery is a context of self-discovery, and the context of justification is a context of self-justification. There is the actualizing of potential, which was met in both physics and biology, but in humanistic psychology this takes the form of the deliberate actualizing of a self.

There are connections here too with the character of life as perceived by the classical religions, in which systems of salvation propose to heal and to help by supplying norms and strength in this struggle. We reach a level advanced beyond that of the physical suffering that proved problematic within organic life. Matter cannot suffer; only life can. But mind can suffer most of all, because mind knows anxiety, despair, hope, culpability. Persons can know affliction beyond suffering. They can know betrayal and loss of prestige, feel alienated, God-forsaken. The person is condemned to self-awareness before death. To persons alone is assigned an inquiry about meaning in a harsh world. There is more than the logic of trial and error; there is the logic and absurdity of trial through tragedy, with all its psychological and existential demands. Indeed, it seems as though where Freud psychoanalyzed away the illusion of religion and behaviorism substituted itself for religion, humanistic psychology almost insists on giving humans something like a religious assignment.

Persons are complex *symbolic beings.* They see beauty or ugliness, fairness or injustice, love or hate. They overlay physical behavior with interpretive categories so as to apprehend and judge what is psychologically there. Persons perceive using cognitive reflection, a perceptive activity that is larger than language but that language makes possible. Language is both catalyst for it and vehicle of its communication and analysis. A maturing self is increasingly directed by, and extended into, abstract symbolic ideals that serve as perpetuated motives and continue to operate even when the primary bodily drives are satisfied. It is precisely these systems of governing meaning that escape much, or even any, causal analysis.

Humans are startlingly advanced over animals in this crucial feature of personality. We have encountered already the use of symbolic systems in genetic sets, where base pair triplets code for amino acids, storing metabolic and structural information in DNA sequences. But in the psyche there is a subjective awareness, together with the astronomical learning capacity and openness of the brain-based mind. There emerges an entirely new level of symbolic, systemic control. Persons are the most advanced cybernetic systems that we know, as is evidenced by their cultural production of myths, models, and paradigms in science and religion. When such persons respond to the world and to each other, it is not the stimulus in itself that produces a response, but rather (as cognitive psychologists have come also to recognize) there is a symbolically mediated interpretation of the stimulus. If this is misjudged, an alternate response will follow the otherwise identical environmental or cultural stimulus.

The personality passes through *stages in development,* more or less successfully maturing by its biological and psychological self-will, but nevertheless according to generalized laws of development. Such stages can be verified by introspection, but also are subject to observation and experiment. Psychologists do not yet agree what these developmental laws are, but there is meanwhile frequent insight in their proposals. Jean Piaget's analysis of four stages—the sensorimotor (up to two years), semiotic (two to seven years), concrete operations (seven to twelve years), and formal (twelve years to adult)—is especially rich in its capacity to blend the rational and empirical components in development. These stages trace how our structures for perception and conception, though genetically based, mature under the stimulus of the environment, the person constituting himself in this transaction.[59]

In another scheme, Erik Erikson traces through eight stages the lifelong struggle for integrity, for instance in the autonomy-versus-security tensions of childhood, the identity crises of youth, the generativity-stagnation ambiguities of the middle years, and the integrity-despair struggles in later life.[60] Abnormal behavior can be interpreted as maladjusted transitions or throwbacks between stages. These general laws by which personality develops do not deny the particular, historical form of an individual's story, or the initiative supplied by the individual. In the work of Lawrence Kohlberg, six developmental stages are traced with particular reference to the development of moral character and ideology. The later, more adult stages take on increasingly deeper religious overtones. Quite contrary to Freud's finding that religion is juvenile and illusory, Kohlberg is inclined to posit a further, seventh stage, the highest and consummately religious one, seldom reached, largely lived toward or approximated as an ideal.[61] Religion is not ultimately childish; rather, moral maturity ultimately requires religion.

There is ample place in personality for *rationality, morality,* and *love,* at least as ideals often incompletely attained. In the defense of the self, an overly assertive ego can prove unreasoning, immoral, unjust, and unloving, quite as authoritarian and destructive as is the behaviorist's environment in determining what the person thinks and how he conducts himself. Our escaping from this involves much ambiguity and stress. Nevertheless, the ego is more or less brought under the discipline of reason, and here we can attend to the deep structures of the mind as rational ones, to logical grammars innate there,[62] a level not envisioned in the Freudian interplay of irrational unconscious drives. The brain has evolved in response to environmental stimuli. But it has gained a logical accuracy in getting around in the world; it learns how intelligently to fit the self into the world in mixed altruism and egoism.

These deep genetic structures operate to organize any immediately supplied environmental stimuli. The intelligent self can deliberate, choose its beliefs, and defend its conduct with reasons. It has gained much autonomy over causal stimuli, although it continues creatively to respond to opportunities and threats presented in the environment. When, in addition to defending self, this response respects integrities outside itself, we become moral and loving. Mature persons are not only well integrated in their self-defense but are well integrated into the community in which they live. The moral life blended with that of love makes this second level of integration possible.

Self-actualizing Persons

In a phrase that comes into evident conflict with the stimulus-response account, the person is and should be *self-actualizing* (not S-R, but SA). We thus reach the fourth main paradigm operating in contemporary psychology. Human life is not a product and resultant of drives in the unconscious mind, nor is it a response determined by environmental stimuli, nor is it just cognitive processing. It is rather lived in personal autonomy. Abraham Maslow calls this "self-actualization."[63] Allport speaks of "propriate functional autonomy."[64] Angyal finds central a "trend toward increased autonomy."[65] Carl Rogers, yet another therapist of personality, finds that the main "curative force in psychotherapy" is "man's tendency to actualize himself, to become his potentialities," a trend that in the healthy is the "mainspring of creativity."[66] "Psychotherapy is the releasing of an already existing capacity in a potentially competent individual, not the expert manipulation of a more or less passive personality."[67]

The individual has dimensions of freedom to construct and maintain her own identity in positive self-regard. Other organisms do this in lesser ways, but humans reach the newly emergent level of personality. This requires that the person is neither a creature of caprice nor one driven by unconscious determinants, but is rather the responsible author of her own personality. Within genetic and environmental parameters, my life must be "up to me." I am free to execute my motivations and, more, free to originate these motivations. Contrary to behaviorism, the world stimuli do not uniquely or overwhelmingly determine the individual. They present to the person nested sets of possibilities. The person selects this and not that from among the stimuli, and thus constitutes a self, and this agentive self can reconstitute the world stimuli in turn. Personality is an *interaction phenomenon,* not just a response to stimuli. We have here an analogue of the organism nursing its way through the quantum states. But there is not a definitive program merely preset in the genes and instinctively defended. We have risen above the organic level to that of a meaningful personal life. The quest is not for information alone, but for self-formation.

My character, my conduct, beliefs, and decisions (though perhaps not my imaginativeness) will be in measure predictable on the basis of my perpetuated motives. We predict such behavior on the basis of common sense quite as much as by any theories discovered by behaviorists or psychoanalysts. A genuinely humane education, if it is a liberal, liberating education, can take place only under natural and cultural stimuli, but it must reach this level of self-actualizing. The preplanned utopias of the most benevolent behaviorism, together with those careers shaped by behavioral modification, now seem subhuman. So far as they succeed, they leave no room for this self-actualizing in which humans approach the image of God.

While my conduct can be increasingly predicted from my character, my character in turn is not merely a predictable social product, not something to be engineered by education. Indeed, by the account of Carl Rogers, it is vital that education and counseling be nondirective, that these make no effort to predict and control conduct, lest in so doing the environmental and cultural stimuli thwart self-actualizing and as a result dehumanize the person. Genuine education leads out (*educo*) the self; it cannot train behavior.

Value Assumptions and Inadequacies in Humanistic Psychologies

Humanistic psychology recognizes the necessity of values and describes how these function, but it *cannot be said to have solved the value problem.* That remains the most difficult issue faced in normal versus normative descriptions of the healthy self. The psychologist can describe what any particular value system does do, and, further, on becoming a therapist, he can allow no value system that is not integrating to the personality. The defense of the self becomes at once a scientific paradigm and quasi-religious criterion. The psychiatrist further recommends that persons be rational, moral, loving, although she will probably be content with reforming a client to normalcy in his cultural setting, while the pastoral counselor may be interested in a parishioner's progressive perfection. Even the recommendation that persons be self-actualizing is a prescription.

Issues that from the philosopher's or theologian's point of view need to be argued over tend to get buried beneath such phrases as "healthy," or "fully functioning person," or "self-actualizing." We are perhaps not quite told, prescriptively, that persons morally ought to live this way. But we are told, descriptively, that the best persons, the type specimens, live in this self-determining way, with the implication that all or most others ought to live like this. Any kind of self-actualizing pattern will do, presumably. The theory is value-neutral about this or that self-actualizing, but only after having imposed a theory about what value is: the subjective choices of the self-actualizing person. The psychological theories do constrain the religious theories up to a point; they demarcate the area of possible normative religious theories.

After that, there may be little inquiry into whether being rational, moral, and loving can in fact undercut, or be undercut by, our self-actualizing tendencies. Especially when learning to be rational, moral, and loving, perhaps we need to say more about the limiting of our self-actualizing drives in favor of our being other-directed. It may be, as we will later hear sociologists and historians contend, that much of value is going on over our heads, that the main dramas are larger than these ephemeral affairs of individualistic self-actualizing.

Nevertheless, after requiring a minimum value set that permits the formation of personality, humanistic psychologists tend to want their science to be value-free. As psychiatrists, humanists do not and cannot judge between those competing ethical systems recommended in the various religions and philosophies, provided only that these systems have been tested for their capacity to supply integration. Psychiatrists describe the *form* of personal life, but they do not prescribe any material *content*, although they proscribe any that are incompatible with this form. They set up life's logic, which, like formal logic, is necessary but insufficient for the rough and tumble of everyday life. We are left encouraged but hanging in the air.

Here a mature religion can supply the actual content for a fully functioning person at home in his world, and this contra Freud, who had pronounced all religion illusion and neurosis. Under these terms, the sense of the sacred, overlooked by behaviorists in their reduction of all to causal stimuli, can be reinstated. The psychiatrist may document the healthiness of meanings found for life or (as so many have) the malaise of meaninglessness in secular life.[68] But neither psychologists nor psychiatrists can by their sciences explain the meaning of human existence in the world. If they try, they become prophets and seers instead.

In humanistic psychologies there is routinely an axiomatic faith in the *person as being innately good.* The human drives are good ones, if suitably released and balanced out, if uninhibited by oppressive parenting, social systems, cultural dogmas. At this point the theologian may return with the criticism of naïveté leveled earlier against behaviorists who wished to enlarge the possibilities for human manipulation. It is rather naïve for the therapist to think that the problem of release or salvation is one of "freeing up the person" to be herself. Fallenness, guilt, repentance, forgiveness, atonement, reconciliation, rebirth, redemptive suffering, divine grace—these are the categories that ethical monotheism has found to be relevant, and they run rather deeper than the humanistic advice to "Be yourself." "Trust yourself!" "Be a fully-functioning person!" "Blossom!" But the blossom theories of humanistic psychology are, to put it mildly, incomplete. They are, to put it frankly, shallow.[69] In contrast to them, there was something congruent with the Christian tradition in the Freudian tendency to find innate, self-destructive tendencies in the unredeemed person. There are evil powers at work within us from which we need to be released, not less than oppressive powers that may work on us from without.

In their formal nature, the humanistic theories are only half-true, and they are empty of significant directions. "Be creative!" "Be human!" "Actualize yourself!" "Have peak experiences!" (Maslow) "Be autonomous!" "Promote yourself!" "Optimize [maximize?] yourself!" All this is well enough as a weapon to defeat behaviorism and to serve as the normal model that the Freudians could not supply. But at the same time, there is a hollow looseness here. Nothing really is critiqued or recommended about directions in this self-actualizing. Too much is left "up to me." Self-actualization is an empty genus without specific contents; and indeed somehow the genus itself as an ultimate paradigm seems suspect and narrow. We want to get past *self-actualizing* to critical *self-appraisal* in the historical context in which one plays his part. Indeed, the religions teach all but unanimously that we ought to seek not *self-actualizing* but *self-transcending.* The one may or may not be compatible with the other. The one may be paradoxically compatible with the other in ways that saints have realized more effectively than psychiatrists can counsel.

The insights of existential psychologists, applied descriptively to the struggling person, will not in every case be superior to those of the humanist and theologian, even though the former are offered in the name of science. To understand the ambiguous human condition or the debilities of pride, should one read Maslow or Erikson? Or can Shakespeare's *Hamlet,* Augustine's *Confessions,* Paul's autobiography in Romans 7, serve equally as well? When science comes to touch the problem of meaning, the tools it has so well honed—observation, experiment, causal and functional analysis—fall short of sufficiency for the task, and the skills of humanistic and religious insight, found in classic and canon, become more requisite. At the critical turnings of religious history, what can be described (in Luther, Ignatius Loyola, or Isaiah) as self-actualizations, or mutant, creative, autonomous visions—what can in behaviorism be called newly emitted operants of conduct—can quite as intelligently be described as revelations and prophecies.

The centered self that is the cardinal frame of reference of humanistic psychology is both asset and liability, just as the environmental determinism of behaviorism is true and half-truth. How far can there be a self sufficient in the generation of its own integrity? Humanistic psychologists can recognize an overly assertive ego, or the need for a supporting community. They describe interpersonal relations; they debate whether the person is or should be an open or closed system.[70] But what can they say about life's big horizons? We still need a satisfactory account of what does in fact lie beyond the individual himself. To this behaviorism gives promise, before it falls into incoherence. We still need an account of what does and does not lie within the individual's capacity and choice. Here humanistic psychology starts, describes a self, but then refuses to prescribe anything, and ultimately of its own resources cannot locate any bearings for the self in search of meaning in the world.

Persons are, or should be—the religions have often insisted—not so much like watches, which are self-contained, self-actualizing, as like compasses, which are not self-actualizing but operate as they take their bearings from transcending fields of force that pass through them, immanent within them though these forces also are. What one wants is not just self-actualizing, but a role in the story, a meaningful place in the sweep of history, directions pointed by the power and grace of God. One wants an answer not just to the "parenting" question dismissed by Freud as juvenile, the question of where we have come from and what forces have generated us; one wants also (and logically connected with this) an answer to the question of what bearings should I take from here, and one needs to feel these cosmic fields of force to take these bearings.

Behaviorism eliminates the self before its overpowering situation; humanistic psychology features a self in search of relationships to, but over against, its environment. The self enjoys an exodus out of nature into autonomous personality. But these sciences supply no ethos interpreting how the person is meaningfully to be joined to her natural surroundings. Humanistic psychology wants a self free within its culture, but supplies no model of how culture is the carrier of meaning to the self, which is otherwise lost and powerless without this education. What is needed is an interactive model, an ecological model that transcends the insights of both sciences under a larger paradigm. But such a unified science, or unity of science and religion, yet awaits us.

Persons in Historical and Cosmic Contexts

The humanistic model needs yet to be made *narrative and historical.* In this sense, if taken alone or as final, it is not only empty of significant directions for life, it is empty of biographically and historically significant categories. If a psychologist abstracts out from human life enough to write a theory of personality, valid over the centuries and around the globe, valid for Gautama Buddha and Martin Luther King, Jr., the detailed biography that makes each life a narrative story has to be eliminated. Again, the psychology catches some of the *form* of life, but none of the *content.* The theory of personality is no story, but each individual personality is a story. Further, every individual human, like every organism, lives beyond himself. Each is set in a larger cultural story, with his role played once only in a singular historical drama. There is nothing in this psychology that can illuminate the historical uniqueness of Abraham setting out from Ur toward Palestine, of Jesus setting his face toward death in Jerusalem, of Abraham Lincoln setting the slaves free in the tragedy of civil war. For, as we will later see in detail, life is irreducibly historical. Before this dimension even the best psychology is incompetent.

There is often a kind of yearning toward a grander theory when humanistic psychologists come to review their work. Gardner Murphy writes in closing a long text on *Personality:* "Psychology . . . has explored by systematic methods only a few aspects of the deeper interindividual unity that is a phase of the man-cosmic unity. . . . In a future psychology of personality there will surely be a place for directly grappling with the questions of man's response to the cosmos, his sense of unity with it, the nature of his esthetic demands upon it, and his feelings of loneliness or consummation in his contemplation of it." Persons "have felt incomplete as human beings except as they have endeavored to understand the filial relations of man to the cosmos which has begotten him, and have tended, in proportion to their degree of seriousness, to recognize the relativity of selfhood and the fundamental unity of that ocean of which the individual personalities are droplets. . . . Our study of man must include the study of his response to the cosmos of which he is a reflection."[71]

Maslow warns, introducing his psychology, "I should also say that I consider

Humanistic, Third Force Psychology to be transitional, a preparation for a still 'higher' Fourth Psychology, transpersonal, transhuman, centered in the cosmos rather than in human needs and interests, going beyond humanness, identity, self-actualization and the like."[72] Both notice that this stage is not yet arrived at, nor is there much suggestion of how this phase will be entered using scientific or even "humanistic" skills and models. Indeed, we have to wonder whether this fuller phase should even be called psychology at all, rather than cosmology and history, philosophy and religion.

Across evolutionary time, the environment has created the person who now acts creatively on it. Even in the personal time span, the individual is a beneficiary of nature and nurture, finding life to be heritage and gift as well as demand. The person needs to be set free in her world without being set adrift in it, free to live under the predestination of the drama in which all take up a part. This perception of a cosmic value set, of the obligation and contribution of the person to her world, and its support of her, cannot be reached by psychoanalysis, by behaviorism, by cognitive psychology, or by humanistic psychology. None of them have enough historical rooting, evolutionary scope, cultural appreciation, or ontological insight. This does not fault what they can successfully abstract from life, but shows them to be incomplete explanations of what it means to live humanly in the world.

What remains is rightly to resolve the Freudian question of our parenting. What is the meaning of this larger natural generation by Mother Nature, of our selection with hand and mind to be free agents in the world, a question not only of origins but of fulfillment in life? What is the appropriate *Response* to this *Stimulus*? Any powers inherent in the self need recentering in our filial transcending by the world. Genuine maturity requires self-actualizing in reciprocal dialogue with the natural and cultural directives out of which we are composed. The person is a particle, an individual, as humanistic psychology sees. But the life he shares is a wave, a series that overleaps the individuals that instantiate it, as behaviorism is aware. But neither sort of psychological science renders a complete account of how the personal quantum of centeredness accords with the cumulative transmission pulsing from past through present to future.

No prevailing paradigm from any psychology deploys well into the larger evolutionary, historical, cosmic picture. In this perspective, persons are not S-R things, not SA things, but they inhabit a person-world ecological unity; they take roles in a cultural-historical drama. In a more searching and more kindly sense, differently from the way Freud thought, we are ruled by voices from our past. In a richer sense than behaviorism knows, we are ruled by forces that are greater than the self. We may want to move beyond even the freedom and dignity of humanistic psychology to admit that our dignity rests in traditional, environmental, and inspirational powers, in the extrahuman and superhuman forces that have supplied our humanity. We need a model in which persons are attuned to the prolife forces of the universe. While better science may help us toward this, such a model will display meanings more than causes, and be as religious as it is scientific.

Further than any socionatural transcendence of the person, religious minds wonder if these phenomenal developments do not signal a supernatural environment, some yet larger prime movement under which the cosmic drama takes place.

Theism has maintained in the Christian doctrine of grace, or in Judaism's vision of a chosen people, or in Islam's predestination, that the autonomous, contained self needs the complementary vision of communion with God. Nor have nontheistic religions, as in Advaita Vedanta or Buddhism, been convinced of the ultimate reality of the self. To the contrary, they have found the phenomenal self to be illusory and have sought to transcend it. Every religion wants to place the self in a much larger environment.

Here the theist detects both in the natural stimuli and the personal response, yielding the rational, moral, loving, free personality, something of the character of this ultimate watching over. The particulate self is thus understood as a wave in God, in whom we live, move, and have our being. Just as we are thrown here by forces that transcend ourselves, so too we are thrown forward by forces that are not merely immanent in the person but transcend our individuality, indeed our comprehension. Creative personality is indeed our calling, but takes place in the divine grace, providence, and election.

To take the Christian account, this providential election leads through sacrifice, pain, travail, and death in a cruciform world. Humans only penultimately bear the image of God in their self-actualizing. They bear the divine likeness ultimately in self-sacrifice, in dying to self and rising to newness of life in God, in self-transcendence and agape love—after the model of the Christ and Suffering Servant. Self-denial, not self-actualizing, propels us toward self-transcendence, toward God. One wants not so much self-actualizing as reinforcement in God, a personal conduct steadily corrected by the divine environment. The quest is not for information, nor even for the inner formation of a self. It is for spiritual formation of the Christ-like image within.

NOTES

1. Paul D. MacLean, "The Triune Brain, Emotion, and Scientific Bias," in Francis O. Schmitt, ed., *The Neurosciences; Second Study Program* (New York: Rockefeller University Press, 1970), pp. 336–49.
2. Eugene P. Wigner, "The Unreasonable Effectiveness of Mathematics in the Natural Sciences," *Communications on Pure and Applied Mathematics* 13 (1960): 1–14, citation on page 3.
3. Michael S. Gazzaniga, "The Split Brain in Man," *Scientific American* 217, no. 2 (August 1967): 24–29.
4. Donald J. Lewis, *Scientific Principles of Psychology* (Englewood Cliffs, N.J.: Prentice-Hall, 1963), p. 12.
5. This dualism is especially discomfiting to the therapist. See Carl R. Rogers, "Persons or Science? A Philosophical Question," *American Psychologist* 10 (1955): 267–78.
6. Sigmund Freud, "A Difficulty in the Path of Psycho-Analysis," *Standard Edition of the Complete Psychological Works,* vol. 17 (London: Hogarth Press, 1953–74), pp. 137–144, quotation on pp. 142–43.
7. Sigmund Freud, *The Future of an Illusion,* in *Complete Psychological Works,* vol. 21, pp. 5–56, quotation on p. 24. *Totem and Taboo,* vol. 13, pp. 1–162, contains also a speculative theory of the historical origins of religion, continued in *Moses and Monothe-*

ism, vol. 23, pp. 3–137. The latter works discuss more complex psychological repressions than does *Future of an Illusion.*

8. *Future of an Illusion,* p. 30.
9. Sigmund Freud, *Psychopathology of Everyday Life, Complete Psychological Works,* vol. 6, pp. 258–59. Italics in original.
10. Carl G. Jung, *Memories, Dreams, Reflections* (New York: Vintage Books, 1961), p. 150. Freud's colleague and opponent develops an entirely different theory of the collective unconscious, one not pejorative toward religion.
11. Sigmund Freud, "On the History of the Psycho-Analytic Movement," *Complete Psychological Works,* vol. 14, pp. 1–66, citation on p. 50.
12. Chapter 1, p. 11.
13. Sigmund Freud, *Introductory Lectures on Psycho-Analysis, Complete Psychological Works,* vols. 15–16, citation at vol. 15, p. 106.
14. Sigmund Freud, *A Short Account of Psycho-Analysis, Complete Psychological Works,* vol. 19, pp. 189–209, citation on p. 197.
15. *Introductory Lectures,* vol. 15, p. 28; *Project for a Scientific Psychology, Complete Psychological Works,* vol. 1, pp. 281–397, citation on p. 295. Freud lived through the introduction of Heisenberg's uncertainty principle into physics and of random mutations into biology, but saw no implications or analogues for psychology.
16. *Future of an Illusion,* p. 54. Freud believes that his method is reputable, using reason and corrigible by criticism and experience, unlike the methods of theologians.
17. For a creditable, though finally unsatisfactory, effort to describe what freedom can mean for Freud, see Ruth Macklin, "A Psychoanalytic Model for Human Freedom and Rationality," *Psychoanalytic Quarterly* 45 (1976): 430–54.
18. Freud's criticisms are in *Civilization and Its Discontents, Complete Psychological Works,* vol. 21, pp. 64–145. For an effort to moderate Freud's criticism, see Ernest Wallwork, "Thou Shalt Love Thy Neighbor as Thyself: The Freudian Critique," *Journal of Religious Ethics* 10 (1982): 264–319.
19. For example, Eric Fromm, *Psychoanalysis and Religion* (New Haven: Yale University Press, 1950).
20. Sigmund Freud, *Civilization and Its Discontents,* p. 74.
21. Bernard Lovell, "In the Centre of Immensities," *The Advancement of Science,* New Issue/no. 1 (August 29, 1975): 2–6, citation on p. 6. See Chapter 2, note 29.
22. John A. Wheeler, interviewed in Florence Helitzer, "The Princeton Galaxy," *Intellectual Digest* 3, no. 10 (June 1973): 25–32, citation on p. 32.
23. For revised, hermeneutic accounts of Freud, emphasizing meanings over causes, see Paul Ricoeur, *Freud and Philosophy* (New Haven: Yale University Press, 1970), or Jürgen Habermas, *Knowledge and Human Interests* (Boston: Beacon Press, 1971), Chapters 10 and 11. By Carl Jung's account the unconscious forces are positive drives toward meaning, which can be satisfied only religiously; see *Man and His Symbols* (New York: Dell Publishing Co., 1968).
24. Philip Rieff, *Freud, The Mind of the Moralist* (Garden City, N.Y.: Doubleday, Anchor Books, 1961), p. 1.
25. John B. Watson, "Psychology as the Behaviorist Views It," *Psychological Review* 20 (1913): 158–177, quotation on p. 166–167, 175.
26. John B. Watson, "Behaviorism—The Modern Note in Psychology," in John B. Watson and William McDougall, *The Battle of Behaviorism* (London: Kegan Paul, Trench, Trubner and Co., 1928), pp. 7–41, citation on pp. 27–28.
27. Anonymous, related by Robert S. Woodworth in Robert A. Baker, ed., *Psychology in the Wry* (Princeton, N.J.: D. Van Nostrand Co., 1963), p. 1.

28. Jay N. Eacker, "On Some Elementary Philosophical Problems of Psychology," *American Psychologist* 27 (1972): 553–565, quotation on p. 564. For a more recent assessment, see Ernest R. Hilgard, "Consciousness in Contemporary Psychology," *Annual Review of Psychology* 31 (1980): 1–26. The cognitive turn, which we recognize below, qualifies Eacker's judgment.
29. B. F. Skinner, *Science and Human Behavior* (New York: Macmillan, 1953), p. 35.
30. Watson, "Psychology as the Behaviorist Views It," p. 167.
31. B. F. Skinner, *About Behaviorism* (New York: Vintage Books, 1976), pp. 30–31, 114–115.
32. B. F. Skinner, "Critique of Psychoanalytic Concepts and Theories," in Herbert Feigl and Michael Scriven, eds., *The Foundations of Science and the Concepts of Psychology and Psychoanalysis* (Minneapolis: University of Minnesota Press, 1956), pp. 77–87, citation on p. 80.
33. Arthur W. Staats and Carolyn K. Staats, *Complex Human Behavior* (New York: Holt, Rinehart and Winston, 1963), pp. 255–257.
34. B. F. Skinner, "Freedom and the Control of Men," *American Scholar* 25 (1955–56): 47–65, citations on pp. 47, 52–53.
35. B. F. Skinner, *Cumulative Record,* 3rd. ed. (New York: Appleton-Century-Crofts, 1972), p. 325. Emphasis in the original.
36. B. F. Skinner, *About Behaviorism,* p. 63.
37. B. F. Skinner, *Beyond Freedom and Dignity* (New York: Alfred A. Knopf, 1971), p. 205.
38. For example, "What is needed is more 'intentional' control, not less, and this is an important engineering problem." *Beyond Freedom and Dignity,* p. 177.
39. *Beyond Freedom and Dignity,* pp. 21, 198.
40. Skinner holds that psychology does not "need the present muddle in the physical sciences." From the fact that electrons are unpredictable, it does not follow that human beings are. A science of behavior seeks "real certainty." See T. W. Wann, ed., *Behaviorism and Phenomenology* (Chicago: University of Chicago Press, 1964), with Skinner quoted in a discussion, pp. 139–40.
41. *Beyond Freedom and Dignity,* p. 211.
42. Chapter 7, p. 322–23.
43. *Beyond Freedom and Dignity,* p. 201.
44. Ibid., p. 158.
45. B. F. Skinner, "Why I Am Not a Cognitive Psychologist," *Behaviorism* 5, no. 2, (Fall, 1977): 1–10.
46. For a current summary, see Anthony Dickinson, *Contemporary Animal Learning Theory* (Cambridge: Cambridge University Press, 1980). Some cognitive emphases have long appeared even in behaviorally oriented studies, such as Edward C. Tolman, "Cognitive Maps in Rats and Men," *Psychological Review* 55 (1948): 189–208.
47. C. Alan Boneau, "Paradigm Regained? Cognitive Behaviorism Restated," *American Psychologist* 29 (1974): 297–309, citation on p. 308. The computer model dominates the survey by Gordon H. Bower, "Cognitive Psychology: An Introduction," in W. K. Estes, ed., *Handbook of Learning and Cognitive Processes,* vol. 1, (Hillsdale, N.J.: Lawrence Erlbaum Associates, 1975), pp. 25–80.
48. John Haugeland, "The Nature and Plausibility of Cognitivism," *The Behavioral and Brain Sciences* 1 (1978): 215–226, citation on p. 220.
49. John Cerella, "The Pigeon's Analysis of Pictures," *Pattern Recognition* 12 (1980): 1–6.
50. Michael G. Wessells, "A Critique of Skinner's Views on the Explanatory Inadequacy of Cognitive Theories," *Behaviorism* 9, no. 2 (Fall, 1981): 153–170, citation on p. 155.

51. For introductions, see Margaret Matlin, *Cognition* (New York: Holt, Rinehart and Winston, 1983), and Roy Lachman, Janet L. Lachman, and Earl C. Butterfield, *Cognitive Psychology and Information Processing: An Introduction* (Hillsdale, N.J.: Lawrence Erlbaum Associates, 1979).
52. R. B. Zajonc, "Feeling and Thinking: Preferences Need No Inferences," *American Psychologist* 35 (1980): 151–175, citation on p. 152.
53. For introductions to personality theory, see Salvatore R. Maddi, *Personality Theories: A Comparative Analysis* (Homewood, Ill.: Dorsey Press, 1972), and Arthur Burton, ed., *Operational Theories of Personality* (New York: Brunner/Mazel, 1974).
54. Gordon W. Allport, *Becoming: Basic Considerations for a Psychology of Personality* (New Haven: Yale University Press, 1955, 1976), p. 37.
55. Andras Angyal, *Foundations for a Science of Personality* (Cambridge, Mass.: Harvard University Press, 1958), pp. 164–165. Italics in original.
56. Gardner Murphy, *Personality: A Biosocial Approach to Origins and Structure* (New York: Harper and Brothers, 1947), p. 270.
57. Allport, *Becoming,* p. 51.
58. Abraham H. Maslow, *Toward a Psychology of Being,* 2nd ed. (Princeton: D. Van Nostrand Co., 1968), p. 163. Maslow finds that 5 to 30 percent become self-determiners in rather limited circumstances.
59. Jean Piaget and Bärbel Inhelder, *The Psychology of the Child* (New York: Harper and Row, Torchbooks, 1969).
60. Erik H. Erikson, *Childhood and Society* (New York: W. W. Norton and Co., 1950).
61. Lawrence Kohlberg, *Essays on Moral Development,* vol. 1: *The Philosophy of Moral Development* (New York: Harper and Row, 1981), pp. 311–72.
62. In the work, for instance, of Noam Chomsky.
63. Abraham H. Maslow, *Motivation and Personality,* 2nd ed., (New York: Harper and Row, 1970), pp. 149–80.
64. Gordon W. Allport, "The Open System in Personality Theory," *Personality and Social Encounter* (Boston: Beacon Press, 1960), pp. 39–54, citation on p. 51.
65. Andras Angyal, *Foundations for a Science of Personality,* pp. 20–55.
66. Carl R. Rogers, *On Becoming a Person* (Boston: Houghton Mifflin Co., 1961), pp. 350–51.
67. Carl R. Rogers, "A Theory of Therapy, Personality, and Interpersonal Relationships, as Developed in the Client-Centered Framework," in Sigmund Koch, ed., *Psychology: A Study of a Science* (New York: McGraw-Hill, 1959) 3:184–256, citation on p. 221.
68. Carl G. Jung, *Modern Man in Search of a Soul* (New York: Harcourt, Brace and Co., 1934); Viktor E. Frankl, *Man's Search for Meaning* (Boston: Beacon Press, 1962).
69. Searching criticisms of self-actualization theories may be found in James Lapsley, *Salvation and Health* (Philadelphia: Westminster Press, 1972), and David Norton, *Personal Destinies* (Princeton: Princeton University Press, 1976).
70. Allport, "The Open System in Personality Theory."
71. Murphy, *Personality,* pp. 919, 923.
72. Maslow, *Toward a Psychology of Being,* pp. iii–iv.

Chapter 5

Culture: Religion and the Social Sciences

The most immediate field that surrounds the self is society. While the person is interlocked with skin, body, and brain, and still recognizably a *natural* product (even granting a unique humanistic factor), human society, by contrast, is more evidently an *artifact.* Culture is by definition an environment built from and overlaid on the realm of spontaneous nature. Society too is in accord with our *human* nature, which has evolved out of nonhuman nature, and there may be basic senses in which cultural processes are themselves governed by natural law. Still, the social world is *made*, while the natural world is *found.*

The natural world is found with constructs made in the social world (including those of science or religion), and the social world is by no means all or even mostly deliberated, for we find ourselves in society when we earliest come to reflective awareness, both as individual persons and historically as a race. Humans are naturally social animals. Nor are humans, primitive or advanced, fully aware of the dynamics of their socializing; these are often unintended and unrecognized. But still, civil institutions, represented by the military salute and chain of command, are different from natural institutions, represented by the pecking order in birds or the food chains into which they fit.

One might think that the sciences of society would be simpler than the natural sciences, since we ourselves are participants with intersubjective access to the workings of our own product. Across some of its phases, society might be simpler than the persons who are its component units, and thus economics can sometimes have a simpler task than does psychology. Abstracted and simplified, the causal web in a regional economic system can appear to be simpler than the 10^{14} circuits in a brain. But the presence of persons as the ultimate social units, their capacities for the diverse production of societies with transnatural elements of humanistic self-definition, and the increased orders of option and complexity make the social sciences rather softer and more open, not in all their component studies, but in their governing paradigms and methodologies. The human being is born and develops in some one of thousands of cultures, supersocieties, each historically conditioned, perpetuated by language and tradition, conventionally established, using symbols with locally effective meanings. Nothing in animal communication approaches this. Here social science is forced to draw near and overlap the humanities, and with this its scientific status becomes more problematic.

Beside the (apparent) concreteness of matter, or the bounded locatability of the biological organism, we find it difficult to say what kind of a "thing" society is, whether indeed there really is any such thing as society over and above its component persons and their relational properties. This intangibility and quasi-empirical character is seen, for instance, when a sociologist remarks that there are "social currents" that carry away the individual in mob behavior, or carry along the person in a religious tradition, or that such "forces" are, in other cases, held in "equilibria" by counterforces. What is meant by differing "social positions"? Have we only metaphors here, or can we analyze rigorously what sort of real but nonphysical, nonbiological "forces," "equilibria," and "locations" are present? A chemical *state* is one thing, a nation-*state* is another, and socioeconomic *status* is something else still. Can one get a scientific hold on "rapport," "social inertia," "social space," or "power"? Even "persons," the ultimate social units, have to be recognized through no skill ever learned in social science, and it does not seem possible to put the kind of operational, quantifiable definition on "person" that the natural sciences have put on "element," "oxygen," "molecule," "water (H_2O)," and the like.

Culture, wrote Edward B. Tylor, an anthropologist, is "that complex whole which includes knowledge, belief, art, morals, law, custom, and any other capabilities and habits acquired by man as a member of society."[1] Culture, says Clifford Geertz, another anthropologist, "denotes an historically transmitted pattern of meanings embodied in symbols, a system of inherited conceptions expressed in symbolic forms, by means of which men communicate, perpetuate, and develop their knowledge about and attitudes toward life."[2]

Subsets of culture have become the provinces of the special social sciences—sociology is the science of society, anthropology the comparative science of cultures and subcultures, formerly the nonliterate or least nonmodern ones. Economics is the science of commerce and industry, politics the science of government. Even history, also among the humanities, is the science of the historic or literate human past. Underlying them all is the hope of bringing culture under scientific analysis. This will inevitably bring science in confrontation with religion. Those social forces that include our knowledge, art, beliefs, morals, laws, and customs, while they need not always be religious, will surely be religious at their wellsprings, characteristically if not inevitably.

The Christian society is the largest and most influential society on Earth. The Jewish society is among the oldest, remarkably small yet powerful and persistent in its witness for monotheism in the world. The kind of life-orienting belief that has classically supported societies has been religious, at least before any social sciences (as in Marxism or Comte's sociology) arose proposing instead to give direction to society. Biology and religion overlap at the word "life." Psychology and religion join at the word "person," and social science and religion meet where each claims to explain the *community*. For when we inquire how we are bound to each other, we begin to understand our freedom and our responsibility.

When culture emerges from nature, there is a puzzling dialectic. In ways that we dimly understand, the evolutionary process fits humans for culture. The hand, the brain, speech, collective emotions, religious capacities, and the long juvenile rearing period evoke culture. Diverse cultures appear as an adaptation for survival, and they are selected for their reproductive capacities. But we are not any longer dealing simply or even principally with genes, and analogies with animal societies (bees or coyotes) are relatively unenlightening. This is true even though the genetic control of society has been under vigorous discussion recently in sociobiology. The novel emergent is *information transmitted and advanced in a heritage.* Moreover, the human maladaptation to nature drives culture in tandem effect with the human adaptation for culture.

Humans are drawn to culture; they find security, fellowship, meaning, freedom, love. But they also are driven to culture (as Freud and others have rightly noted) because they cannot survive or prosper alone. Human life could never have evolved within a totally hostile nature, nor does it really seem satisfactory to say that nature evolved humans by chance. Nature is in some sense our parent. The evolutionary process had this potential and this heading. On the other hand, in a nature wholly irenic to human life little culture would have evolved. If the human adaptation to nature had been perfect, culture would have provided no benefit to be selected for. The need for food, which nature sparsely provides, for the comforts of shelter and security, the care of the young and the old, the benefits of reciprocal cooperation —such pressures force humans into culture. Over time, they lure increasing cultural development. Once again, what seems evil in nature is paradoxically productive of a good; here too there is a suffering through to something higher.

In the sections to follow, we look (1) at concepts of society as these characterize social science, moving (2) to society as a meaning system that exceeds causal scientific analysis and demands a kind of interpretation approaching religious levels. (3) Against this background, we criticize attempts in social science to explain religion (away) as a causal social projection, to be displaced by science, a form of explanation exemplified in Émile Durkheim. Finally (4), we press the question of value-free and value-laden social science as this bears on religious belief.

1. SOCIETY AND THE INDIVIDUAL: MODELS, LAWS, CAUSES

"Theory" in social science never reaches the sort of consensus found in quantum theory, atomic theory, and genetic theory after Newton, Einstein, and Darwin. It tends to mean "schools of thought," often unsettled and competing, often attached (as in philosophy and theology) to the name of a master, denominational rather than commanding the entire field. There are virtually as many live options in kaleidoscopic social theory as there are major creeds in contemporary religion. Sociology is a multiple-paradigm science. Less charitably put, it is a muddle of models.

This prevents a unified science, but it also lets us mix the models if we do so with the humility that becomes a soft science. At the same time, social theory can sometimes be pretentious and dogmatic, partly in compensation for its juvenile state. Early theories in social science tended to be ambitious, quite hostile to religion. But with the maturing of social science these have become less sophomoric, more open to their own incompleteness. We here begin with such supertheory, and then move to consider some main models of society, as this leads toward religious issues.

The Law of the Three Stages

Auguste Comte, a principal founder of sociology, held that societies follow determinate stages, beginning with primitive societies in the *religious stage,* where nature and culture are imagined to be governed by gods. A society subsequently moves to the *metaphysical stage,* with explanations in terms of abstract forces, timeless principles, essences, forms. Finally, on becoming modern, society moves to the *scientific stage,* with explanations in terms of causal law. "*Social Physics* . . . occupies itself with the study of social phenomena, considered in the same light as astronomical, physical, chemical, and physiological phenomena, that is to say as being subject to natural and invariable Laws." By a threefold and "necessary chain of successive transformations the human race, starting from a condition barely superior to that of a society of great apes, has been gradually led up to the present stage of European civilisation," the course "rigorously determined" "in accordance with a law as necessary as, though more easily modified than, that of gravitation."[3] This is verified by the secularization of modern societies.

This explains religion, and explains it away. The model (we shall call it a myth) is still found on the opening pages of Don Martindale's leading text in modern social theory. "Sociology is a part of that great evolution of thought in Western civilization that passes from religion through philosophy to science."[4]

We have at once to notice anomalies in the model. Comte's social series is also a liberating intellectual development. Although other causal factors, such as population growth and division of labor, enter in, he singles out "intellectual evolution as the preponderant principle."[5] But how can a movement be rigorously causal, like gravitation, a kind of *social physics,* and simultaneously an intellectual maturation? We are wise to suspect even biology when it treats all informational advancing as so much physics, and we may further suspect any advancing social-intellectual development conceived as a sociophysical mechanism. An end has become only a consequence. How can the Comtean sociologist be genuinely educated if his knowledge too is the inexorable result of a determinism? Comte insisted that all earlier religious and philosophical stages were error, and he was glad for their replacing by science, his reasoned truth. But if the last stage too is automatic, by what exemption can the finalists hold their beliefs any more autonomously? What one wants, nonnegotiably, is rather a *logic* in the evolution. Are there good reasons by which religion is displaced by philosophy, with that in turn displaced by science? It is not sufficient to be told that this must take place, or that it has taken place (oddly, only in the recent West), until we know whether it logically ought to take place.

Martindale leaves unaddressed whether science can really do the job of its classical predecessors at the level of explanation and providing value that sustains a society. If the religion, found early, and the science, coming late, should prove to have differing competences, if science works best with causes and religion works best with meanings, then perhaps the evolution here predicted is ill-conceived or even incoherent. Martindale simultaneously wants to claim (contrary now to Comte), that his science is value-free. Sociology "seeks maximum freedom from value suppositions." "Humanism [philosophy, theology, or literature] is a system of values and modes of conduct designed to secure them; science is evaluatively neutral pursuit of knowledge, renouncing all claims to prescribe what ought to be." Social science has "cast off its ideological moorings, for sociology is neither revolutionary nor reactionary, neither liberal nor conservative; it is a science—an objective enterprise in empirical knowledge."[6]

If so, by what logic can this sociology as science, in the third stage, give guidance in the place of the religion and philosophy, which it has replaced from the first and second stages? Either it cannot, and the great evolution of thought will leave us directionless. Or, if it can, it proves itself a nonscience, an ideology in disguise, and thus by its very success disproves the supertheory.

Society as an Organism

One trouble with such brash predictions lies in unresolved accounts of how society works, and only after we examine this can we intelligently project the course of religion within it. Sociology developed under the conservative theory that society is a quasi-organism. The model is adapted from biology, where (from the skin in) the organism is a model of cooperation, with highly functional parts cooperating in a whole. The analogy is at least as old as the biblical organic image of the church,[7] but when sociology followed biology as a developing science, the organic model had a renewed appeal. Much in society demonstrates a part-in-whole cooperativeness, the more so as societies advance and differentiate, with labor becoming specialized and with communications, interdependencies, and roles highly developed. Physics, chemistry, geology, and astronomy do not involve this level of organization, but it has undeniably appeared in biology.

So it seems likely that by extrapolation from the organic model we can begin to understand society. Herbert Spencer, another founder of sociology, put it this way: "All kinds of creatures are alike in so far as each exhibits co-operation among its components for the benefit of the whole; and this trait, common to them, is a trait common also to societies. Further, among individual organisms, the degree of co-operation measures the degree of evolution; and this general truth, too, holds among social organisms."[8] Societies as a whole have "needs" and "face problems" and are "self-regulating." A group acts like an animal body.

But a good many puzzles and disanalogies immediately appear. When one moves outside bloodstreams and neural circuitries, it is difficult to say just what sort of causality, if any, is the underpinning joining these cooperations. Further, while the liver, finger, or cell seems appropriately subordinated to the organism, in human society we are not just dealing with biological organs, not with a somatic physiology, but with persons who have a self-reflective life, with those self-actualizing subjects that humanistic psychology has described. The individual who is a subservient member of society is also in some ways superordinate to it.

A society has no center of experience, no subjective inwardness; only individual persons do. A society has no will or feeling, no autonomy, perhaps no rights, except in some derivative and associative sense from wills, feelings, autonomies, rights in personal lives. A person has a unified center of experience, as the parts in organisms do not, and it is by no means obvious that the person is, or ought to be, by organic analogy "nothing but" a contributor to her social system, since the centers of experience remain in the parts, in the persons, and are not found in the whole, in society. Persons may be rather more or less free in, or even free from, their systems, self-actualized within and even determiners of those systems, rather than being determined by them.

Yet what humans feel, choose, and will, what rights humans own, are deeply culturally conditioned, indeed, some say, entirely so determined. We have no truly separate self, but always one that is enmeshed in, a product and instance of, the society that has made us what we are. We each carry on within ourselves the work of numberless generations, and we receive daily the products of the work of our contemporaries. So perhaps individuals are organic-like parts-in-wholes, particles determined by their cultural waves at least enough for us to write a social science that analyzes this process. Social models that give the system more priority over the individual are termed *conservative;* those that give the individual more priority over the system are termed *liberal,* though even the latter may recognize a basic priority in the system. Both have something of a tendency to prefer what they find, or at least to trim their values accordingly.

Society as an Equilibrating System

There is a tendency already in the organic model to think of society mechanistically, just as biology was reducing the whole organism to the interplay of its parts. Sociology has theorists who pull the organic model in the direction of making it mathematical, operational, empirical, positivist, by finding laws in measured causal correlations. We are no longer dealing with an organic-like system, for organisms have more centeredness, more cooperation of parts and coherence; we are dealing with a network of equilibrating forces, devoid of centeredness, but laden with push-pull interactions, checks and balances. Society is not a centered working together, but an individualistic working for one's own benefits, pitted against those of others, with only secondary cooperation in enlightened self-interest. Society is more like a market (which is indeed one kind of society) than an individual organism; there is no control center, much less any center of experience. There is only a web of interacting forces.

Comte wanted a "social statics" and "social dynamics," and others have posited the hydraulic interplay of diverse social forces that arise from the drives of individual actors. Persons seek to maximize their money, status, or power. Given the motivational sets of the social actors, the dynamics of society can be understood and predicted on the basis of mutual attractions and repulsions, resultants of social vectors. Stability results from the equilibria of counterthrusting currents; change results from disequilibria. These social "positions," "forces," "powers," "pressures," "fields," suitably analyzed and measured, will give us complete access to social movements.

Vilfredo Pareto, another founder of sociology, illustrates this in a form that quite dismisses the efficacy of belief systems. Sociology is to be modeled after thermodynamics. Persons are "molecules" driven by deep-seated, nonlogical springs of action. They attract or repel each other on the basis of unconscious "residues," that is, instinctual drives producing social circulations that can be lawfully described. This analysis undercuts all belief systems as superficial. Hence Pareto flails all metaphysical, religious, and moral systems that pretend rationally to control actions. The most liberated and modern of them are as vacuous and mythical as the most savage magical incantations—all of them, that is, except his own science, by which he has finally uncovered the causal rather than the ideational principles of conduct.[9] (His theory too, of course, gives us every reason to distrust reason, whether ours or his. It leaves Pareto and his enlightened sociological elite as "residues" of another sort, unexplained residues.)

Even among those who later and with increasing sophistication wish to emphasize how mental factors too are causal, there remains this tendency to look to physics for axioms. Robert MacIver, a social theorist, asserts, "We postulate a social law roughly corresponding to the physical law of inertia, to the effect that every social system tends to maintain itself, to persevere in its present state, until compelled by some force to alter that state. Every social system is at every moment and in every part sustained by codes and institutions, by traditions, by interests. If a social order or any social situation within it, suffers significant change, we think of some insurgent or invading force, breaking up as it were this 'inertia,' the *status quo.*"[10] Neil Smelser agrees: "All [social] systems of action are governed by the principle of equilibrium."[11]

Individuals here may be the propelling units, on the basis of their beliefs or not; but still each individual is caught within an interactionary system where the currents flow and equilibria result from pulling and hauling between individuals, variously arranged in their social groups. Any atomic individual in his volition is quite subordinated to the larger social movements. His society may be something of an organic whole, but it is more an organization of mechanisms, a distributive system of social forces. Further, it is a system that governs the individual, statistically if not in detail.

Society as a Structural-Functional System

Contemporary sociology has sought some distinctive model of its own, not organic, much less mechanical, but peculiarly social and yet at the same time lawlike and scientific. The leading such paradigm is named *structural functionalism,* rather unfortunately using two words so widely employed elsewhere in science that they are rather nondescript in their sociological use. But the intention is to portray a genuinely social system with communal structures and functions, again in a scientifically describable equilibrium.

Another prominent sociologist, Talcott Parsons, like McIver and Smelser, finds the "basic paradigm" to be an interactive system with a "tendency to self-maintenance, which is very generally expressed in the concept of equilibrium." A society is "a huge moving equilibrium."[12] Yet, unlike a merely equilibrating system and surpassing an organism, a society is a self-maintaining community with intermeshed corporate structures and functions. Unlike individual organisms, societies have no finite lifetimes or fixed anatomies and morphologies. A society need not age and die, it can elaborate and transform its structures and functions indefinitely. In the organism, structure dictates function; in society, function requires but can dictate structure. Social movements, having causal priority over the individual, are system-determined and system-maintaining activities that satisfy societal needs and fulfill societal "purposes."

This model fully accepts the reality of the societal system, prior and primary to the individuals who are embedded within it, a whole greater than its parts. The social process unites interdependent individuals, induces their particular behavior, and persists over time with the replacement of those individuals and the education of new ones. Dysfunctional behaviors that threaten the social welfare appear from time to time; but, rather like mutations and anomalous perturbations, they are deviations that will either be edited out by social selection pressures or converted to functional form and retained. Many apparent dysfunctions may really be serving covert functions that clever sociologists can detect (as with crime or deviant behaviors). Some dysfunctional trials regularly ventured in the ever-changing social environment provide the variation introducing emerging social trends.

Still, in its lawlike structural-functional processes, this system "locates" individuals, gives them "status," "roles," "meaningful orientation," "values," "ideas," "symbols," places them in "institutions," informs their "drives" and "emotions." Social "tensions" constrain individuals and motivate them to adapt and adjust to the system. Persons maximize their status; but since their status is given by social approval, society in fact determines their behavior, and induces such behavior as is socially functional. There is really little self-actualizing (SA) in any deep sense, little personal originality or authenticity, but mostly an S-R effect again. But now the S stands for Society, the corporate Stimulus that governs human nature in a social mold.

To understand what is going on in society, we do not have to understand individuals so much as their organizations: for instance, the dynamics of the small groups that make so many of the decisions of large latent groups, or the self-perpetuating tendencies of agencies and clubs, or the nature of bureaucracies, the relations between superiors and their subordinates, or how sluggish communities need rejuvenation by charismatic leadership. We notice how heritages are transmitted over generations (American, Jewish, or Christian ways of life), and so on.

But rather soon such a theory, being relatively sketchy and loose, begins to permit interminable revisions, adaptations, and adjustments within itself and becomes a hotchpotch that is difficult to test. Like natural selection with its non-negotiable positing of some survival value for any persisting organismic change, like Skinner so sure of a stimulus in the reinforcement history for every response, or Freud's dogma that a suitably explanatory unconscious mechanism must be there driving every behavior, so Parsons is overconfident that all social behavior, on statistical average, is structural functional. This includes religious belief, and, more recently, scientific belief. A way will be found to interpret everything so, and anomalies will not be allowed to challenge the theory. After all, it is not very insightful or surprising to theorize that persisting social patterns must somehow work; in one way or another this seems almost inevitable, although we have to be careful about endorsing what *is* as what *ought* to be, especially where social powers over the individual are found.

Society as a Cybernetic System

Cybernetics has given the promise of a social systems theory that is neither mechanistic nor organic, but is informational.[13] Society is not part-in-whole (organismlike), not an equilibrating network of causal forces, and the really important element in its structures and functions is how society is a cognitive network. It is the power of a society to transmit and use information that is novel over anything known in biology, even cognitive biology. While the macrosocial theorists have failed to put any one paradigm in command of the field, the microsocial analysts have with increasing skill and success been able to document statistical covariances between many of the social forces. But they, in order to get any correlated variables into focus at all, had to bring them into too flat a focus. As these correlations accumulate and begin to interweave and overlap, it begins to be possible to move from microsocial observations or regularities back at least to the mezzosocial level, now to construct at more depth a theoretical model of the social process. This will be less global and coarse, less formal and empty than the classical theories, but now supplying a reliable multidimensional map of parts of the social landscape.

Computers are electronic machines that overlay cognitive processes on causal circuits, with great scientific respectability, and perhaps the cybernetic social forces can be computer-analyzed or -modeled. Recent social systems theory has become more sophisticated, using computers, moving away from linear cause-and-effect sequences to interactive circular webs with feedback loops, stronger or weaker hookups between variables, step variables or buffer effects that allow for delayed regulations and potential buildups, or lag and slippage effects as from educating a next generation. Ecological and resource constraints can be included. Computer models, as of a national economy, or presidential election, or military operation, or social preferences, or demographic trends, seem so respectably scientific, and they are not merely causal but also cognitive. They are also metric. We are forced to put measures on social forces, to write equations about them, and we can make predictions and test our theories. We can do computer analogues of that "huge moving equilibrium" which Parsons could only vaguely describe.

The abstractive simplicity of even the most advanced of these models is recognized; correlation coefficients are low, and only modest claims are made. All science abstracts and simplifies for analysis. But impressive as all this is, any really rigorous and extensive theory, as opposed to piecemeal bits of analysis, that explains social dynamics with few or no exceptions remains a promise and not a reality. In this fascination with computers (which we earlier saw also in biology and psychology), the ultimate hope is something more than just an electronic simulation of social forces. It is to begin to understand a social cybernetic system by its analysis on an electronic one.

But the problem here is the same as the promise—the mathematics and pressure toward exact specification. Society is often less logical than we could wish, and even when it is logical, it is so by a more informal, subtle logic than that of the game theories or economic formulas that do computerize. Game playing is rather algorithmic; there are canonical rules that govern all that one can do on a chessboard. But a new canon in society may emerge from moral criticism of the existing order. The mix of deception and cooperation in games is really quite different from that in society. Life is not one game but many incommensurable games; interacting with a business opponent who is also a fellow churchman is like playing polo with the right hand and checkers with the left, and there are no rules for the trade-offs in concentration and allegiance required.

The effect of religion on politics goes poorly into syllogisms, set theory, or feedback loops. One can take the really fertile analytical tools of sociology—what it means to be "legitimate" or "heretical," concepts such as "role," "status," "authority," "anomie," "consensus," "bureaucracy"—and abstract something out for the computer run. But this only mimics on electrical, causal circuits what is a very much richer reality. One has the skeleton, not the flesh.

The really distinctive sociological elements here are those hardest, even impossible, to put into hardware or formalism. Gestaltic relations of part to whole may not be mathematical, not even logical in a scientific sense, but only in a meaningful sense. Society is a fuzzy system; it is an adaptive learning system, and one finds it hard to preprogram into models or computers what the next move in constitutional or canonical interpretation will be. Unlike game simulation, the rules themselves are an open set. We short-circuit the inwardness, the subjectivity, the agency, that the social actors in fact enjoy. The reality we study stays soft, and thus even when our science seems to become hard, it actually stays a soft science, only loosely comprehending the reality it models.

Society in Historical Conflict and Change

None of the previous models handles conflict and change very well, and society is rampant with both. Organicism explains the cooperations and self-regulations, but, from the skin out, evolutionary and ecological biology add conflictual and contingent elements, as well as further fittedness and cooperation. Various persons, groups, and classes are in competition, in which the fittest prevail, rise to the position of dominance, fill a functional role in the survival of that society, but are later ousted by newcomers. The whole is not a quasi-organic cooperation, not a structural functionalism, not a cybernetic or equilibrating system, but a struggle for power.

Marxism attempts to predict the inevitable course of class struggles, without notable success. All such historicist predictions of what conflicts will arise, how they will be resolved, and what will be the direction of change have proved to be projections, not predictions. They are laden with provisos and surprises, always with many trailing "ifs" and "assuming thats." It seems almost logically impossible for a social system to know, ahead of time, what new discoveries, paradigm revolutions, moral convictions, or other unexpected truths will supplant the existing order. One can, of course, believe in certain likely directions of development, but there is a difference between causal prediction in natural systems and social prophecy. When we try to look ahead we quickly get over our heads.

The notion of an equilibrating social system allows for disequilibria upon the appearance of "some insurgent or invading force," but it must treat this as an anomaly which it cannot predict, which it explains only ex post facto as a counterthrusting given. It has little theory for the long-range trends and resolutions of conflicts, little for the evolution and revolution that is always upsetting the equilibria. No social science could predict the appearance or effect of the railroad in nineteenth-century America; none anticipated the transistor in twentieth-century America. None could foresee the emergence and resolution of conflicts in the churches about women's ordination or homosexuality. None anticipated the electronic church, depending as this does on computerized communications.

Structural functionalism explains how social dynamisms work once they are in place, as biochemistry explains the morphology and physiology of an organism. It explains how the cogs turn, but what explains what generates the past and future variety in these turnings? Any cybernetic theory stalls here precisely because it lacks the information it needs. It cannot prespecify conceptual innovations, technological discoveries, the power they will have, or the moral response to them. We cannot predict the movements of the spirit. Peter Berger, a sociologist, reflects, "I have often thought that even a person equipped with all the tools of modern social science would have been hard put to predict the Reformation, say, at the onset of the 16th century."[14] But few events have affected history as dramatically.

In short, that problem returns, now compounded, which we have already met in evolutionary biology with its mutations and survival contests, in quantum theory with its superposed and indeterminate quantum states. The system is an open system. The events to be studied are no longer matters of nature but are matters of mind, where one generation is unable to anticipate the mental life (the new theories, the reformed sense of justice) in subsequent eras.

Individual Agency in the Social System

Sometimes persons appear (as did Luther) who refuse merely to internalize the prevailing social norms. Social theorists divide over such individual agentive capacities as they make a difference in the course of society, or in the course of one's own life within that society. Social-action theorists will give the individual some priority over her system, some significant options within it, while social behaviorists will tether her closely to it. A sociological liberal theory will believe that persons somehow, sometimes are in command of their systems, or will become so with education and time. At this point organic, equilibrating, structural-functional, and even conflictual models can be oppressive, so far as the individual is but a cog in a machine. One is forced into crime, or conformity. But if we can replace the mechanical, coercive, oversocialized "forces" with a richer metaphor, that of a person in a community, this situated freedom may be welcomed over more mythical and destructive extremes of freedom.

So far as one finds or desires options and openness, these seem to make social science impossible, at least in any hard causal or naturalistic sense. It is not the closing of the feedback loops that makes teleology possible, but rather their staying partly open, for point insertions of mutation, choice, learning, innovation, genuine action. Jon Elster, a philosopher of social science, writes, "The basic postulate from which I start is that *the goal of the social sciences is the liberation of man.*" "Man should somehow be able to *choose himself.*" But echoes of the old tensions remain, as Elster puzzles over the requisite determinism and its impossibility. "I do not argue that social facts are inherently less subject to causal determinism than natural facts, only that we must at present treat them as if they were, constructing our theories so as to admit more degrees of freedom than is usual in the natural sciences. On a more speculative level I think that in some cases the process *must* be treated as if it were inherently non-deterministic, even if we know that in 'reality itself' it must be completely determined."[15]

May Brodbeck, another philosopher of science, expresses the ardent hope for and promise of a lawlike, causal social science. "Two centuries have passed since the vision of a science of man first fired the imagination of the great social critics of the Enlightenment. Seeing man as part of the natural order, they envisaged a science of man and society, modeled on Newton's explanation of heaven and earth, by whose application the potentialities of man could be realized to form a more just and humane social order. . . . In essence the vision has withstood all challenges."[16]

But, to the contrary, retorts Alan Donagan, a philosopher of history, "Far from withstanding all challenges, I submit that it [a causal social science] has not withstood any serious challenge at all: its supporters imagine it to have done so, because they have failed to notice that they would consider nothing a challenge to it. . . . What challenges could be offered to the Enlightenment ideal except the two that have been: that history, and successful research into human affairs generally, does not accord with it; and that no example of a complete explanation according to the Enlightenment ideal exists?"[17] And none exists because every historical, social, and biographical explanation is incompletely causal, being perforated with noncausal decisions and quests for meaning.

The issue is partly how far society is dominantly influencing the individual, or vice versa. But the issue is further how that influence is to be conceived. Many sociologists have wanted to make this *causal* in such a way that the actor's meaning and intentions are epiphenomenal. There is a second main myth in traditional social science, which lingers because it is an occasional and rudimentary truth. It is closely related to the first myth, Comte's evolution from religion to science. This second myth is illustrated in Durkheim's principle "that social life should be explained, not by the notions of those who participate in it, but by depth causes which escape conscious awareness," with his corollary that "these causes are to be sought principally in the manner in which the associated individuals are grouped."[18] He insists, with emphasis, "*The determining cause of a social fact should be sought among the social facts preceding it and not among the states of the individual consciousness.*"[19] Here we investigate social reality by using concepts not available to members of that society, by invoking causal explanations of which the agents themselves are unaware. The reasons they give for their behavior and the causes the sociologist finds are mutually exclusive.

But this sort of hard causality has been progressively softened in later social science in favor of more volition, more agency, more freedom. Ideas, beliefs, information make a difference, sometimes *the* difference in the dynamics of social systems, but with this we pass over from causes to reasons, to "causes" in some nonnaturalistic, teleological, Aristotelian sense. Here the concept of "causes" is difficult to use without some overlay of "reasons"; they coinhabit the term "function." But whether that cohabitation is licit or illicit is never resolved. The social system has in part its causes, but it also has its logic. That was already true even in biology, for those scientists who are sensitive to information in noncultural, organic forms. It now becomes overwhelmingly evident. The system is a *semantic* one. The place of the individual in his society and the ultimate nature of the social powers of persuasion have to be cast in a different light.

The culture that an individual faces is not an all-determining and closed-determinate thing; rather, to a considerable extent society presents the individual with nested sets of possibilities, before which the individual may actualize himself in this way or that, at his option. But as he does so, he is not only actualizing himself, but further actualizing his culture, perhaps seminally, perhaps only incrementally. In this respect, we have an analogue of what we met in biology—the organism nursing itself through the superposed quantum states, actualizing this and not that outcome. We do not have merely a one-way influence, but interaction between individuals and their societies, each presenting the other with possibilities that are as significant as any precedents and causal functions.

2. INTERPRETIVE SOCIAL SCIENCE

Without denying that mechanical, organic, functional, and cybernetic subroutines are present in society, we advance to what matters most to the community—its meaningfulness. The peculiar feature of the human social system (made possible because of the psychological capacities of the persons who compose it) is that society carries meaning. These purposes can be handled on social circuits, use feedback loops, and are overlaid on causal hookups. But to understand the bare "path analysis" here is not yet to entertain these meanings, any more than to map computer circuits is to understand the program that is being run on such hardware. Nor is it simply an issue of information handling, of the communication of signals and skills, the distribution and diversifying of labor or know-how and the like. A further sort of information is still more fundamental. We move beyond technique to *worth,* beyond fact to *shared judgments of value,* beyond science to wisdom.

Communities of Shared Meaningfulness

Organic models of society immediately notice that a society has no skin. This is not really a problem of soft boundaries, of its being difficult to tell where one society leaves off and another begins. The problem is what *binder* joins the social units together, since no physical or biological metabolism keeps components in contiguous touch. A frequent answer is that language serves this "function," but even a highly developed common language is not enough. Nor is it a matter merely of *organization,* for that word, however beloved in sociology, remains too structural, too functional, too organic. What binds a society together, after economic, geographic, and other cohesions have been recognized, is *shared meaningfulness.* The members of a society may not share these meanings evenly, and these may not be a consistent set; there may not be a consensus about all these meanings, but dispute and tolerance at many points. Or some in power may impose their meanings on others. Nevertheless, if there are no governing meanings a society will not long persist. What persons try to maximize is not merely money, status, power, but, most of all, meaning, and yet they draw their meanings from the societal meanings.

If we now try to speak of this as the "cause" of social cohesion, connecting this use of the word "cause" with what has gone before in physics and biology, we find that the word "cause," infiltrated with meanings, has been taken under a new control. By "functional" analysis we may learn first the simple covariances between social events, and after that the contribution of parts to a whole. But we need still more explanatory work. Intersocial and intrasocial contests and cooperations are not so much equilibrating forces as they are roles in stories acted out. What society "seeks" is not survival, much less equilibria. Nor is "information" an adequate category. *A society seeks meaning.* Humans do not live by bread alone, but they live for and off ideas and ideologies.

The coming of mind makes possible that emergent sort of information transfer and storage, sharing and evaluating that we call a cultural heritage. Humans are not always, sometimes not often, rational; and their rationality is flickering and fitful, a matter of degree. But they live by meanings, worse and better ones, but meanings still. Eventually we reach what Talcott Parsons called the "highest cybernetic level." Systems high in meaning, even though low in energy, will dominate, direct, and replace systems that are low in meaning, even if high in energy.[20] That is, in the historical evaluation of societies, success is not so much a matter of technique, power, or industry as of having a more religious steering, a better sense of ultimate orientation in the natural and social worlds.

But sociologists will find it difficult to identify high-meaning systems, against low-meaning systems, without empathic use of some value sets of their own. They will do all this poorly on any computer run, and with no skill that is merely scientific. Indeed, the highest social-meaning system, says Parsons, is one that reaches for "the ultimate ground of meaning."[21] Such a social selection principle, passing beyond but continuing the upslope of natural selection, is certainly (contra Comte and Martindale) a pressure to be religious, almost a lure toward something like God. At this point, quite contrary to Freud's finding of monotheism to be the juvenile individual's projection of an illusory heavenly Father, an unconscious defense against hostile nature, Parsons finds quite appealing the way in which Judeo-Christian society has interpreted life as a divine gift and demanded an appropriate response in the recipients.[22]

What has to be entered into are the mentalities that permeate a culture. Advocates of *Verstehen,* understanding, from Max Weber through Karl Mannheim, Wilhelm Dilthey, and more recently Peter Winch, all leading social theorists, have long maintained that understanding a society requires not only an *objective analysis* of social forces, conflicts, equilibria, functions, and the like, but an empathic interpretation of the *subjective Weltanschauung* of its actors. Meaningful behavior cannot be exhaustively described by external study of what pulls and pushes the actors but requires a shared inwardness, one consciousness appreciating another consciousness. We can discover empirically, by observing behavior, that x values y. Perhaps we can even metricize this. But we cannot gain nonempathic access to the meaning of y to x.

When sociologists work mostly within the confines of the cultures in which they have been reared, these may be so obvious as to be uncontested and even invisible. But when anthropologists seek to know why the Hopis dance before the rain—to cause it? to beg for it? to praise it? to get ready for it?—the sequence of production and response intended or latent in the ritual will, until found, leave any explanation incomplete. In insisting on something more than empirical structural functionalism, something more than cybernetic social systems, interpretive social science in sociology and anthropology forms the parallel to humanistic psychology in its insisting on something beyond behavioral science, beyond causal or even cognitive psychology.

The Scientific Study of Religious Prejudice

Take, for example, studies of the role religious convictions play in forming and overcoming various kinds of prejudice, by which individuals are predisposed on the basis of their belief systems to regard others favorably or unfavorably in disregard of relevant facts. These prejudices will surely have to be understood in order to model, much less to predict or control, the dynamics of a society. One might wish to do this in a strategy for a just and meaningful society. We find that it is possible, although difficult, to test for these concepts, even to scale their covariances, and to portray (and even graph) a picture of the following type. There are differing institutional levels—ecclesiastical officials, the general clergy, the laity—and the higher up, so to speak, one goes in the ecclesiastical institution, the less the prejudice in principle and in practice, with some lag between principle and practice, while the more rank-and-file members have greater amounts of prejudice. We find that prejudice increases (though not much) with conservatism. The official pronouncements of a denomination about racial, ethnic, or religious prejudices are about half shared, about half ignored or rejected by the laity.

We find that theological exclusivism (my form of Christianity is the only way) breeds religious prejudice, often unawares. Theological individualism (each person is individually in control of and responsible for her own destiny) breeds racial prejudice. For example, one believes that underprivileged Negroes or persecuted Jews get what they deserve, and are not merely the victims of prejudiced social forces. Theological universalism (God wills and works graciously to save all) promotes tolerance. Prejudice strongly corresponds with religion where group dominance is emphasized (consensual religion), but negatively correlates with intensely and personally held religion (committed religion). A careful description of all this may enable a sociologist to describe about what effectiveness the churches do have and will have in combating racial and religious prejudice.[23] We might say that these relationships appear to be causal in some way, since beliefs are indisputably correlated with prejudice. One seems to be producing the other.

But the sort of causality here involved is something else from anything met in physics or biology, where neither religiousness nor prejudice appears. The entire inquiry is infected with meanings. Notice, first, the measuring process (as distinct from what is measured). In the survey research, the researcher has to work with language. He must himself make discretionary judgments as to whether the general concepts to be measured—"conservative belief," "liberal belief," "prejudice," "fidelity to a faith"—are in fact measured by the array of specific questions assembled as empirical verifiers of the presence of that concept, whether the split half of a test really double-checks the same concept. If, for example, a Jewish mother discourages her son from marrying a Gentile woman, does this count as "prejudice"? If a Protestant clergyman disapproves of ordaining homosexuals, is this "prejudice"?

Respondents will understand questions with varying meanings, and, especially if they are informed about the character of the research, as the American Sociological Association Code of Ethics requires,[24] the researcher will have to trouble about how much deception, if any, is ethical, how much is being suspected by respondents, how to avoid sensitive questions (those that offend the respondents), and yet how to get the information he desires. He must give assurances that there are no right or wrong answers in order to get honest answers, preventing normative answers (when respondents answer what they believe the researchers want, or what respondents idealize). He must overcome a respondent's reluctance to admit to proscribed behavior, and get cooperation by making the survey seem worthwhile and legitimate. He must worry about the effect on the study of the authority or lack of authority of a sponsor, about his own seeming religiousness or lack of it.

The researcher has the problem of seeking permission, of diplomatically "getting into" the denominations he researches, and of getting some of their religiousness into himself as observer, of achieving rapport and credibility. He has to estimate and discount reactivity between the group and the observer. He has to worry that conviction and prejudice register poorly on a thermometer scale of 1 to 10, that better-educated believers will have more linguistic skills on open-ended questions, and that lesser-educated ones will need boxes to check, often oversimplifying their replies. Questions about the meanings of words, sentences, concepts—prejudice, grace, personal responsibility—soon spill over into questions about the meanings of events, actions, and of life style.

Meaning Systems and Causal Law

If we think, secondly, of the character of what is actually being measured, religiousness and prejudice, sociology wants to maintain, not deny, that social facts are emergents not present in physics or chemistry. But the character of these dynamically unique forces requires us to ask in what sense any *causal* account, even one of a sociocausal sort, can be complete here. What is novel is that we are dealing with *beliefs*. The lawlike order to be studied here involves a logic and illogic of belief.

This can be readily seen if one moves from the internal validity of such research, in a studied sample of Protestants, Catholics, and Jews, to ask of its external validity when transferred to Muslims, Hindus, and Buddhists. They live with the same physiology as do we, but their social world is different, and hence we are willing to project few of our Western conclusions about religiousness and prejudice into the Middle or Far East, not at least until we have examined all relevant obstacles to cross-cultural transferability from one meaning system to another. Is the Hindu caste system also prejudice, associated with the doctrine of *karma*, and does it work like the doctrine of theological exclusivism?

If a "cause" includes any "bringing about," whether efficient or teleological, if "cause" marks any link between variables, then we can call the connections "causal."

We say that T (belief in a nonexclusive, gracious God) is the *cause* of O (unprejudiced, caring behavior) in the sense that the appearance of *T makes O more probable,* but what is the sense of this handy but treacherous and equivocal "makes"? It is not physical or biological necessitation, as in "the sun makes the corn grow," nor any formal logical necessitation, as in "2 plus 2 makes 4." It is a motivational, an obligative "makes." It is much nearer to our sense in which the social world is not "found" but "made." Given this belief, the agent resolves to "make" her world less prejudiced. Her creed urges this conduct. The presence of will, deliberation, obligation, and valuation, of acting on judgments of meaningfulness, advances this activity far beyond any mechanical, organic, or functional notions of causality.

The sort of *law* that operates in a society, reflected for instance in its moral codes with their sanctions, must not be confused with the sort of *law* that operates in a biochemical system. The Ten Commandments, written not so much on tablets of stone as in the canon of Israel, have the effect they have on the behavior of those they control because of the meanings they have. These meanings are not empirical facts. They cannot be "observed." This does not merely involve the type of nonobservation that occurs in natural science, where neutrinos or the primeval origin of life cannot be observed either. It is a matter of how these laws operate to bind by traveling through the participant's theories about the world.

Nothing of this kind is found in genetics or in geology. This emergent sort of *law* yields rule-guided behavior where the rules are in part, though not in whole, consciously followed, wholly different from the "obeying" of "rules" in natural science. This sort of being has an *ought,* beyond an *is,* prescriptive behavior beyond descriptive behavior. Humans can form and maintain a "policy"—to use a model from political science—a policy, for instance, to reduce prejudice against blacks, women, Jews, outcasts, based on considerations of meaningfulness. By contrast, it makes no sense to ascribe policy decisions to animals, plants, or the nonbiotic materials of spontaneous nature.

We do not wish to deny causes among the social determinants. Crowding and summer heat cause race riots. Things happen that nobody intends. What is rational for the individual may cause dysfunctions in society. An anthropologist may help us to understand what happens when money is introduced into what was previously a barter economy; an economist may help us to understand what happens when too much money inflates a modern economy. Further, a sociologist may sometimes gain insight into the intentions of his subjects of which they themselves are unaware, or only opaquely aware. Where the *norms* of society remain subconscious, we may yet speak of causes, as when your nearness infringes upon the norm of personal distance in my culture and "causes" me to step back.

But as these norms increasingly enter our awareness, tacitly if not yet explicitly, these rules come to govern aimed behavior. When we label a person as deviant or competent, saint or atheist, and notice the effect of this labeling on the person so labeled, or on those who label others so, these effects are "caused" by the labeling, but the labeling itself is a value judgment, so that the "cause" is nonmaterial, nonbiological, not so much something that can be objectively observed as part of the "outlook" of a subject. In social relations, especially religious ones, language is loaded with words and phrases whose meaning is indexed to socially shared under-

standings. Here meanings are also conveyed by gestures, inflections, rituals, and mannerisms that are culturally embedded.

Latent and Manifest Social Functions

We may now employ the distinction between manifest and latent functions. "*Manifest functions* are those objective consequences contributing to the adjustment or adaptation of the system which are intended and recognized by the participants in the system; *latent functions,* correlatively, being those which are neither intended nor recognized."[25] Latent functions may have *ends;* but only manifest functions have *ends-in-view,* although insightful social science can bring into view the ends of latent functions. An adequate account of the social process (of ancestor rites, for instance) is going to require, first, an explication of the ends-in-view, which can be had only by empathically sharing the view of the actors, and, second, an explication of the latent functions, as these fit into the ends-in-view, sometimes in phase, sometimes out of phase with them, even though the participants may be unconscious of these latter perversions. Nor will it be possible to discern the latter without checking the actual against the presumed intentions of social participants. Access to inwardness is quite necessary, if also insufficient, for describing latent functions.

At this point, we cannot discount the insider biased by her covert culture while favoring the sociologist and anthropologist on the presumption that they have no covert culture that is slanting their judgments some other way. There is certainly no reason to prefer the interpretation of a culture in terms of the covert dispositions of foreigners over the covert dispositions of natives, merely because the former are foreigners, or merely because they are secular and scientific. Those who feel the latent forces are quite as likely to be able to read the behavioral cues, to probe the motivations or rationalizations, and so to uncover and make the tacit explicit, as are those who cannot sense these forces. One may need to experience these esoteric, subconscious forces to know whether one is dealing with a meaning or a cause, with something complementary to or competing with the exoteric culture.

In description here, the word "cause" remains the more helpful the more one stays on the latent end of the spectrum. Even here behaviors are tacit symbols, needing interpretations to be ferreted out in the context of the ends-in-view. Latent causality is woven into a system of meaning. Moving to the manifest end of the spectrum, however, the sociologist must deal with "reasons," because his subjects are dealing with reasons. Here they do intend and debate their policies and creeds.

As social processes become fully manifest, one can keep the word "cause" only by using it nonscientifically and half-accurately, as in everyday conversation, where, with Aristotle, a cause is any explanatory factor, including meanings, purposes, reasons, such as a political or religious "cause." But to call this a hard cause has become wordplay. What is required for full accuracy is not more causal explanation at all, but judgments of rationale, the capacity to think like one's subjects of study, and indeed to overthink and second-guess them, to see deeper into their own meaning system than they themselves may see.

In the broad sense, the deductive model is still valid (if T_s, then O_s), for sociologists argue out of their theory (T_s) to explain and predict the behavior they observe (O_s). Such a theory is capable of corroboration or of falsification, and there is regularity enough to handle it systematically. The novelty now is that the sociologists' theory is, in crucial part, about the theories of social participants (T_p), as these "determine" their public behavior (O_p). There too one can make inferences, as do the social actors themselves, from the theoretical beliefs (T_p) held by such actors to their observed behavior (O_p). Our meaning systems motivate us. We presume logic and volition such that: if T_p, then O_p. We can appreciate connections and often predict the future.

But the influence under study is now not *causal* in the same way as in physics or biology, where the objects of study hold no such mind-sets. Paleontologists hold theories about dinosaurs, which behaved as they did without benefit of language or theory. But paleoanthropologists hold theories about the Neanderthals, who behaved as they did by benefit of theories, made possible by language. We are looking at what they looked through, viewing world views. The social subject is acting on beliefs, perhaps inherited, perhaps recently made, perhaps endorsing, perhaps criticizing her society, and this not all the time, but at least at decision points. Such a meaningful symbol system is the fundamental social field, a reality beyond the scope of earlier, objective notions of causality because it now involves  judgments of value and plausibility. There is a world outlook, a logic-in-use, and decisions made in faith.

The dependent-independent variable format is here often inadequate. Causally, a researcher can hold constant all variables but one, vary that, and see what happens to the dependent variable. The production is linear and sequential. But where all meanings are holistically interwoven, if we vary one factor the whole belief system can be strained. The gestalt distortion, with the resulting change in the life narrative and behavior, is too complex to be described merely as a dependent variable. It nevertheless follows as an inference from the revised character of the meaning system.

Historical Self-definition and the Logic of Social Science

We find elements of self-definition in the face of options in an ongoing historical consciousness. Selection from these options is by reference to prevailing systems of meaningfulness, which partly constrain choices made. These choices are not and ought not to be capricious, nor can we do anything we want. Our society may not have a coherent or feasible goal set; there will be conflicting claims, a pluralism of interests and creeds, a society of societies mixedly cooperating and clashing. But nevertheless there are genuine decision points, if only with incremental dimensions of freedom, and with failure as much as with success. Once again, but now at a higher level than in physics, biology, or even psychology, it is not just the causal necessities that impress us, or, on the other hand, just the contingencies and openness, but the mixing of the two so as to make a place for a genuinely creative personal life in interaction with a historical culture.

By initiative, by education and innovation, by conspiracies, revolutions, and elections, the social actors opt for this path and not that one, as surely as they are also induced to follow paths begun before by others, whose endings are out of their sight. Leaders in groups sets new paces, and new groups interact with others old and new. There is emergent behavior, which may not be entirely the result of any single leading persons or groups, but it is greatly influenced by the pacesetters. When social scientists describe this capacity of a culture for self-definition, they have been brought to the borders of their competence. From here onward any description of a society must remain partly indeterminate. Elucidate a people's symbol system though they may, interpretive social scientists can go no further. Here sociologists, anthropologists, political scientists, and economists, *qua scientists,* have no powers for prescription of ends as distinguished from a description of means to such ends.

We are encountering mutations again, only this time not as atomic or genetic variants, but as cultural change points before superposed possibilities of culture. The plot has thickened. Before, in physics, statistical regularities for the most part would mask atomic indeterminacies. In biology, the notion of efficient causality still worked half-plausibly, however overtaken by a prolife vitality that began to grope for ends, though not yet ends-in-view. But in culture, when we conceive of social forces mechanistically, organically, functionally, or even cybernetically, we cannot explain the social turnings at their most critical junctures, where a people come to options on the cutting edge of a historical self-consciousness. Social scientists can only propose alternative scenarios, and then join in debate with their fellow citizens in the choice for this reform against that one.

Sociologists and anthropologists, it has sometimes been remarked, are good at studying their "inferiors"—army privates and Arunta chiefs—but falter at studying their equals. Philosophers and theologians study their equals, or, better still, their superiors, the masters and saints. This insight does not mean that social scientists are less wise than humanists, but only that the causal and tacit forms of explanation work better with less self-conscious social phenomena. Many social scientists abandon all judgments about "inferiors" and "equals," but at least literate persons in advanced societies are equally as self-consciously reflective about their behavior as are the social scientists who study them. By contrast, the less literate are less critical, perhaps not less complex in their societies, but less deliberate about them, and therefore more amenable to causal scientific study. But after latent meanings have been brought up into view, brought up for review, after functional covariances have been described, causality becomes increasingly useless.

Finding oneself among equals and superiors, what one wants is sympathetic and critical interpretation of the intersubjective culture and the valuing process of which one is socially a part. The more educated and self-conscious a modern society becomes, the more deliberately it may seek to reform itself. Thus, at length, Charles Taylor, a philosopher of social science, concludes, "A study of the science of man is inseparable from an examination of the options between which men must choose."[26] Before such an agenda, sociologists and anthropologists have no expertise that qualifies them over any disciplined humanistic thinker.

It is, then, partly true that the social scientist needs a different logic of inquiry from that used in natural science. In the broad sense, all critical thinking, whether in science or in theology, has a single logic of inquiry. The logical form of social science does not here differ from that of natural science or even of the humanities in the forming and testing of theory as this is referred to experience. On the other hand, in the narrow sense, the many sciences all use their differing techniques. Bacteriologists look through microscopes; meteorologists send up weather balloons; social scientists distribute survey questionnaires or consult public welfare records. Their methods of inquiry may be entirely different. But in the middle scale of methodology, social science does have a different character of phenomena under study, one that cannot with entire success be handled under the causal and empirical model, because it requires another kind of access, a meaning model. A differing logic of explanation is in order.

This is not merely in the context of discovery where social scientists use empathy to come up with theories about belief and behavior. It extends to the context of justification, where no explanation can be complete with only formal, repeated covariances between stretches of behavior. Here explanation always further demands penetrating *insight*, beyond mere *sight* of the behavior, appreciating the experienced form of life and its logical and volitional coupling with manifest behavior.  There is always an interpretation that goes with the facts. This interpretation, in the first instance, is intersubjective among the social actors, although not merely in the private agents but inlaid into their institutions and traditions. But just for this reason the interpretation becomes a *fact* objective to the social scientist, although a funny kind of fact, since this "fact" is in turn a symbol system itself representing the facts.

Society as a Text to Be Interpreted

Those who work in the intersubjective arenas of social science have not borrowed their paradigms from physics or biology. They have looked to quite another extreme,  to theology: *a society is a text to be interpreted.*[27] The logic needed is nearer to that of a theologian exegeting a biblical text than it is to a physicist watching billiard balls. The science of *hermeneutics* originated in wrestling with explicating the meaning in a canonical text.[28] This involves recovering meaning as contained in scriptures now often archaic and problematic, relics from an exotic culture, as well as how to reform such meaning for present society. This turn toward theology and comparative religion is no accident, however, because the religious texts of any society are repositories, *par excellence,* of meaning. A canon is the epitome of rule-oriented behavior. Even where there is no text, we study some text analogue, some artifacts, or behavior in which these humans have expressed or betrayed their thoughts about meaningfulness.

Social science at this level is discovering that the form of logic required is not  mechanical, not organic, not functional, not even cybernetic. It is dramatic and historical. "Sociology," says Martindale, "seeks abstract knowledge not biased by the normative patterns of local time and place. . . . It aspires to be a knowledge of the general rather than the unique."[29] But if so, then it cannot tell us what we most want to know: the meanings of our case history. In society, as in chess, the meaning of the piece is its role in the game; and more: the value of the person is his role in a particular story.

The logic that we eventually want, whatever truth may lie in these simpler laws, is hermeneutic, one that honors the significance of a distinctive society. Here science looks to religion for instruction; it becomes more religious in its methodology. Even the natural sciences cannot remain outside of this development, for our experience of nature is not really disjoined from our experience in culture. Rather, all our experiences of nature, including those we have in natural science, pure and applied, are mediated by these interpretive symbol-systems that are so cultural, so religious, historical, and philosophical at their cores.

In this broad sense, especially at its religious and philosophical foundations, a culture is a propositional set. Impressive as the "propositional set" that we found inlaid in the genes was, there is a more marvelous element here. There emerges from the beginning of the human race, increasing over time, deliberative power in governing what we have made. On the present cutting edge, we are called upon to remake ourselves. But a culture is not the sort of propositional set that goes into formal logic or game theory. It is that kind of propositional set which forms a *story.* What we need is not theory, not model, not law, not part in organic whole, not push-pull forces in an interactive network, not cognitive processing, but rather paradigm in the *dramatic narrative sense.* Parts of the story have scientifically to be described, but the narrative plot of the story has interpretively to be explained, its meaning discerned. Indeed, we find that we are actors in this story, charged with a prescriptive task at the end of our descriptive labors. We are left with an *is* that requires the choice of an *ought.* We are called to decide between good and evil. But that choice does not belong to any science.

3. RELIGION AS A SOCIAL PROJECTION

"From the end of the nineteenth century to the middle of the twentieth century," writes Daniel Bell, a leading sociologist, "almost every sociological thinker—I exempt Max Scheler and a few others—expected religion to disappear by the onset of the twenty-first century."[30] Like early Christian predictions of the end of the world, this predicted end of religion has, to say the least, been greatly delayed, even debunked. But it will be instructive for understanding the ongoing relations of religion and social science to see by what theories sociologists made such predictions, and where they went wrong. We have already met such a theory in Comte, echoed in Martindale, but we may turn to Émile Durkheim for more detail of argument, with also some revealing hedging.

The major thesis here is that religion cannot be inconsequential, for it is too widespread, indeed, universal in all premodern societies. It must therefore serve some crucial function in society. On the other hand, religion cannot be what it evidently claims, for the diverse religious accounts are incredible and conflicting, superstitious and nonscientific. Religion is something else than it pretends to be. So the task of social science is to discover why people behave like that, why they believe all those diverse and incredible things. The religious actors are being traversed by causal processes that they do not understand. They are in the dark about what is really happening, which is that religion is a social projection required for the functional maintenance of their society. But sociology illuminates all this, brings to light the latent function of religion, whereupon the puzzle of religion stands solved and religion dissolved. Once these hidden origins and workings are exposed, we understand religion's power and universality on that account alone, and have no further need for continuing with the manifest account that takes the question of religious truth seriously. What counts is the social force; everything else is epiphenomenal and contingent. This discovery applies alike to the Australian Arunta and the European Protestants. Sociology here is antitheological, not merely nontheological.

But this leaves the sociologist in a quandary, since Durkheim simultaneously finds religious doctrines to be untrue and indispensable. Religion of some sort provides the social cement that glues a society together, and yet all past religion is really phony. No religion can survive the social science analysis of its function. For once believers are let in on what they were unwittingly doing, they lose their earlier motivations for doing these religious things. They are not really dancing to make it rain, or serving God. They are rather generating social cohesions. The study of religion is its obituary. Although the analysis is different, Durkheim's treatment is like Freud's in that a discovery of the conception of religion results in its abortion. One cannot pretend the old religion, since it so greatly overshot the truth. But what will now generate these social cohesions formerly provided by religion? That question much troubled Durkheim, and it has yet to receive a satisfactory answer. Durkheim hoped that some science of morality or education, some social substitute "religion," could do the job.

This sort of social science thus predicts that religion as it has so far been known will disappear, and works toward causing this to happen by explaining it away, but offers nothing very definite to fill the vacuum. Perhaps the best we can make of this is to see the whole program as a test, an experiment. Sociology here proposes the explaining away of religion, and we have now to see whether religion, explained so, does in fact disappear, and whether anything can substitute for it. If not, then perhaps more is going on than such sociologists have yet discovered.

The Collective Function of Religion

Notice, says Durkheim, the congregational nature of religion. There is no religion without a church, in the broad sense. Religion is "an eminently collective thing."[31] It reproduces itself by cult and ritual, by sacrament, festival, service, ceremony, and dance. It always contains an ethics, which enshrines the community's value commitments, sanctioning these with sacred authority. Israel is rescued from Egypt, charged to keep the Passover and distinctive ceremonial laws, and to worship the God who has chosen them. The Ten Commandments are urged in the name of Yahweh, thundering from Sinai, and provide the social constitution for the fledgling Israeli people, protecting the family, the person, life, children, property, truth, and ordering a Sabbath to seal the whole contract. As revealed by the etymology of the word, religion is essentially a *binder* (Latin *ligo,* to bind).

Judaism splendidly illustrates how religion is the bearer of culture, but that was likewise true of the Greek, Roman, and Egyptian gods, true of Shinto, Buddhism, Hinduism, or even (to notice subcultures) Puritans or Mormons. Religion is everywhere a means of social control; there are rites of birth, death, puberty, marriage, and counsel in time of crisis. One's station, place, and duties are defined by caste, or a work ethic, or a duty to God and country, or the divine rights of kings. Religion has a strongly conservative nature; it allies itself with the prevailing economic and social interests. Prayers and oaths add solemnity in government.

The establishment of religion (as in the Anglican Church) joins church and state, but even where these are separated there runs a civil religion that nondenominationally uses God to sanction our form of life.[32] "God" is a power greater than the individual, on which each depends, imposing authority and commanding respect, invigorating, comforting, mysterious, sacred. But society is all these things too, a power that overleaps generations, penetrating and organizing itself within us, calling us into being and sustaining us. Without this, we fall into Durkheim's *anomie,* a sense of confusion and meaninglessness.

Religion does indeed have a foundation in fact, concludes Durkheim, but "it does not follow that the reality which is its foundation conforms objectively to the idea which believers have of it." To the contrary, none of these face-value explanations are adequate, supposing as they do their mystic realms. The truth is screened off from believers. Religion is based on a reality that is scientific and secular. "This reality, which mythologies have represented under so many different forms, but which is the universal and eternal objective cause of these sensations *sui generis* out of which religious experience is made, is society. . . . It is obviously necessary that the religious life be the eminent form, and, as it were, the concentrated expression of the whole collective life. . . . The idea of society is the soul of religion. . . . Religion is the product of social causes." "God" is nothing but society conceived symbolically, an "effervescence" of the collective life. "Religious force is only the sentiment inspired by the group in its members, but projected outside the consciousness that experiences them, and objectified."[33]

In sum, writes Durkheim, "I see in the Divinity only society transfigured and symbolically expressed."[34] Religion is reduced to sociology.

Durkheim has undoubtedly gotten into the roots of much religious life, but it does not follow either that he has hold of the main root or that he has uprooted the classical religions in this dimension that he has identified. Christians who share the Lord's Supper know they are binding themselves together socially; that is part of the meaning of the ritual. They believe that this is God's will to be done, and that they gain God's help to do so. Forgiveness is an excellent social cement. Christians ought to be the salt of the earth. Social cohesiveness is part of the good news. But it is not all of it.

Even where the practices are tacit, not explicit, the social unconscious is neither religion's sole origin nor its chief engine.

Nonsocial Dimensions in Religious Life

Religion is generated, for instance, in confrontation with nature, as when enjoying a sunset, or fearing a storm, or awestruck under the night sky. Often there is here a sense of the numinous, the supernatural, the ultimate. Religion is generated in the crises of personal life, when confronting birth and death, or disease and tragedy, or guilt and despair, or in mystic trance or quiet meditation. Durkheim overlooks the dimension that Alfred North Whitehead recognized in his aphorism that "Religion is what the individual does with his own solitariness."[35] There is an important sense in which one's "sacred space" is not the reflection of society at all, but a place for distance from society, space for privacy and aloneness, for contacting realities that outlast society, for freedom from social forces, for submission to God alone. These other generators of religion—nature, supernature, the personal life—do indeed, as we have maintained, have a social coefficient. What we experience is theory-laden. Our encounter with nature is mediated by culture.

But Durkheim has simplistically reduced many roots to one, and prejudged all the claims. There remain these nonsocial and presocial factors, present to us through cultural lenses, but encountered as something other than society, and before which we become religious. We do so not simply to establish solidarity, not secretly to keep social actors civil. To recall the arguments of this book, when modern humans face the unsettling revelations of physics about matter-energy, or assess the evolutionary sweep of life and its cybernetic upslope, or the emergent psyche enjoying its mental and moral life—in all the terrain that we have surveyed in earlier chapters—religious issues arose repeatedly outside of social contexts. We have really no reason to think that in this study we have been dupes of society, unwittingly trying to generate coping behavior, nothing more. What is mostly proved by such ongoing religious

inquiry is not that religion is everywhere a social projection but that sociologists must lay aside their Durkheimian spectacles and see what is going on in other ways.

Universal and Prophetic Dimensions in Religion

Nor is religion, when it does touch culture, so culture-bound as Durkheim would have us think.[36] In the history of religion, tribal religions have tended to die out and have been replaced by universal religions. Christianity, Islam, and Buddhism have proved transcultural. They bear cultures, to be sure, but they transcend the cultures they bear. Even a religion that remains rather ethnic, as does Judaism, may transcend diverse cultural situations, as has Jewry. The leading elements in it may be shared internationally, as has the ethical monotheism that emerged from Judaism. Indeed, Judaism rather carefully distinguished between what in it is ceremonially Jewish, preserving the covenant people, and what it offers as the divine message for the Gentiles, ethical monotheism. But how could it do this if it were only another social reflection? Whence the universalism, and the capacity to articulate this from the ethnic elements?

The classical religions spread from culture to culture. They may have arisen within one historical culture, but they interacted with others to evangelize them, if also to be modified by them. Here religion, at its best, is not only the carrier of worth in a culture but also culture's critic. It tells a people's story; it incorporates individuals into that story, and so constitutes a community. But it lifts them to a higher drama. It contains an *ought* with its *is;* it speaks to and does not merely reflect a culture. It may speak out for freedom, for love, for justice, against the evils of racial prejudice, or the industrial revolution, or the Vietnam War.

Here we must recognize the initiatory power of individuals—Moses, Jesus, Muhammad, Buddha, Martin Luther, Martin Luther King, Jr.—charismatics who reshape a culture quite as much as they are themselves shaped by it. Nonconformists are called to and call others to new directions. The anomic may not be the suicides, dropouts, and criminals; they may become the seers, the innovators, the newborn and reborn creators of what is to come. They are not necessarily the misfits but are often the mutants. They do not internalize their social norms; they externalize their newfound norms upon society through religious or political enterprise. Sometimes, the most effective critics of society are those who say, "We must obey God rather than men."[37] They hardly seem to be projecting their society, nothing more. They seem rather to have a vantage point that has priority over society. Such prophetic vocation is as powerful as the priestly tradition, and ultimately even more in need of explanation. Here one finds the emergents, the inspired ideas, that prevail over previous ones, because they are in some sense more functional, and also perhaps truer and more noble than the creeds and class consciousness they supplant.

These founders generate new enclaves, at first small, novel, and not merely reflecting any preexisting society. They see in vision some revised ideal, and become centers of propaganda, rather like a novel set of biological mutations become propagating centers. These seeds expand, grow, conquer, as did the primitive church, the Buddhist Sangha, or the émigré Muslims. Analogously, the reforming genius in science does not merely reproduce prevailing paradigms, but criticizes and may even overthrow them. Such reformations are often dysfunctional in the short range, causing revolutions and social upheavals, though they become socially functional in the long range. Religion generates much of the ambivalence individuals may have to the rules and values of their society, and we cannot, without begging the question, redefine "their society" until that subclass coincides with all and only those individuals who have identical values.

Durkheim knows such facts, but is determined to subordinate all such evidence to his paradigm. The critical and the charismatic, the prophetic and the idealistic powers of religion are all hidden in that "effervescence" that, he thinks, must be a social bubbling over from which God arises. But one can argue away large ranges of religious creativity by ascribing this to a mysterious effervescence, and if one is convinced enough that this must be a merely social effervescence, then one cannot seriously consider whether sometimes this enthusiasm might be a revelation of God.

What we have in him is not scientific fact, not even any longer a scientific theory, but what we earlier called a "blik."[38] Here is a monocausal presumption fixated and run riot, with all evidence cut to fit it, discounting religion of every kind as nothing but society writ large, adjusting its account of religion and the scope and character of society to make each coincide with the other. We have but another instance, now in sociology, of what we have previously met in Darwinian biology, in Freudian psychoanalysis, and in Skinnerean behaviorism: a hard-core paradigm become exempt from falsification.

It may well be that the forces of culture and of human society determine in some general way that every people will have some religious or religious-like beliefs, although they leave open which life-orienting views they must have. Humans must have some religious beliefs, as a library must have some books, but what books and beliefs is an open question. The social forces make necessary some "sacred canopy,"[39] but they are sufficient for no particular form of religion.

But such beliefs are not all and only determined by the social system. They are not latent causes that maintain a society by deceiving and coercing the social actors, making them its religious puppets. These belief systems are posits that explain why there is a world, what life is, what it ought to be like. They offer meaning to society because society needs an identity in the world, some grounding of human existence. Society needs such meaning not for its blind self-maintenance but because humans have logical, epistemological, and metaphysical drives. They have religious urges; they want explanations. They want to be responsible to whatever powers are responsible for their being here.

All this must be social, in that humans do these things with inescapable togetherness. These meanings traditionally come from outside any one individual; they enter, uplift, and constrain her. Any sacred rituals and practices are likely to be such as contribute, in their color, tradition, and detail, to group loyalty and experience. But that there are group values here does not mean that there is no universal intent. What believers are doing is not merely, not by disguise, establishing their togetherness. They are figuring out who they are and where they are, and so religious symbols are not merely symbols *of society;* they are symbols *of the whole world* and the powers that move us, including, but not limited to, the human place and role here. Religions are indeed constitutional of society, but in far more than a hidden social sense. By them the believer is constituted in his world.

If such opinions serve society, they also transcend it in their scope and concern. We do not assess these beliefs at all if we notice that they are socially functional in one or more societies, concluding that because all are functional, more or less, none can be true, more or less. The finding of a socially functional religion is only the first, and very trivial, bit of evidence that a religion might be true. If there is a God (G) of the kind Judeo-Christian monotheists allege, who calls a chosen people as a witness to the world, who calls into history a church community, who desires love of God and neighbor, then one would certainly expect, resulting from any revelation, a religion that is functional (F) in society, a religion with both priestly and prophetic capacities. If G, then F. Where F is observed, G is corroborated. That is but weak, backtracking evidence; still, it is exactly the logic of all science. It may well be that other theories, such as the natural and social selection of religiously stabilized societies, will also yield this observation. One would not expect dysfunctional religions long to persist.

After that, we shall have to choose between theories on further grounds, or perhaps to find elements of truth in both. Past functionality, everything remains by way of examining the cognitive adequacy of a faith. Can it stand its own ground in the competitive marketplace of religions? We still have to look critically at truth claims in any religion, in the light of our best knowledge of the world and of human character.

Primitive religion, like primitive science, is full of error, but even here we may want to acknowledge that such beliefs had other referents than society, and that, though mistaken, the Arunta genuinely held the beliefs they did, not merely hiddenly for latent functions that escaped them. Their theories will have now to be discarded, but often one can appreciate worthy elements in them, given the time, place, and limited areas of human experience before which they were formed. We do not now condemn our forefathers for believing in phlogiston, the flat earth, or the six-day creation, nor the Arunta for their totemism. All were functional theories in their time.

With an enlarging observational and experiential base, with better theories that interpret such events, these myths fall. So also in part will what we now believe true in our science and religion. Comparing the Arunta in Australia for ten thousand years and the Americans over a few centuries, it is even possible to ask whether some elements in the primitive world views, in which we no longer believe, were not more socially functional, more environmentally sound, than the secular, scientific, technological mind-set that has given us the emptiness of the modern era, the threat of nuclear war, an ecological crisis—and social science!

The Religious Problem in Secular Society

When the sociologist leaves Australia and comes to assess the religion of her contemporaries, she finds that Durkheim's explanation, having assessed the believers and not their beliefs, leaves both believer and sociologist with the religious problem still on their hands and completely untouched. "We must," says Durkheim, "discover the rational substitutes for those religious notions that for a long time have served as the vehicle for the most essential moral ideas."[40] In religion's claims about the world, life, society, in its advocacy of realms of the sacred, in all cognitive truths, religion will "progressively retire" before the science that, as Comte predicted, will replace it. Nevertheless "there is something eternal in religion which is destined to survive all the particular symbols in which religious thought has successively enveloped itself." That something is these "collective sentiments" that keep us "closely united to one another."[41]

So Durkheim, having found God an illusion, wants (like the Marxists) to reconstitute Society with a capital S, to make it sacred enough to take up the room of the God and gods confused with it. Yet he finds nothing in social science that even begins to have the kind of affective power he needs, nor can he find it elsewhere on the modern horizon. He can only cry out, alas, that "the old gods are growing old or already dead, and others are not yet born." What Durkheim wants to do is to naturalize society, and yet, since society now worships nothing else, make it worthy of worship itself. But his aspiration enormously outstrips his achievement, his promises fail because his premises are misguided, and in the end he can only lament "this state of incertitude and confused agitation" in which he finds himself and the modern world.[42]

His is a theoretical impossibility from which he desperately needs exemption. His science finds that society needs religion, at the same time that it keeps on dissolving religion by oversocializing it. He hopes for some binder that persists through this dissolution, but cannot locate this postreligious savior for Society.

Believers have an ideal society—the Kingdom of God, the fellowship of believers, the Church, Israel, the covenant people—which is worthy of veneration and can be effectively life-orienting. But that society, with its ideals, its union of the dead, the living, and the yet unborn, its shaping of the *is* by the *ought-to-be,* is no scientific object. It is both distinct from and yet partially present in any real society; it is on the frontier. It does not operate causally in any naturalistic sense or only tacitly on believers. The laying on of hands, for instance, is the handing on of meanings, and all that is happening here may indeed escape the believer; but it can as well be more, not less, than he thinks. These things operate by compulsions of worth, this perhaps hiddenly, but not, as Durkheim thinks, deceptively. Here society operates, if you will, spiritually, not scientifically.

No science can predict, control, explain, or critique this society exhaustively. Among these "collective forces" that exalt and inspire us, there may indeed be present secret winds of the divine Spirit, a spirit also inspiring the nature that, like culture, surrounds and bears us up. This ideal society is arriving, coming but yet to come, one that "in God" we might make and that makes us, as we are drawn to and by its coming. But is this sacred, social field "only a projection"? Or could it not with just as plausible a diagnosis be considered rather a supersocial vision, a call to and by something beyond society, a call to responsible love and freedom in which

God may be present? This religious power would then yield society as its derivative, and not the other way around. At the depths and in the emergents, these social currents are not so much forces available for causal, scientific analysis as they are spiritual forces.

Passing beyond the social functions of religion, if sociologists ever step into the substantive debate about the soundness of a belief system, seeking to choose, say, between Christian theism, Advaita Vedanta, and Zen Buddhism, they will find that there is in social science no particular skill, given its limited phenomena of study, that qualifies them to make judgments about the cosmic claims of religious belief. Social scientists are, from their discipline alone, undertrained for this task (as is every other scientist). The mysteries of birth and death, of being and nonbeing, good and evil, the worth of redemptive suffering, the tragedy of meaningless suffering—all the core problems remain quite inexorably there after anything tacit and sociologically functional in religion has been exposed and reviewed. The questioner's puzzlement does not vanish.

This is why religion has not disappeared with the arrival of social science. It is why, further, social science can give us only an incomplete answer to the question of, and questions in, religion. In this sense and in the end, the question of religious origins is not a scientific question.

4. VALUES IN SOCIAL SCIENCE

A Value-Free Science?

Few watchwords are more often pledged in social science than that it ought to be value-free; indeed, one suspects that this is repeated so frequently because it is never very clearly said just what this means, or very convincingly said whether and how this can be so. Already announced by Martindale, this theme deserves closer inspection, especially since we have seen how some social science predicts and contributes to the dissolution of religion, while at the same time discovering the need for values that it cannot supply. We may be dealing with another myth, the objectivity of the social sciences, or with a confused half-truth on which we must get clearer if we are to understand the relations between social science and religion.

"Sociology," Robert Bierstedt insists, opening a textbook, "is a categorical, not a normative discipline; that is, it confines itself to statements about what is, not what should or ought to be. As a science, sociology is necessarily silent about questions of value; it cannot decide the directions in which society ought to go, and it makes no recommendations on matters of social policy. This is not to say that sociological knowledge is useless for purposes of social and political judgment, but only that sociology cannot itself deal with problems of good and evil, right and wrong, better or worse, or any others that concern human values. . . . It is this canon that distinguishes sociology, as a science, from social and political philosophy and from ethics and religion."[43]

Bierstedt's judgment itself has a good deal of *ought* mixed with his *is.* He is proposing what sociology should be, not what it is invariably is. But even as an ideal the question of a sociology scrubbed free of values is more complex than first appears. By factoring out various dimensions here we can expand an account of the presence and absence of values in social science. We can allow at the start the presumption that social science itself is good, as well as foundational assumptions about the worth of human rationality and free inquiry. Past these background values, what is the actual or necessary connection of social science with ideologies?

Value Presumptions and Consequences in Social Theory

We judge a theory in part by the light and shadows that are cast over it from outside social science, and this is not a mere psychological matter, but a logical one. There are, of course, questions of value that arise within a discipline. Is this a reliable questionnaire? A poorly grounded hypothesis? Here what counts as better or worse sociology is a proper value judgment for sociologists to make, although we need not reserve these judgments exclusively to sociologists, since colleagues in other sciences and in the humanities may from their perspectives be able to spot flaws invisible to the sociologist.

But even judgments within sociology about the adequacy of this research instrument or that microtheory are in some degree keyed to what one thinks society is in its fullness, and hence keyed to judgments about what is getting emphasized or slighted in the adequacy of a description. We are here dealing not merely with values as they enter into decisions about what problems are chosen when we start a study, but with their continuing in the solutions selected over the course of the study.

Here the scientist's beliefs about the nature of society must sooner or later be value-laden, for she herself is social, and these overbeliefs feed back into one's estimate of the acceptability of this or that piece of sociological work. There may be (as psychoanalysts have taught us) tacit motives behind the reasons, undermining the explicit reasons given for this or that conclusion, a failing that can beset sociologists as well as theologians. But which theories one ought to promote within the discipline, especially which big theories, will be and ought to be overcast by one's larger experience in the world, not just by one's sociological experience. Ulterior interests of this kind have every right to be brought into a unified theory. By such wide-ranging ulterior beliefs, we can best lay off all the blinkers and frames that restrict us into seeing too little because we are looking with too limited a paradigm.

In judging, for instance, the adequacy of Durkheim's account, one may find that there is too much in religious phenomena that he overlooks and underweights, and here good theology or religious studies may be helpful in detecting bad sociology. How one thinks in substantive intellectual problems bridges over disciplinary lines, and we judge the smaller, sociological units out of a gestalt over the whole in which values are present.

Elements somewhat like this are present in natural science, but we reach compounded and emergent difficulties in social science. Now the objects of study, persons in society, are themselves the holders of value sets that figure among the phenomena studied. The scientists who pursue such studies themselves inescapably hold value sets as members of a society, not less than their subjects. Thus, scientists and their subjects alike are in the valuational stew. Something is likely cooking between values in the scientists and values in their subjects, no matter whether the two live in the same, in alternative, or in competing societies. Something is at stake in the controversy.

In a limited but important sense, sociology needs to be free from governmental, institutional, commercial, or religious control, as such forces, which are themselves social forces, may impinge on how sociologists do their work. There ought to be no values that are driven from without into sociology. Society ought not to prejudge which conclusions sociologists can draw, and it ought not to hamstring the directions in which sociology is permitted to go. Yet, on the other hand, sociology is licensed and financed from outside itself, and it is unlikely that social institutions will support sociology where they are not in sympathy with its emphasis and focus.

Indeed, it is unreasonable to expect them to do so. If one is working within a science, one ought to seek truth in the likeliest directions. If one is outside yet supporting science, one ought to support those phases of science that give the most promise, as best one can judge these. Proportionately as one believes that a science is seriously wrong-headed, wasteful, or gone up a blind alley, one ought to withdraw support. This is true with any science, but sociology differs from chemistry again in that what sociologists are studying is their supporting matrix, directly or indirectly, and this self-referential problem is not similarly present in the natural sciences. There are stretches within sociology where sociologists can deal descriptively with social practices, prescribing nothing, and here they can keep their supporters, their science, and their subjects all at arm's length. But proportionately as sociology focuses on the core institutions in which the scientists, their subjects, and their supporters are enmeshed, there will be inevitable tension. In sociology it is a lot harder to keep these political or economic, religious or antireligious, liberal or conservative forces at a distance and to conduct a cool, cognitive discussion, since what is under discussion are these heated forces that drive life. At times it can be irresponsible not to consider their strength in the light of what one has personally experienced of their power.

Decisions about facts and theory ought to be made in an unconstrained atmosphere, but they will not be made (as noted earlier) by sociologists liberated from their own value sets. Supporters have every right to be aware of this, and quite as much right to use their own value sets in deciding whether to trust what they receive from sociology. Social science has a right to responsible autonomy, but this will remain in tension with the right and responsibility of its supporters to invest in what they deem to be sober and relevant science.

Values, though different from descriptions, may hang on them. Thus, when descriptions change as a result of studies in sociology, values may have to be revised in the light of the revised description of processes. If Max Weber has shown a closer link than was earlier known between the Protestant ethic and the rise of capitalism,[44] one may need to revise one's valuation of Protestantism, capitalism, or Catholicism and socialism as a result. If indeed certain creeds generate prejudice, then our evaluation of such creeds, or of what we are willing to call prejudice, may alter. To whatever extent latent functions in religion are convincingly argued for, depending somewhat on their character, one may have to revise one's estimate of the manifest functions in religion.

By altering our descriptions of the socializing processes, sociology forces revisions of value. In this sense and with such topics, no one can say in advance whether sociological inquiry is going to be value-neutral or not. That depends on what is found out. This can be also true in natural science. The study that yielded evolutionary theory was not value-free in its consequences. It rather forced dramatic revisions of meaningfulness and estimates of good and evil in nature. But the problem is compounded in sociology, owing again to its self-referential nature.

The physical and biological sciences study natural events—matter, life, ecosystems—and their discoveries may enter into the content of a religious belief by revising the description of the objective event. But the human sciences have religion itself among their objects. They take as content our value systems and behavior (as chemistry, physics, even biology do not—sociobiology excepted), and in this sense the threat they pose from discoveries that they may make about the workings of value systems is correspondingly higher. One needs more empathy, or more cool, or more courage, or more passion just to describe what is so, because of its import for what ought to be. There is an inevitable overflow problem, because social scientists, more than other scientists, are called to stand in an affective relation with what they study.

We judge theories in part by the light and shadows they cast, and where less or more unfolds from theories we are slower or faster to accept them, depending in part on our values. We want better evidence or are sold on less evidence. This is clearly so in applied sociology. We do or do not want to put the theory to work yet, depending on the risked good or harm we think may be involved. This will be true when a panel decides that the responsibility for violence in a community, evidenced in riots or juvenile delinquency, lies in unresponsive institutions and not with the lawless individuals who commit violence. How much evidence do we need to decide whether school busing has failed to improve the education of underprivileged black children? Or we may worry over the finding that pornographic materials do not contribute to sexual deviancy in adults (as some social scientists claim), noticing how this claim weakens before more equivocal data about the effect on youth. Has enough research been done on single-parent adoption to recommend this equally with two-parent adoption?

But this will not be only in applied matters. Our readiness to accept the pure science is wired up to the impact we think the theory has; we choose more cautiously if the costs of mistakes are high. This is so in all science, seen in our readiness or reluctance to accept theoretical explanations about genetic effects involving radiobiological hazards. A different measure of statistical insignificance is used in medical research from that in nonmedical biology. But the distance between our value sets and our evaluation of theory narrows in social science.

This was true in our earlier considering of Durkheim's theory that religion is only a social projection. It is true, for example, in appraising Kingsley Davis's theory that social classes are inevitable and functional, especially in advanced societies, rather than being inegalitarian, immoral social flaws. A consequence of this theory is that there neither can nor ought to be a future classless society.[45] It will be true in theories of alienation, or of human irrationality or determinism. What we are ready to accept in sociology depends in part on how hard this hits our value sets. This is not only in the application of a theory, but in any revision of our hopes for society that may be tied into a revised social theory. No social (or other) theory ought to be suppressed because it is dangerous, but the standards of proof ought to be higher for dangerous doctrines. One rejects (as opposed to suppresses) such theories sooner and longer and insists on more evidence, especially in applied science but to some degree in pure science. Moreover, deciding whether a novel theory is dangerous can sometimes be as difficult as deciding whether it is true.

Naturalistic Tendencies in Social Science

Certain sorts of presumptions, especially those of the monocausal, "nothing but" kind, do contain covert judgments, which are often prejudgments, about the essential nature of the social practice studied, and hence about the value of it. The most widespread of these is the conviction that religious phenomena must be completely amenable to scientific study, that religion can be explained without residue in terms of the categories of social science, terms outside of those of religion itself. J. Milton Yinger, a sociologist of religion, says, "Science inevitably takes a naturalistic view of religion."[46] But, depending a little on how open one's concept of the naturalistic is, and how much one expects by way of completeness in any naturalistic explanation, such a presumption can be inevitably to misview it.

The spectacles with which we look determine what we can and cannot see. An advance decision to force-fit religion into causal scientific categories is not neutral but sinister to religion. The methods used in a science, like the presuppositions held, can rig the substantive conclusions found. The strategy used makes it inevitable that religion will be some sort of by-product or have some derivative status. We have already met such presumptions in Durkheim and in Comte, who judged all the traditional faiths to be ripe for replacement.

We recommend to social science the endeavor to explain the phenomena as far as it can. Sometimes in its account of religious events there will be conflict between what participants think is going on and what sociologists and anthropologists say is going on, between what are called *emic* and *etic* explanations. What *neither* side can *presume,* however, is that the other side cannot be right. Neither side ought so to adjust its mode of inquiry that it prevents or short-circuits seriously evaluating the cognitive claims of the other. That the other side is wrong may be a conclusion, but it cannot soundly be a conclusion precontained in the assumptions and methods brought to the study, for that is to operate under an incorrigible blik, a closed ideology that cannot hear contrary truth.

Hence, for instance, owing to his "scientific" methods, Durkheim is no more prepared seriously to entertain the religious claims of the Arunta or of European Protestants than the Arunta or European Protestants, owing to their "religious" convictions, are prepared seriously to entertain the claims of Durkheim. Ian Hamnett, a sociologist, says of this sort of sociology, "The sociology of religion is, in fact, the sociology of error."[47] It assumes beforehand that believers' accounts are spurious and looks for the "correct" account to causal forces of which believers are unaware. In so doing, such sociologists have taken sides on the truth question before they start. These strong-armed sociologies and anthropologies cannot be value-free. They cannibalize religion. The sociology of religion here makes a religion of sociology, that is, sociology becomes the principal orienting explanatory and valuational scheme, the ultimate criterion of explanation and value.

But there may be more modest, weak sociologies and anthropologies that are willing to acknowledge their incompleteness in explanatory power, and, having pressed explanations as far as they can on their own terms, are then willing to proceed to the further, extrasociological, extra-anthropological debates about the adequacy of belief systems, as these may now be revised in the light of whatever findings of social science remain convincing in the full debate. Within these limitations, as nearly as may be after recognizing our earlier constraints, social science ought to be value-free. The valuational debate, when it occurs, draws social science outside its boundaries, with the social scientist as a participant with other inquirers, one among equals in a larger philosophical, humanistic, and religious forum. Here we may anticipate that most primitive belief systems will fail in the modern debate. But what will happen to the classical and universal religions, or to newly originated ones, cannot be prejudged out of sociology or anthropology.

Therapeutic Tendencies in Social Science

We must also recognize a therapeutic impulse in social science, one which is typically absent from the natural sciences, although present in psychology. We study celestial systems, oceanic currents, or ecosystems with no urge to fix them, even when we employ facts discovered there to our own advantage in navigation or agriculture. But when sociologists study societies, especially their own, it is difficult, even inhumane, not to form some salutary opinion, beyond what *is,* about what *ought* to be. Lee Benson, in a presidential address to social science historians, advocates "changing social science to change the world." Social scientists ought to develop "credible empirical theories about human behavior highly useful to human beings struggling to create a better world."[48] Even anthropologists are likely to be drawn to the primitives they study and to wish their betterment, or perhaps to judge them well enough off and to seek their protection from the inroads of modernity.

A textbook in social psychology begins with a mixed description of and prescription for social science: "A revolutionary idea is affecting man's thinking about social problems. He is hopefully pursuing the notion that the sciences of man will soon be effective in preventing the hateful turning of man upon man in racial prejudice, the bitter conflicts among people of different economic and political ideologies, and the awful obscenity of war. And the urgency of this hope grows as he contemplates the even greater evils now made possible by the science of things. The social scientists encourage this hope as they engage in a bewildering variety of activities. They can be observed studying the effectiveness of polar exploration teams and PTA conference committees. They sit with management and labor at the bargaining table to discover the sources of misunderstanding and conflict. They climb stairs, push doorbells, and interview the citizenry as they seek to measure public opinion. They experiment on groups in the laboratory. Nor do they limit themselves to observing and measuring and experimenting. Many of them engage in action designed to achieve a tolerant world, a peaceful world, a better world."[49] However worthwhile all this may be, social scientists have here become saviors, bringing in a secular version of what theologians have called the Kingdom of God.

Mere descriptions of the relationship between religion and prejudice or of the social origins of religion seem pointless unless one is prepared to act on these findings. Hence, though descriptions may stand independently of prescriptions, some of them are nevertheless likely to lean toward some duty, or to suggest options between which the members of a society might choose. In this light we understand Karl Marx's impatience with any science or philosophy that undertakes merely to understand the world and not to change it. There is a bit of preaching, of meliorism, even of utopianism lurking near most theories in social science. We tend to get led into, or on from, our "problems," not by pure theory but by social needs. A problem in sociology is likely to be a problem in society, and a solution in the former likely to suggest one in the latter.

Good and Evil in the Cultural Community

Yet, rather paradoxically, social science of itself, after describing these "problems," cannot tell us what we must do. The social scientist, as social scientist, describes parameters within which we must work, perhaps describes what cannot be abandoned, or what must be, sets forth alternative routes, but the obstinate question of good and evil is never decided merely from within sociology and anthropology. There is always an exodus out of social science into philosophy, ethics, and religion. What we find out in social science is going to need supercharging if we are to get final directional guidance and motivational energy out of it. Up front, on the cutting  edges of moral, political, and spiritual decisions, we can use input and advice from the social sciences, but we always need more to clinch a decision. Confronted with choices between meaningful stretches of life, we need an act of faith.

Humans are in part self-defining animals. The more they learn, the more powers of revision they have. Even the finding in social science, so far as it is true, that systems have in the past had priority over the individual, need not be true in the future, not at least in the same degree. For once we awaken to the power of social currents over us, we immediately gain some measure of possible autonomy over them, increased capacities for reasoning about them, and we thereby face the value-laden choice whether we ought to continue in, or rebel against, the tradition that binds us. We cannot learn from the past unless we can receive tradition and be persuaded by it. But we cannot critique it without some autonomy of judgment.

Such autonomy of mind is equally essential in religion and in science, but in religion we add now an *ought* to an *is*. In this sense, after the most convincing accounts of how religious traditions operate over the individual to produce solidarity within society, after all this is consciously recognized, social science increases rather than decreases our religious options. But at the same time it is powerless to help us choose between whatever religious options do transcend the analysis of the social sciences. What is happening here is that the set of sociological premises about how religion operates is being revised in the very drawing of conclusions from these premises. Perhaps religious life cannot afterward be the same, but neither can the sociological study of it.

One encounters within the social sciences a good many schools of thought, which may go under the names of critical, reflexive, radical, or polemical sociology, where sociologists do propose revisions of society on the basis of their scientific findings.[50] It is well here to remember that such sociologists have crossed over, sometimes without showing any passport, into social philosophy. They are no longer within the realm of science, either hard or soft, but, additionally to whatever science is used, are operating out of commitments to a world view, partisans of what social scientists call (in both good and bad senses) an ideology.

The notion that social science is value-free is thus a myth, but, unlike our earlier two myths (about an inevitable social evolution away from religion and about causal explanations entirely supplanting intentional ones), this myth is a useful one. Value-free social science, to change the metaphor, is an ideal type, one that in real form is inevitably constrained by concrete entanglements. But just this accentuation helps us to separate science and nonscience, to find leaks in the science, and to measure the spillover between what *is* and what *ought* to be. Value-free social science is a heuristic criterion, a regulative maxim that helps us to locate the semipermeable boundary between science and values, and to recognize mixtures of the two. It thus operates, as do ideal laws, helpfully but only partly to describe the real, since the real is more contingent, more contaminated, more confused, quite as subject to deformation by other processes as to formation by the ideal type.

Like so many other analytical constructs that we overlay on the world and society, this one too is a somewhat artificial mapping device, yet it does help us to map the terrain between science and religion. This ideal thus guides as well as describes. By it we clarify how our decisions even on crucial scientific theories overreach mere science and require religious, ethical, and metaphysical resources. By it we learn that whenever we endorse meanings, embracing creeds that sustain society and make sense of nature, we are nearer to religion than to science. By it we learn that when a particular sociology on its own resources has judged a religion false, we have been told as much about the sociology involved as we have about the religion judged. We have learned that such a sociology itself has in fact religious content, and this of a sort competitive with the religion that it has cast out.

The human sciences have been all too anxious to borrow from the natural sciences both their paradigms and their associated methods. Meanwhile, the two main branches of the human sciences have in fact unique paradigms of their own. Psychology has the category of *mind*, the psychic unity. Social science has the paradigm of the *cultural community*. They both sell themselves short by groping for mechanistic, organic, causal, equilibrating, functional, or even cybernetic models, since mind and community are richer processes than can be fully illuminated by lesser models. In their effort to make social science legitimate, they may in fact make it illegitimate by making it oversimple and inadequate. Indeed, it is sometimes more productive to read the higher model back down into the lower sciences, as for instance when ecologists borrow from the social sciences the notion of a community, or when biochemists borrow for DNA and RNA the notion of a language, first known to us from the domain of mind.

There is, further, no particular warrant for thinking that either human minds or their communities are the exhaustive paradigms with which we must interpret culture, set as this is in natural history. Nothing learned in social science forbids asking whether there is something transcendent to the human world, something sacred exerting its pull over society, and out of which the human and natural worlds may be derived. What if there are some challenges and conflicts that a society can solve only religiously? We can even agree with Durkheim that religion and society vary together, but put between these two a quite different account of their correspondence. The fact is not so much that religion calls for an explanation outside itself in society. It is rather society that calls for an explanation outside itself in those realities to which religion points. Society, not just religion, is the effect.

En route to this more comprehensive explanation, we recall how human culture is the end product of the incubating natural matrix. Transcend this though society may, nature, the generating field, nowhere disappears but ever accompanies social development, and forms an overarching assignment demanding interpretation. Add then to *culture* that other, equally great effect that needs explanation, *nature,* and let the two merge in the full historical drama. We are ready to examine the credibility of more robust explanations regarding the nature and society in which we seek meaningfully to participate. We know well enough that life is in society, but we know too that life is only partly fulfilled by society and needs pointing not only beyond society, but even beyond nature and history. That pointing beyond is, ultimately, to Presence.

NOTES

1. Edward B. Tylor, *Primitive Culture,* 4th ed., vol. 1 (London: John Murray, 1903), p. 1.
2. Clifford Geertz, *The Interpretation of Cultures* (New York: Basic Books, 1973), p. 89.
3. Auguste Comte, *System of Positive Philosophy,* vol. 4, (New York: Burt Franklin, 1967; London, 1877), pp. 599, 557, 560.
4. Don Martindale, *The Nature and Types of Sociological Theory,* 2nd ed. (Boston: Houghton Mifflin, 1981), p. 3; cf. pp. 7, 14.
5. Auguste Comte, *The Positive Philosophy of Auguste Comte,* vol. 2, trans. Harriet Martineau (London: George Bell and Sons, 1896), p. 307.

6. Martindale, *Nature and Types,* pp. 6, 37, 131.
7. 1 Cor. 12.
8. Herbert Spencer, *Principles of Sociology,* vol. 1, part 2 (New York: D. Appleton and Co., 1898), p. 592.
9. Vilfredo Pareto, *The Mind and Society,* vol. 4 (New York: Harcourt, Brace, 1935), p. 1442.
10. Robert M. MacIver, *Social Causation* (New York: Ginn and Company, 1942), p. 173.
11. Neil J. Smelser, *Social Change in the Industrial Revolution* (Chicago: University of Chicago Press, 1959), p. 10.
12. Talcott Parsons and Edward A. Shils, in Parsons and Shils, eds., *Toward a General Theory of Action* (Cambridge, Mass.: Harvard University Press, 1967), pp. 107, 226.
13. Walter Buckley, *Sociology and Modern Systems Theory* (Englewood Cliffs, N.J.: Prentice-Hall, 1967).
14. Peter Berger, "From Secularity to World Religions," in James M. Wall, ed. *Theologians in Transition* (New York: Crossroad Publishing Co., 1981), pp. 21–28, citation on p. 28.
15. Jon Elster, *Logic and Society: Contradictions and Possible Worlds* (New York: John Wiley and Sons, 1978), pp. 158, 162, 183. Italics in the original.
16. May Brodbeck, "General Introduction," in Brodbeck, ed., *Readings in the Philosophy of the Social Sciences* (New York: Macmillan, 1968), p. 1.
17. Alan Donagan, "Can Philosophers Learn from Historians?", in Howard E. Kiefer and Milton K. Munitz, eds., *Mind, Science, and History* (Albany: State University of New York Press, 1970), p. 234–250, citation on p. 243.
18. Émile Durkheim, "Review of Antonio Labriola's *Essays on the Materialist Concept of History*" in *Revue philosophique de la France et de l'étranger* 44 (1897): 645–51, citation on p. 648. Translated from the French.
19. Émile Durkheim, *The Rules of Sociological Method* (1895) (Glencoe, Ill.: Free Press, 1938), p. 110.
20. Talcott Parsons, *Societies: Evolutionary and Comparative Perspectives* (Englewood Cliffs, N.J.: Prentice-Hall, 1966), pp. 113, 28–29.
21. Talcott Parsons, *Action Theory and the Human Condition* (New York: Free Press, 1978), p. 382.
22. Ibid., pp. 264–99.
23. Rodney Stark and Charles Y. Glock, "Prejudice and the Churches," in Glock, ed., *Religion in Sociological Perspective* (Belmont, Calif.: Wadsworth, 1973), pp. 88–101; Bernard Spilka and James F. Reynolds, "Religion and Prejudice: A Factor-Analytic Study," *Review of Religious Research* 6 (1965): 163–68; Russell Allen and Bernard Spilka, "Committed and Consensual Religion: A Specification of Religion-Prejudice Relationships," *Journal for the Scientific Study of Religion* 6 (1967): 191–206; Frederick L. Whitam, "Subdimensions of Religiosity and Race Prejudice," *Review of Religious Research* 3 (1962): 166–74. See the summary of research in this area in C. Daniel Batson and W. Larry Ventis, *The Religious Experience* (New York: Oxford University Press, 1982), pp. 256–81.
24. The ASA Code of Ethics states, "All research should avoid causing personal harm to subjects used in research." In most interpretations, deceiving subjects about the true purpose of research harms them. "Toward a Code of Ethics for Sociologists," *American Sociologist* 3 (1968): 316–318, citation on p. 318.
25. Robert K. Merton, *Social Theory and Social Structure,* enlarged ed. (New York: Free Press, 1968), p. 105.

26. Charles Taylor, "Interpretation and the Sciences of Man," *Review of Metaphysics* 25 (1971): 3–51, citation on p. 48.
27. Paul Ricoeur, "The Model of the Text: Meaningful Action Considered as a Text," *Social Research* 38 (1971): 529–562; Clifford Geertz, " 'From the Native's Point of View': On the Nature of Anthropological Understanding," in Keith H. Basso and Henry A. Selby, eds., *Meaning in Anthropology* (Albuquerque: University of New Mexico Press, 1976), 221–37.
28. See Hans-Georg Gadamer, *Truth and Method* (New York: Crossroad, 1975), and earlier in Wilhelm Dilthey. See H. A. Hodges, *The Philosophy of Wilhelm Dilthey* (London: Routledge and Kegan Paul, 1952).
29. Martindale, p. 16. See also George A. Lundberg, *Foundations of Sociology* (New York: Macmillan, 1939), p. 140.
30. Daniel Bell, "The Return of the Sacred: The Argument about the Future of Religion," *Zygon* 13 (1978): 187–208, citation on p. 188.
31. Émile Durkheim, *The Elementary Forms of the Religious Life* (1912) (Glencoe, Ill.: Free Press, 1947), p. 47. For a more recent statement of Durkheim's position, see G. E. Swanson, *The Birth of the Gods* (Ann Arbor: University of Michigan, 1960).
32. For a study of American society following "the Durkheimian notion that every group has a religious dimension," see Robert N. Bellah, "Civil Religion in America," *Daedalus* 96 (1967): 1–21, citation on p. 19. Unlike Durkheim, however, Bellah does not presume that a sociological analysis of religion as a social binder dissolves the question of its truth. "I would argue that the civil religion at its best is a genuine apprehension of universal and transcendent religious reality as seen in or, one could almost say, as revealed through the experience of the American people" (p. 12).
33. Durkheim, *Elementary Forms,* pp. 417–24, 229.
34. Émile Durkheim, "The Determination of Moral Facts," *Sociology and Philosophy* (Glencoe, Ill.: Free Press, 1953), p. 52.
35. Alfred North Whitehead, *Religion in the Making* (New York: New American Library, 1960) p. 16.
36. See, for instance, H. Richard Niebuhr, *Christ and Culture* (New York: Harper and Row, 1951). Niebuhr, a theologian, traces historically five different relationships between Christianity and culture, including Christ as a transformer of culture. All the relations he finds between religion and society are more sophisticated than any portrayed in Durkheim.
37. Acts 5:29.
38. Chapter 1, p. 10.
39. To borrow a term from Peter L. Berger, *The Sacred Canopy* (Garden City, N.Y.: Doubleday, 1967).
40. Émile Durkheim, *Moral Education* (New York: Free Press, 1969), p. 9.
41. Durkheim, *Elementary Forms,* p. 427. Durkheim can give religion some left-handed praise while with his right hand he is exorcising it. "Religion will survive the attacks of which it is the object. As long as there are men who live together there will be some common faith between them" ("Les Études de Science Sociale," in *Revue philosophique de la France et de l'étranger* 22 [1886]: 61–80, citation on p. 69). But this "common faith" will henceforth be radically unlike all previous religion: it will be secular and scientific. "Our principal objective is to extend scientific rationalism to human behavior. It can be shown that the behavior of the past, when analyzed, can be reduced to relationships of cause and effect. These relationships can then be transformed, by an equally logical operation, into rules of action for the future" (*Rules of Sociological Method* [1895], Chicago: University of Chicago Press, 1938, pp. xxxix, xl).

42. *Elementary Forms,* p. 427.
43. Robert Bierstedt, *The Social Order,* 3rd ed. (New York: McGraw-Hill, 1970), p. 11.
44. Max Weber, *The Protestant Ethic and the Spirit of Capitalism* (New York: Charles Scribner's Sons, 1930).
45. Kingsley Davis, *Human Society* (New York: Macmillan, 1949), pp. 366–78.
46. J. Milton Yinger, *The Scientific Study of Religion* (New York: Macmillan, 1970), p. 531.
47. Ian Hamnett, "Sociology of Religion and Sociology of Error," *Religion* 3 (1973): 1–12, citation on p. 1.
48. Lee Benson, "Changing Social Science to Change the World: A Discussion Paper," *Social Science History* 2 (1978): 427–441, citation on p. 427.
49. David Krech et al., *Individual in Society: A Textbook of Social Psychology* (New York: McGraw-Hill, 1962), p. 1.
50. For example, Alvin W. Gouldner, *The Coming Crisis of Western Sociology* (New York: Avon Books, 1970); Alfred McClung Lee, *Sociology for Whom?* (New York: Oxford University Press, 1978); Paul Connerton, ed., *Critical Sociology* (New York: Penguin Books, 1976).

Chapter 6

Nature and History

It is sometimes quipped that with God the problem is whether God exists, given what we suppose such a being might be like. But with nature the problem is reversed. We know that it exists, but what do we suppose nature is like? Is it even spatiotemporal, ultimately, much less good or evil? We think we know what God would mean, if God existed. (Ah, but do we?) We know that nature exists, but what does it finally mean? What nature is like and whether God exists are overlapping questions, and, driving toward the latter, we will start by surveying nature after science. Then we ask whether this calls for a world view that is nontheistic, such as may be found in hard or soft naturalisms and in some Eastern religions. Past this, if God can be supposed to exist compatibly with the findings of natural and social science, what can God be like? That will lead, in this chapter and the sequel, from nature through history toward God.

1. NATURE AFTER SCIENCE

Immensity, Diversity, Unity in the Natural World

Nature is *immense* at orders of magnitude that stagger the imagination. A light ray would take twenty billion years to cross the known universe, and we do not know whether the universe is infinite in extent. The farthest edges we see are receding at two-thirds the speed of light. This spatial immensity is matched by a temporal one. The time frame that we share at the molar level, though perhaps only relative and not absolute, is twenty billion years old. We do not know whether anything preceded the big bang, or when our epoch will end. As much time may lie ahead as behind. Within this spatiotemporal immensity, there is a further largess both of bits and potential. A vast number of infinitestimal particles (wave clouds) have yet more terrific orders of magnitude of possible combinations into diverse things. There are 10^{22} stars in 10^{9} galaxies, with a wholly unknown number of planetary systems. We do not know whether the rarefied simplicity of interstellar and intergalactic space shields our access to multiple worlds as rich as our own. In our juvenile phases of space exploration, the heavens have proved anything but dull. In quantum mechanics, the number of possible superposition states is presumed infinite, only a few of which actualize. The few that do materialize result in the construction of the exuberant array of kinds we next notice.

Nature is *diverse.* We have hints of this in astronomical nature. Each natural locus is a separate inertial reference frame with quite limited possibilities of communication with any other, owing to the finite speed of light. There is no universal now, only intersecting historical lines. In the physical world we find quarks and gluons, black holes and ringed planets. But the principal evidence for nature's plurality lies immediately at hand, where the rarefied simplicity of astronomical nature is replaced by the dazzling display of earthen fauna and flora. There are between 5 and 10 million extant species, with only about 1.6 million identified. Nature must love these myriad and bizarre creatures; she has made so many of them—from aardvarks to loons, from microbes to men. The present species constitute less than 1 percent of those that once were but are now extinct. But the trail of extinctions has steadily increased the diversity and complexity of life, which is now richer than it has ever before been (apart, alas, from human-caused extinctions).

Nature takes a plural when we restrict our scope from usage with a capital N, *Nature,* covering everything that exists, to refer in a more limited way to the *natures* of things, to essential structures and behaviors of particular systems and natural kinds. Different things have different essential natures, in distinction from which there can be abnormal or adventitious properties. Any property is natural that is tributary to a local integrity; and unnatural properties will be unstable and collapse, be stillborn, edited out by evolution, or remain vestigial, monstrous, futile. Nature incessantly experiments to explore what further properties can be natural, and selects some successes from many trials.

She gives every biological individual a share of this experimental heritage, a funding from a gene pool, and yet also some novelty and an organic self that typically can separate itself from billions of others. The more complex these individual items become, the more they rise to a further sort of unbounded richness, one of particular kinds in their exploits. At the middle ranges of existence as we know it, humans bring a new kind of infinitely extended openness. Owing to their astronomical number of brain synapses, humans may set up an inexhaustible variety of circuits, and think new thoughts forever. With all this development, we can see no end of affairs and goings on.

A *unity* nevertheless counterpoints this diversity. Astronomy, geology, and plant and animal taxonomy have multiplied the natural kinds, but microphysics, chemistry, and biochemistry have found a relatively few tectonic materials. Landscapes are built from rocks and soil, where three dozen minerals dominate, and all the hundreds of minerals known fit into thirty-two crystal classes. All living forms employ two dozen amino acids, structural motifs mostly of carbon, hydrogen, oxygen, and nitrogen. The less than a hundred natural elements are formed from protons, electrons, neutrons. These are elementary in some first-level sense; they are common and stable, although of course by hitting them hard enough we can annihilate them and re-create new assortments of particles. The neutron and the proton may be versions of the same particle, and all the "particles," diverse though they are, are wave functions crinkling a great plasma-ether. There seem to be only four sorts of forces, two of which are evident only at nuclear levels and two of which (electromagnetism and gravity) account for all the bindings and motions we know at macroscopic levels.

If one traces the now so greatly expanded universe back in time everything shrinks to one originating point, a big bang, from which the emanating parts are strikingly fine-tuned to cooperate in the development that has resulted, and in few other universes, perhaps no other imaginable universe, could there have been matter, life, and mind. Such fine-tuning extends to parts of the universe so remote that, at least since the originating naked singularity, they have not (by any laws known to us) been in contact with each other.

If one asks whether nature after science is simpler or more complex, the question is best answered: Yes! On the one hand, the multiple sciences spell out detail in a nature everywhere enriched, never impoverished. On the other hand, unifying relationships run through the theatrical whole, often by singular laws that embrace multifarious phenomena. Consider the two greatest paradigms of twentieth-century science: relativity and natural selection. $E = mc^2$. The fittest survive. Are these formulas simple or complex? Einstein concluded, "Our experience . . . justifies us in believing that nature is the realisation of the simplest conceivable mathematical ideas."[1] Edwin Taylor and John Wheeler, theoretical physicists, say, "The principles of special relativity are remarkably simple. They are very much simpler than the axioms of Euclid or the principles of operating an automobile."[2] The atomic table is easily mastered by college freshmen, and by its means thousands of compounds and reactions are explained. Likewise in the citric acid cycle, with DNA replication, in phylogenetic relationships, with the equations of mechanics and thermodynamics, or with electromagnetic radiation across a spectrum of wavelengths, we find simplicity beneath complexity. There is a richness of results yielded by simple laws.

This simplicity has proved a bit deceptive. The number of elementary particles has multiplied so as to leave physics in confusion, although many of these particles are not found in the ordinary matter of Earth, stars, and space, but are only made up in high-powered exotic manipulations. Even here physicists are trying to show how these in turn are composites of half a dozen or fewer kinds of quarks. Proteins are formed from amino acids, and form further into enzymes, which may be keyed with receptor sites using the rarer elements, which makes the active difference biologically. We hardly dare to generalize from biology and Earth's only known case of life to laws elsewhere. Plural sorts of life may be possible.

Still, the basic themes and fundamentals are there, undergirding the variations. All life here is kindred, and consubstantial with the physicochemical Earth. We do generalize from physics and chemistry to the solar system and beyond, believing that physicochemical laws apply uniformly. Gravity and electromagnetism everywhere obtain at appropriate structural levels. Matter is everywhere electronic. Six basic magnitudes—the mass of the proton, that of the electron, the charge unit, the velocity of light, Newton's gravitational constant, and the quantum of action, together with atomic number and weight, which are built on these—are fundamental to all else that happens. Nature is "simple—but subtle."[3] There is sophisticated simplicity!

Some have thought that nature is irretrievably pluralistic. There is no *Nature-as-a-whole,* because Nature is not a whole but only scattered natures locally conditioned from a chaotic foam without ultimate form. But science wants unity in nature, quite as much as does religion. It loves to make links over far distances and diverse levels, more and more, and to find how natural things in the plural point back to Nature in the singular, how diverse properties arise from one grounding matrix. It wants kinships. The *nature* of a particular thing is the *Nature* expressed in that thing. No theory of Nature can yet predict these particular natures, but we do hope retrospectively to illuminate them when they are found. The pieces first seem separate and disarranged, but some of them form a jigsaw puzzle. One wants complexity, of course, for this is enriching, and yet one does not want it without the ordered patterns that also make for simplicity. Suitably integrated, they both supply meaningfulness.

Energetic, Formational, Informational Nature

Nature is everywhere energetic, sometimes as pure energy, sometimes incarnated into material forms. Nature's projects are thrust through incessant change. Astronomical nature may be inert on the lunar landscape, but the overwhelming mass of it is constructed into nuclear power plants, stars, yielding both radiant energy and —what presages so much of interest to come—all the heavier elements on which everything else is built. The immensities of space-time and the incredible energies out there provide production sites for making all the ingredients used in the subsequent story. The heavens are the starry firmament indeed, foundational to everything else. We have astral bodies after all! The celestial dust swirls into local worlds. One planet, under stellar irradiation, and given those anthropic constants about which we have learned in astronomy, is the scene of adventures up a ladder. This earthen energetic process becomes genetic and then nervous. Pure energy cultures itself into a sentient history by some blending of evolution and openness for which we lack any satisfactory model. *Being* is always a matter of *becoming,* and structure is never separable from process. The outcome is, at the present moment, persons thrown and drawn toward the doing of science and religion.

The story occurs first at *formational* and later at *informational* levels. There is a proliferating of both structure and sophistication. An ontological ascent results during a historical flow, roughly diagrammed in Figure 6.1. Stages, genera, phase transitions, emergents may be passed through, but all the strata are conserved in the maturing overlays, continuing to coexist both in the life pyramid and as more or less independent phenomena.

Some say that the physical phenomena that come early, into which complicated things may even yet be dissolved, must be more real than the later biological, psychological, or social arrivals. Yet on closer analysis there is equally as much reality in the compositions, whose configurations are maintained and developed over time. The higher levels take over to modulate, in part, the lower levels. Identity in a life form outlasts the parts and energies that life absorbs and excretes, and nothing in the substrate proves permanent or unchanging. The individual uses and passes through particles, but the individual is itself a particle over which the wave carries on, a long-lasting wave just because it is transient over individual parts. This is true from the lowest levels up to the transcending of the individual persons by society and history. Death occurs, but of individuals far more than lines, and the death (extinction) of lines is steadily supplanted by new, novel, and often more complex lines. Death seems a necessity if the genetic process is to mature. And not only in

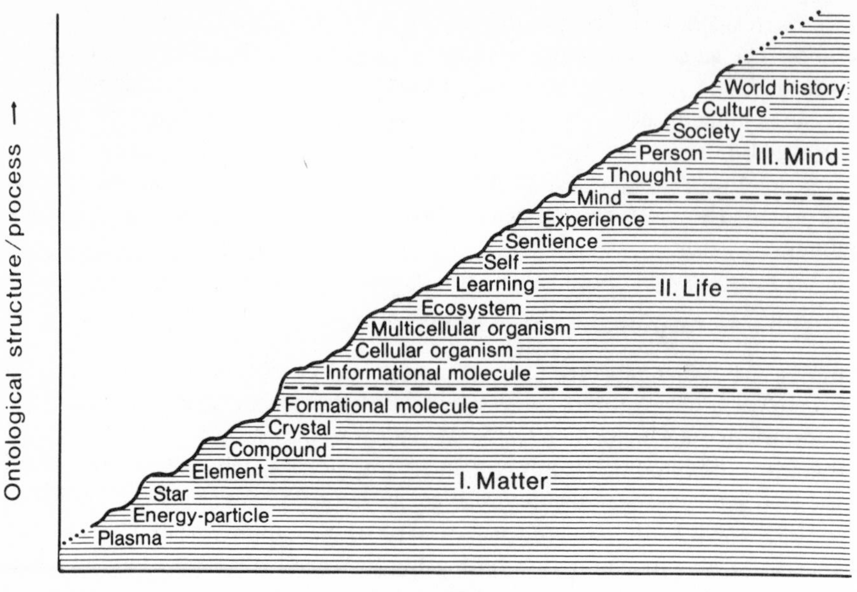

Figure 6.1 The Earth story

biology but also in culture does the death of individuals and of cultures pave the way for advancement and cultural proliferation.

It is true that living and thinking things can be collapsed into material rubble, but then it was just this rubble that erected itself into life and mind in the first place. Life and mind are nothing beyond their activities, but then neither are atoms and molecules. Nowhere is there any inert stuff distinct from goings on. Seemingly ancient matter is not so much material substances once made in stellar furnaces as it is energetic configurations into which nature was once put. These have since continued with many electronic inputs and outputs, and with those modulating overlays that make up more and more story. Little suggests an absolute beginning or end; nothing about the levels of being-becoming implies that we are at the floor or ceiling, although in several directions we face limits to our knowing.

We may naïvely say that cognitive processes arrive only with human awareness. But the life adventures have from their inception been steered by information, imprinted early in the genes and later in neural structures. The classical debate about mechanical versus teleological processes in nature has been bypassed when we realize how the crucial line is crossed when abiotic formations get transformed into loci of information. The *factors* come to include *actors* that exploit their environment. There appears, and remains, a kind of objective knowledge without subjective awareness, passed along in provision without prevision. While we have theories about the incremental assembly of parts or about replication, even about the molecular symbolisms used, we have little or no theory about how the formational levels are pumped up to informational levels, and none whatsoever about how objectivity passes over to subjectivity.

The Newtonian themes of "matter" and its "motion" have to be revised to "mass-energy" (= "particle–wave clouds") and "information." In their multiple phases these are the phenomena that need to be accounted for. Matter has been reduced to energy, and energy is basic to everything. But information seems quite as real. Norbert Wiener, the founder of cybernetics, warns: "Information is information, not matter or energy."[4] In short-scope runs or at simple levels, mass and energy are predictable with high probability. But information, though likewise often predictably conserved, has also a future that is regularly more than the past. Matter and energy obey the first and second laws of thermodynamics, with the conservation of mass, energy, and increase of entropy. But the informational trend, though subject to destruction, erasing, and even tragedy, has proved a fire surprisingly resistant to quenching, a wave that does not subside but ever swells. This tracks through constantly changing environments by deploying themes conserved by their very elaboration into ever more novel ideas. Mass-energy and information are the scientific equivalents of the Platonic-Aristotelian matter and its forms.

But neither are mass-energy and information the end of the drama. A third phenomenon emerges still more dramatically: "consciousness," an "inner-formation" past information, when subjective awareness appears overlaid upon and superintending objective matter, not less real for its late-coming appearance. Out of these principles-processes is spun the ultimate phenomenon, neither mathematics, nor matter, nor energy, nor information, nor even consciousness, but *story*—the storied achievements within our historical universe.

Causes and Meanings in World Events

Science has disenchanted nature and found there a *secular integrity* couched in the concept of *spontaneous causal order.* Primitive peoples (being also great tellers of stories) gave animistic explanations of eclipses, lightning, storms, droughts, diseases, birth, death. They feared mana, worked magic, and peopled their worlds with gods, demons, spirits. But now, having gained scientific explanations for these phenomena, persons believe (at most!) in one God or Ultimate that orders a lawful, constant, natural economy. The correct residue in Comte's belief that we pass from religion to science is that we replace animistic with naturalistic explanations. Only he confused this with whether religion can override this shift and coexist with science, with whether it can show how nature, though disenchanted, is still sacred.

Despite this secular autonomy, there remains a *haunting incompleteness* to scientific explanations. True, there are always causes behind effects, but these nevertheless have surprising effects that the causes never seem completely to specify. The stream steadily rises above its source. Some engine carries life steadily upslope. The effects over time, whether probable or improbable, initiate events the likes of which have not been seen before: life, learning, joy, suffering, resolvedness. Like a sleepwalker, nature does valuable things unawares. If autonomous, she is also pregnant and bears her children. We know something of how this occurs in the organic mother, where the embryo formation is steered by parental genetic information, but we can scarcely envision how this waking up occurs when no such information preexists. We have hit no rock bottom in nature, and in the developing evolution we keep getting more out of less. The best causal accounts do not tell us how there is something rather than nothing, why simple things became complicated things, why uninformed nature seeks information. The creativity in nature is poorly authorized by science, which gives us rather piecemeal descriptions of these events, and little or no imperative for the commanding drama. Even if we did glimpse how this happens, could any *causal* account ever supply the *meanings* in these events? That all this is God's creation may be doubted, but that it is a *creative* universe can hardly be denied.

Yet the world does become a theater of *meanings,* beyond a sphere of *causal* sequences. As soon as life appears, values and disvalues appear in such things as prolife information and environmental resources. Perhaps indeed value was already present in the incubating processes that prefaced life. World events can be better or worse, as viewed by organisms with a good-of-their-own. These valuational themes are conserved and deployed over time, eventually to flower in human life. Nature is an intellectual, aesthetic, moral, and religious theater. Nature contains its imperatives about meanings, beyond a mere chronicle of events. Humans have opportunity and endeavor urged upon themselves, superimposed upon the causal sequences. Nature thrusts at least one kind of its creatures into responsibility, and this call to morality refuses to dissolve before the reductions of science. Science has hardly given us enough imperative even for the causal assembly that has taken place, but science is much harder pressed to give any account at all of the moral and religious imperatives that are thrust upon us, concluding the chapters of causal assembly.

Nature is hostile on some readings, indifferent on others, and hospitable on still others. In the upshot, fine-tuned for development from the start, if also mysteriously random in later crises, she does let life come in precarious vulnerability, tolerating life on this one planet at least, and that not grudgingly but exuberantly. She assigns to humans, who are the richest expression of this life, an intellectual and moral role. In John Keats's poetic phrase, not dislodged by science, the planetary place is a "vale of soul-making."[5] If this is nature's chief beauty, then any elements of ordeal in nature have to be assessed against their contribution to soul making. Everywhere, the ladder of ascent is climbed by problem solving. Nature produces a thesis, then an antithesis to it, and the result is a higher synthesis. Upward on the scale, joy and success come in counterpoint to agony and failure. The outcome is the dramatic quest for meaningful life.

Intelligibility and Mystery in Nature

We continue to keep terms in yoked antithesis by judging nature to be at once *rational* and *mysterious.* Everything studied proves to be intelligible. Formed things have their passive intelligibility when probed by the human mind. Living things further have their active intelligibility. They are informed; they behave in ways that make sense, given their circumstances and endowments. They replicate this information by a proto-linguistic process. Life forms are manifest fits, cleverly adapted to their environments by a trial-and-error process that bears some, though not all, of the marks of human intelligibility, some marks even of the scientific method. Yet this physical and biological intelligibility never banishes a sense of mystery. J. B. S.

Haldane, a geneticist, after a lifetime of study of evolution, concluded that its principal marks were "beauty, "tragedy," and "inexhaustible queerness."[6] Fred Hoyle, an astronomer, exclaims, "No literary genius could have invented a story one-hundredth part as fantastic as the sober facts that have been unearthed by astronomical science."[7]

The world can be penetrated by thought, but we are left in awe both that the world is so built and that we are so built, that nature is comprehensible and that we, the evolutionary product of nature, are up to its comprehension—only in part, but nevertheless far beyond anything that we can claim any right to know, far beyond what our ancestors ever dreamed of knowing, far beyond what evolutionary theory predicts that we should know. Reflecting over this, Einstein commented, "The eternal mystery of the world is its comprehensibility. . . . The fact that it is comprehensible is a miracle."[8]

We climb one summit to see further peaks, a "mountain range" effect. The area of the island of knowledge grows only to enlarge the shoreline where it touches the unknown, a "coral reef" effect. Victor Weisskopf, a theoretical physicist, writes, "When we know more, we have more questions to ask. Our knowledge is an island in the infinite ocean of the unknown, and the larger this island grows, the more extended are its boundaries toward the unknown."[8] Our beams probe farther out, only to confront more dark sky. There is an explosion of knowledge, but it ignites an explosion of mystery. Science removes the small mysteries to replace them with bigger ones.

Even in the known, mystery remains. Where there is light, is the light any less puzzling? We cannot see the other side of a radically inexhaustible nature; we cannot see even what we do see rationally enough to evaporate the mystery. So the question persists whether this secular autonomy veils a sacred presence. So far from proscribing the sacred, good science does not profane its object. It may first manipulate it, but later it becomes captivated by nature, submits to this overarching omnipresence, and longs properly to appreciate this womb from which we come and which we really never leave.

Discontinuity and Continuity in the Human Place in Nature

The human place in nature is one of *discontinuity in continuity,* and here we resort again to dialectic and ambivalence. *Humans* are cognate with the *humus;* they arise from the mud and yet become spectacularly informed, a last-minute but dominant species. Humans do not escape their ecosystemic grounding, and there are animal roots for many human skills. But the display of personality is nowhere approached in the animal world. We come to command nature quite as much as to obey her. We ourselves combine both this rationality and this mystery as nowhere encountered before, dwarfed in size by the stars, yet astronomical in our own complexity, and, compared with animals, strangely unconstrained in this complexity.

Though genetically controlled in part, still the human being is born very unformed when compared with other mammals. Our species is nonspecific in almost a biological sense, not an exact natural kind, because our specific kind is added culturally, transferred from genetic to cultural control, from nature to nurture. By our prolonged infancy, nature requires our filling in by an education; she permits our self-fulfillment. The enormous potential in the hand and brain is left unspecified and perhaps unspecifiable. Any normal human genetic set can take expression in any culture, and this remains true even though some cultures in minor ways have coevolved with certain genetic sets.

The animal takes an interest in its own sector of the environment, for which it is specialized, and largely ignores all else. Humans continue but exceed these utilitarian interests. They can focus on any sector of their world or abstract to the character of the whole; they can know nature's diversity and simplicity, rationality and mystery. Humans are only part of the world in a biophysical sense (as science maintains), but they are the only part of the world that can try to comprehend the whole of it. In that lordly sense, they are divine (as religion maintains). Their unique status lies not in their metaphysical stuff but in their metaphysical possibilities—only they can "do metaphysics." But their mastery over the world is not a human feat as such, but the culmination of the long evolutionary struggle in which they stand as heirs. Their dominion takes place as they enjoy *standing on* an ecological pyramid. Humans are *standouts* in the world. They are in the world, but they are the only thing that in this global sense can see the world they are in, and therefore the only thing that can attempt to fit their own stories into the cosmic story.

The human sciences have not integrated well with the natural sciences, nor have they found convincing or singular paradigms of their own. Their successes have proved toeholds, capable of explaining a little, but only a little, of those human affairs we most want explained. Owing largely to stumbling over the human presence, we do not yet have in sight even a convergence of the sciences, much less a convergence of the religions. But if we turn solely to the natural sciences, they only glance toward this, the most remarkable phenomenon of all. Physics and astronomy tie matter and energy together, but factor out the person—at least pure physics and astronomy do, despite some puzzlements about the anthropic principle. Biochemistry and evolutionary biology trace the threads of life, but they do not touch inwardness; and, in their present state, they offer such a halting account of our arrival and destiny.

Humans are late and rare, but there seems no reason why late and rare events cannot be important events, nor why they cannot teach us about the ever-present potential and working of the whole. A physicist may examine ten thousand photographs of nuclear events and select one as revealing some previously unknown secret of nature. A biologist does not think that mammals are less important than microbes, although they come later and in far fewer numbers. There are fewer of everything as one goes up the later-coming levels, fewer atoms than electrons, fewer molecules than atoms, fewer cells than molecules, fewer organisms, fewer animals, fewer humans, fewer societies, fewer cultures. Indeed, in their actual historical narratives, the things that interest us are not merely rare; each is one of a kind.

Natural History

Nature after science is *historical* to the core, more historical after than before. Yet just this historical element transcends the skills of science. We will later return to this dimension in more detail, and need here for our initial orientation only an advance recognition of the historical element, especially of how unique events make history possible. These singular features are there in the fortuitous physical constants that make this particular universe possible, there in the interaction rates during the early seconds of the big bang, and in the continuing "fortunate" constants that make possible the subsequent "fortunes" of the universe and of our earthen world. The universe is *one,* and it happened *once,* certainly once like this. It is a particular, not an instance of a species or class. But likewise Earth spins out its particular story.

What is "first" (primary) in nature is history, that is, nature's historical sequences are its most evident and undeniable feature, and the feature most in need of an adequate explanation. What physics finds "first" (primitive energy, particles), essential and impressive though this is, does not really offer any satisfactory explanation of this primary historicity, especially not when we consider the later chapters of life and mind.

In its sophomoric days, science supposed that nature was ever the same, repetitive and nomothetic. But the fuller truth is that nature is never the same, always idiographic. To push the point to an extreme, even at quantum levels, no two electrons in the universe can be in the same state, that is, have the same values for all of a complete set of quantum numbers. At advanced levels of structure and story, there is but one Everglades, but one Grand Canyon, but one Jesus Christ, but one Abraham Lincoln, or one individual instantiating your proper name in the way you do. Each of the instances of the natural kinds is itself one of a kind.

There are similarities in the differences that make it half-true that nature is lawlike. Without these there would be no intelligibility, no science, no religion. Many of the differences are trivial in kind, as when one is inspecting aspen leaves. We excise these differences when we bring nature under laboratory control and general theory. But then too we tamper with what is in the field and get from science only leads, not a full specification of the historical phenomenon in its neighborhood. Such analyses will never quite integrate into a comprehensive synthesis.

There are ever new recombinations of old materials, with the manner of their recombination sometimes but not always predictable or understandable in retrospect. There are myriads of intersecting but unrelated causal lines, which are much too fortuitous and messy for science ever to write laws or theories about. There are, even more dramatically, genetic mutants and innovating geniuses, in which the possibility of novel history arises. These are troubling causally and they classify poorly, since they cannot be reduced to their precedents and they break out of the old classes. Yet if we miss these particular events we may miss much of the beauty, half the meaning, and most of the drama. There is no science, not even astrophysics, that does not find that nature has a startling history.

2. HARD NATURALISM

We plan to move critically through a half dozen philosophical and religious accounts of nature and history, beginning in naturalistic camps, then looking at nontheistic modes of accounting for nature and history, and finally (in Chapter 7) at monotheism as it revises or retains more or less supernaturalistic accounts. These position sketches are not exhaustive of possible theories, and the representatives we portray are skeletons that can be variously fleshed out. We do not survey all the philosophical and religious schools, any more than we have surveyed all the theories of the various sciences, but we inquire of leading candidates in terms of their party affiliations. We use large-scale maps, overlooking local detail. If this is oversimple before the diverse creeds, it will nevertheless give us ample material with which to argue over what creeds are adequate to explain nature and history. These are locations on a spectrum, or, to change the figure again, six families within which any specific type will need to be described in more taxonomic detail. In the midst of this review, the narrative features of history, just noticed, will become more urgent, as will the place of suffering.

An Economical and Scientific World View: Nature without Supernature

Of available theories, what we will call *hard naturalism* is pressed as the most economical, scientific, and hard-nosed account. Its advocates do not think of themselves as being religious at all; they rather reject all concepts of the sacred. Their view is nevertheless life-orienting and in some sense takes up the room of religion. Despite protests to the contrary, this account includes elements of belief and is a faith position, whatever its reasonableness. The articles that summarize this creed are:

1. Nature is all there is; nothing supernatural exists.
2. Elementary nature has always existed; nature is its own eternal necessary and sufficient cause.
3. All events have necessary and sufficient causes. No aspect of any happening is uncaused. Determinism is true. If some allowance must be made for microscopic indeterminism, this is negligible at macroscopic levels, where statistico-determinism is true.
4. Nature is fundamentally nonpersonal; human are epiphenomenal. Mind has evolved from matter but is nevertheless eccentric to it.
5. Nature is essentially value-neutral. Human values are real yet nothing more than human values, our own creations. They neither have nor need any explanation outside themselves by grounding in natural or sacred values.
6. The scientific method is the only route to truth; every other supposed method is myth and emotion.[10]

The first principle of naturalism, says Ernest Nagel, is "the existential and causal primacy of organized matter in the executive order of nature."[11] The scientific method, continues Herbert Feigl, is the only hope of truth; the phrase "scientific knowledge" is a tautology, for there really is no other kind.[12] Bertrand Russell puts it quite bluntly: "Whatever knowledge is attainable, must be attained by scientific methods; and what science cannot discover, mankind cannot know."[13]

This position, while austere, has a superficial attractiveness. It proposes to take with complete seriousness the facts of nature as scientifically described, to be purified by a reckless disregard for what humans may psychologically wish to believe. The human race, resolves Russell, "must struggle alone, with what of courage it can command, against the whole weight of a universe that cares nothing for its hopes and fears."[14] Hard naturalism proposes to be rationally coherent and simple, to promote and utilize science to its fullest, and to defend human values for their intrinsic worth, all without excess metaphysical baggage. It proposes a solid foundation in experience and observation. It is modern, realistic, liberal, secular, humane, positive. It has pruned away all dead beliefs and has rationalized every belief that remains.

Difficulties in Hard Naturalism

But these appearances of positivistic naturalism considerably soften under critical pressure. Let us begin by testing this set of claims for internal coherence, trying to fit the final claim against the others. Many features in the scientific method we have already commended as shared also by good religious inquiry. But in this creed, which is typically hostile to theology and metaphysics, something more restrictive is meant, consistent with the other creedal elements. The scientific method is naturalistic and causal, featuring observation, detachment, objectivity, repeatability, prediction, and control. But has this sort of scientific method in some way demonstrated that only nature exists, and nothing more? Perhaps such a method is competent only for working among naturalistic phenomena? We look through a viewfinder that is focused so as to miss all else. We might be stalled, spinning around on limited premises.

Has the scientific method verified that persons are epiphenomenal because science has found them to be rare and recent on the cosmic scale? Has the scientific method shown nature to be eternal, its own necessary and sufficient cause? If anything, science rather finds a nature always contingent on antecedents; the causal chains back up without cessation. Nowhere is any natural thing found to be its own necessary and sufficient cause. Yet this attribute is nevertheless posited for the whole. Has the scientific method established that human values are nothing but human values, with nature value-neutral? Many others believe that science is, or ought to be, value-free at the central decision points in life.

Indeed, to turn the final article on itself, can one establish by the scientific method that the scientific method is the only rational and legitimate method of obtaining truth? As a matter of fact, it is not clear that *the* scientific method is one hard and fixed method, or that we know exactly what it is. It too is revisable, changes over time, is "soft" rather than "hard," and we cannot always say in advance just what will count as being "scientific."

Each of the six claims exceeds what the methodological claim can deliver.

Persons as Epiphenomenal in Deterministic Nature

If we try to mesh the third claim with the fourth, the strong causal hypothesis overlies an epiphenomenal concession. The evolutionary sequence is entirely one of necessary and sufficient causes (statistico-deterministically, perhaps), and yet out of it comes an anomalous by-product, or epiphenomenon, a fluke. One's main theory, causal materialism, is surprised by an odd upshot, personality. First you say that it is causally predictable (in principle) and then you say that it is atypical and adventitious. This is a very peculiar explanation for a very peculiar result. Is this, or is this not, a *non sequitur,* a break in the interpretive scheme? Mind emerges from life, life from matter. In each case something arises that did not previously exist, more from less, caused by but eccentric in its matrix.

Mind appears from absent-minded nature, and when it comes is incidental to nature (epiphenomenal). Mind is hardly more than apparition, a freakish bit of inspirited matter. Is what goes before as a sufficient cause really a sufficient explanation? If so, why the surprise? The causal premise wants a tight explanation, while the epiphenomenal premise relaxes it. Life and mind may be rare, but they are the stubbornest facts of all, and nothing is gained by way of an explanation that makes anomalies (epiphenomena) out of them in a nature that, on the whole, is physically, causally, even mechanistically conceived. On the contrary, such explanations suppress the most storied events in nature as mere epiphenomena, freakish accidents that have no meaning there.

Neither does the determinist claim square well within the descriptions of science. Almost every science has progressively qualified determinism before a recalcitrant, open, evasive nature and human nature. Physics has discovered quantum uncertainty, interpreted by most as objective indeterminism in micronature, by others as a permanent limitation upon our capacity to decide the point. At the cosmic level, there is little evidence that the universe, or any part of it, is a closed system. We hardly know where the vast energies are coming from and whether there is still more incoming. Limited in our looking power, we cannot see the present world edges, or get information out of black holes. We cannot know in local systems what visiting influences may arrive. Even in closed systems physics puzzles over irreversible thermodynamics and climatic indeterminisms.

Biology posits mutations that are partially nonpredictable and random. Indeed, the evolutionary *advance* is said by some to be entirely a matter of accident and probability. In game theory we find that the best survival strategy sometimes preserves elements of randomizing, especially when one is up against a superior opponent, as are prey before their predators. Humanistic psychology boldly affirms self-creativity in the personal life, while S-R psychology fails to disconfirm it. Sociology cannot predict with any confidence the innovative societal movements. Everywhere science rests content with more or less open systems. Of course, determinism is still a useful working hypothesis in science, for it enables us to find all the causes we can, but by now we have failed to find causes often enough to cease to assume their omnipresence. Determinism as an absolute presupposition of, or conclusion from, science is a broken-backed hypothesis. Causal chains disappear, run out of our reach, grow rusty, wear thin, and snap.

We have yet to puzzle over the relation of history to science, but we can here anticipate how history describes a one-time narrative in which the future is never logically deducible in detail from the past, especially not at its crucial turnings. Hard naturalism is tempted to bring every new event under a covering law, à la natural science, and thus causally to explain it. But, as we presently see, such explanations strike others as being thin and trivial in human affairs. All naturalism is pushed to do well at handling much history.

In other quarters, from a logical point of view, philosophers continue to have trouble getting a careful hold on causation, especially on necessitation and tight specification of analogous cases. Further, they cannot reconcile how, if determinism is true, one can select a position rationally in the face of options. They stumble over the by now-familiar catch that advocates of causally rigorous systems have trouble including themselves within their systems. Determinism might have been true in some other world. But in this world the theory refutes itself when anyone *tries* to assert it, that is, to choose it rationally in the face of options. The assertion provides a counterexample to the theory. When minds change, as they do, is this mere causal efficacy come home to inquirers? Or is the new opinion more logically defensible than its rivals?

Here we have only to notice what is going on in order to refute it. When it is asked whether determinism is true, we face a self-destroying question. If the premise is true that every result is causally determined, arguments in its favor lose their effectiveness. Rational appeal before options in belief is precluded. The hard naturalist answer destroys the question, for if the answer is yes, we cannot entertain the question seriously. The determinist claim vetoes itself, and we do not know how, on its own terms, to decide that hard naturalism is true and the other views are false. All deterministic thinkers lock themselves into prisons of their own making.

As if this were not enough, Kurt Gödel's discovery of formally undecidable propositions in something as relatively simple as ordinary arithmetic counts against the possibility of our ever making watertight historical explanations or predictions.[15]

Some will plead that natural and human systems are very complex, and the sciences still in their infancy. The maturing sciences will become more precise; logicians will overcome these difficulties. But this moves determinism over to the realm of promises, not observations. The faith position becomes more obvious, but it is very hard (as religious critics know) to refute expectations and hopes. Meanwhile, let no one say that religion, with its claims to meaning, has to suffer enormous amounts of meaningless noise in the world that it seeks to interpret, but hard science, and hard naturalism based upon it, faces no such noise, finding causes at every turn. This is manifestly not so. To the unbiased ear, every science is full of noise in the background, if also of causality in the foreground.

Human Values and Natural History

By the fifth claim, values are a freakish thing, cosmically absent although terribly important locally. Hard naturalists have often been admirably humanitarian, willing to concentrate on this life (and not the next), on material improvements (and not spiritual consolations). But is it really so logically satisfying (omitting entirely whether it is psychologically satisfying) to posit human affairs as a supremely valuable phenomenon so lost and lonesome, so ephemeral on and disconnected from a larger scene starkly impervious to meaning? Human culture is a tiny island of value in a valueless natural ocean. Such a position has pushed many less courageous souls, who reason just as well, into nihilistic existentialism with its exasperation and despair before sterile nature. This isolation easily converts to alienation; the world becomes absurd. How do humans come to be so charged up with values, if there has been nothing in nature charging them up so?

To suppose that nature is its own prime mover is not enough. One needs not just a first cause; one needs a *prime meaner,* something logically adequate to explain these meanings that trail onto the landscape, both in some start-up push and in nature's traveling on toward meaning in our "epiphenomenal" phases of its history. Even nature as its own sufficient cause would be insufficient to account for this, since chains of *causes* never add up to *meanings,* any more than long chains of zeroes add up to one. Despite all the hard causality, there is here too much luck and serendipity. Something more does need to be said about meanings. In this sense, though hard naturalism functions in the room of religion, it fails to integrate meaning in human life into the cosmic matrix, and this just so far as it succeeds in being what it wants to be: a nonreligious account. Find what causal connections it may, it cannot fit human values into the natural void. Its isolated, privileged humanism fails to make sense in the scheme of things entire.

Our broader claim remains true that no nonreligious scheme can deliver sustained and systematic meaningfulness into life. Should a naturalism begin to pass over and find nature also meaningful, it will no longer be merely humanistic, but religiously naturalistic, as we shall see with soft naturalisms. We may begin now to suspect that the reason why, in the fifth article, human values are not related to any natural values is not that hard naturalists do not wish this could be so, but rather that they see no way to do it. They find values to be a bizarre gift in a valueless world, found close at hand but absent on the cosmic scene. The local and cosmic perspectives cannot be harmonized. The positive affirmation that nothing more needs to be said for human values is really overlaid on something negative, a lack of any theory that can unify these incommensurables.

Is hard naturalism so simple and economical as advertised? Nature is all there is. But one finds it very hard to say exactly what this nature is—not an inert stuff, not material, not something permitting us to have in science more than nonrepresentational analogues of what nature is really like at the substrate levels. It is spatiotemporal at grosser levels, relatively but not absolutely so, and is perhaps not spatiotemporal at all in the fine grain. It is an energetic gauzy haze, bafflingly productive of value in our rare, human case only.

In the nature become human nature, what can we say is going on? What happens when a woman changes her opinion about racial prejudice? She behaves differently because there was some variant reinforcement, some new learning, selected for by socioenvironmental stimuli. But in the understories, her behavior is, like that of those who stimulate her, simultaneously caused by altered synaptic neural connections, caused in turn by enzymatically controlled, DNA-based biochemistries. Seen at still smaller scales, these are the shifting around of molecules and electrons, an affair of wave packets having an affinity for this structure and not that one. Do we here have a simple, economical explanation either of how nature comes to produce the values we enjoy or of how a woman enjoys liberation into a new value set? The proposed theory is too flat, physicalistic, too causal to account for the produced and enjoyed dimensions of depth.

And when one quits in physics, this is not because the explanation at that level is very complete, or logically (much less psychologically) satisfying, but only because one has run out of sciences. Quite counter to the claim that nature is its own necessary and sufficient cause, C. W. Misner, a theoretical physicist, judges that "Physics does not even appear to be approaching an understanding of the Universe that would make its existence necessary."[16]

Hard naturalists propose a view that, since it removes the supernatural, presents us with nothing miraculous. We can certainly favor the principle that one ought to believe as little miracle as possible. We should take the simplest adequate hypothesis. But we wonder whether belief in a hard causality that assembles nature out of itself and lets it erupt at the rarest of points into exquisite meaningfulness portrays any less a preposterous feat than do competing accounts that posit more by way of meaningful precedents in nature, or accounts of supernatural forces at work creating nature and eliciting meaningfulness from it. The magic stays, even when God is banished.

Even more than simplicity and economy, one wants that theory which most adequately corresponds to all of occurrent reality. One ought be determined to hear *nothing but the truth,* but one also ought be skeptical whether *the truth as "nothing but"* is really the whole truth. Hard naturalism does not give us one world, scientifically and most economically. Rather, it gives us half a world, and that too cheaply. Its downfall is its simplicity. It borrows the prestige of science and tricks us by carrying this over to realms of belief where science does not apply. Shorn of such illegitimate prestige, hard naturalism is quite as susceptible to being tied up in mental knots as are any of the positions to follow. A world view requires hard thinking, but thought ought to be, if hard, also solid, that is, well grounded in adequate explanatory principles. Hard naturalism is a view so economical that it no longer goes down deep enough to support the evident world of nature and history that we experience in the superstory.

3. SOFT NATURALISM

Science seeks to give us objectivity, prediction, quantity, and control; but nature offers also subjectivity, surprise, and quality, and requires our submission. Hard naturalism weights heavily the first four factors, but what we call *soft naturalism* is prepared to emphasize the latter four, finding these features in and overleaping science. The adjective "soft" here has multiple senses. It finds hard naturalism inelastic and severe, supposing an exact, metric, mechanistic nature. Soft nature is rather an open system, before which we need to be humble, wondering, and often agnostic in quasi-religious confession that nature is greater than we can know, certainly greater than hard science can specify. Nature is as organic as it is mechanical. Biology and even psychology teach us as much as does physics about the ultimate structure of things, although they too fall short of the whole. While hard naturalism is always a little embarrassed by emergence, and would like to take everything back to physics and chemistry, soft naturalism delights in emergence, enjoys the organic sciences, and yearns for holism against reductionism. Hard naturalism keeps an emotional distance from nature, despite its thoroughly intellectual naturalism. It is alienated toward nature and laments the cold, hard universe. But soft naturalism loves nature and wants communion with it.

Creative and Transformative Nature

We can gather these convictions, somewhat loosely, around a half-dozen points.

1. Nature is all there is; there is no *super*nature, but nature is *superb* in a sublime, mysterious, even a divine sense.
2. Elementary nature has always existed; nature is uncreated, but rather itself contains a creative, transformative principle.
3. Nature is causally ordered but not entirely so. There is significant randomness and novelty. This combination of causation and openness, together with the transformative principle, permits freedom and directedness increasingly in the higher evolutionary forms.
4. Nature is simple and nonpersonal across great ranges, but locally and at complex levels becomes personalized. Persons stand in essential continuity with nature. Both the physical and the psychical dimensions of nature are keys to its understanding.
5. Values are not all human values; there is intrinsic value in natural things.
6. The scientific method can teach us much but not all about nature. Philosophical and religious judgments are required positively to evaluate its meanings.[17]

Energy is driven to form matter, and, driven over matter, proves seminal and creative. Here nature contains two trends. One is entropic and disorganizing, but in dialectic with it is a tectonic trend that is negentropic and organizing. Against the *chaos* there is a *logos.* The latter trend is more constitutional than the former, and in it much mystery hides.

Anthony F. C. Wallace, an anthropologist, writes, "The central theme of the religious event is the dialectic of organization and disorganization. . . . On the one hand, men universally observed the increase of entropy (disorganization) in familiar systems: metals rust and corrode, woods and fabrics rot, people sicken and die, personalities disintegrate, social groups splinter and disband. On the other hand, men universally experience the contrary process of organization: much energy is spent preventing rust, corrosion, rot, decay, sickness, death, and disillusion, and, indeed, at least locally there may be an absolute gain of organization, a real growth or revitalization. This dialectic, the 'struggle' (to use an easy metaphor) between entropy and organization, is what religion is all about."[18]

By this dialectic things are formed, including the planet Earth, which is subsequently formed more and more toward life. Living things become informed, some of them eventually reaching cognitive self-consciousness. This build-up is all built into the nature of nature, but not simply as a causal unfolding. There is a creative urge in nature that is steadily self-transcending; nature's physical side has a nisus for the emergence of psyche. Here a little of the idealism that (as we later see) in theism becomes a dominant theme begins to enter, as does considerable awe.

Soft naturalists can be ecstatic about nature; they have numinous feelings before it. There is no God overhead, but nature is bottomless. We are to look *in* nature, not *behind* it, and yet nature does not wear her identity on her sleeve. The empirical surfaces of things have to be perceived at more depth. There is only Nature (which needs in such moments to be spelled with a nineteenth-century capital), but there is enough on and beneath the surface of phenomenally evident nature to call forth love and reverence. Even worship is appropriate. Soft naturalists may bow before nature on one knee (perhaps not two). They offer what Roman Catholics call high veneration, not divine worship. They will not be found in church, but they may prefer the field to the laboratory, the wilderness to the city. Even when they are social, they may see in society much carry-through of the wisdom of nature.

Soft naturalists dislike the technological mastery and control over nature sought by nonreligious naturalists with their nature-as-mere-resource mentality. They dislike a similar dogma in Judaism and Christianity that disenchants nature and places humans in dominion. Either view results in the same arrogant pride. Neither creed is ecologically minded, neither is respectful enough of the mysteriously generative nature in which we live, move, and have our being. Nature is not so much to be mastered as it is to be endorsed and followed. Nature is only secondarily a resource; ultimately it is our Source. There is no Father God, but some of his properties are retained in Mother Nature. "I do not believe that there is a god separate from the ocean out there," said Gregory Bateson, an anthropologist, watching the Pacific. "On the other hand, I do have a sense that the ocean is alive. Is that . . . religious?" Increasingly he found a "growth of his own personal identification with nature." He found himself "wondering about the sacredness of nature and the nature of the sacred."[19] "Call the systemic forces 'God' if you will."[20]

Sanborn C. Brown, a physicist, gives us the following creed: "I believe in the forces of Nature, the forces Almighty, creators of Heaven and Earth, and in human beings, not as Nature's only sons but as beings who must fit into the vast and interrelated universe which formed us and controls our destiny."[21]

Causality is not here denied, but hard determinism is hollowed out and replaced by a softer transcausal principle. When life and mind appear, we cannot say that efficient causes of a necessary and sufficient kind were actually and previously present in inert matter. We do get more out of less. A genuine potential has "materialized." This engineering power in nature is real but supercausal. It is demonstrated in what has actually taken place, but an analysis of it escapes the clutches of quantitative science, showing up as randomness, spontaneity, emergence, mystery, irreducibility, learning, decision—all that world-building quality which is so paradoxically the principal thing to be explained and so poorly explained by existing science. In each of the upstrokes we are getting more than causal recombinations of what was there before; we are getting potential actualized, novel quanta of worth. Nature is its own Prime Mover, both at the start and all along its ascending course. This brooding fertility is what some will call its divinity. It recalls the sense in which we earlier heard Ernst Mayr concede that "virtually all biologists are religious, in the deeper sense of this word."[22]

Thus, the simple causal principle of hard naturalism is replaced by a tectonic, valuational principle, the ground of meaningfulness. Access to it is transscientific, for our judgments about good and evil, beauty and ugliness belong to no science, although science in its redescriptions of nature is a relevant prerequisite for making mature judgments here. R. W. Sperry, a neurophysiologist, affirms, "The grand design of nature perceived broadly in four dimensions, including the forces that move the universe and created man, with special focus on evolution in our own biosphere, is something intrinsically good that it is right to preserve and enhance, and wrong to destroy or degrade."[23] That sort of judgment certainly depends on science, but exceeds it in finding value dispersed through the universe, culminating in humans. The ultimate good lies in the grand thrust of nature. Human values, eminently real, are yet surrounded by natural norms. All value appears under the influence of a valence in nature toward value.

Difficulties in Soft Naturalism

The openness of this position makes it at once commendable and difficult to criticize. Perhaps the principal point of attack is the very softness that is also its virtue. Soft naturalists complain about hard naturalists that they have strained away the mysteriously constructive elements in nature, and reduced all to mechanical causality. But when it comes time for them to catch hold of this dimension and say what it is, their sieve lets it leak away too. They may not feel liable, or able, to give much account of the central creativity here, unlike the theists who will find this a sign of God. We are told that there is no unknown beyond the natural, but we are also told that the natural itself is quite largely unknown, especially at its creative cores. The softness in this creed reflects a frequent weakness in science: the seminal principle is positively missing, only quietly, mysteriously there. The account is vaguely reasonable so long as it is kept reasonably vague, but looked at more closely it is ignorance presented as belief.

What first looks like light in the creed—the emphatic belief in generative nature —is, looked at more closely, really a series of troublesome dark spots. The openness is really agnosticism about how nature comes from itself, why it climbs constructively upslope, how an innovating combination of randomness and natural selection yields such impressive advances, how physical nature actually is in continuity with psychical nature. The affirmations we have outlined are all seriously but also rather loosely believed, for the model is only suggestive. It is not enough that we are told that all this is possible because of a potential in nature. That is only explanation by wordplay. We want to know how it is possible, by what "power" there can be a fine-tuning of an expanding universe resulting in the constructing of life and an arousing of experience. We are not helped by naming this an "evolutionary" or "emergent" power.

Soft naturalists do not face this demand squarely, but they hide what hard naturalists had obscured in the word "epiphenomenal" by using softer words like "generative" and "potential." Since they cannot get any energetic input or controlling force out of God, soft naturalists must derive all this from an anonymous nature itself, which is all that there is. They need from science and natural philosophy alone both the grand sweep and the executive power adequate to the spectacular results. Nature must hold itself together. In its deepest currents nature must be singular and self-explanatory, however heterogeneous it is on the surface. It is not easy to envision such a unified theory without approaching a divine nature. Without more clarity, this can become only a "mystic chant over an unintelligible Universe."[24]

Persons as Children of Nature

In its upper reaches, this generative principle falls short of supplying enough anthropic principle, both in the matter of origins and to instruct our present conduct. Humans are "children of the universe."[25] So we are assured by two astronomers, and the phrase has some warmth. But just what does it say about how we are related to the universe and how we are, in consequence, to behave? We are assured that the psychical dimension of nature is a key to its understanding, assured that humans lie in full continuity with nature. But soft naturalists are not theists. However mysterious, organic, or even divine nature is, it has no conscious center of experience, no deliberative, moral, or even cerebral qualities resembling those that the last of her offspring does have. She is inferior to her product. Nature is a fertilizer that is requisite for, but not so marvelous as, the plant she grows. But if there is any universal unifying principle, this cannot (like gravity, electricity, or vitality) be simpler, less spirited than we are, for that would make us too complex for it. We should be disconfirming anomalies to the theory. Nature would be primal but primitive, original but afterward secondary.

This is what has kept theism a vital option, with its insistence that whatever divine Ground lurks in, with, or under nature, it must overreach the personal level. Else we cannot have the results we see manifestly delivered in ourselves. Nature has to be up to its own cognition, since we ourselves provide evidence that this has happened. But how can it happen that nature comes around to such capacity for *experience* unless there is all along in the natural potential a principle adequate to this achievement? Any other explanation is too soft, too weak, and we suspect more in this Power that breeds something greater than itself. We may not always need a like cause for a like effect, but we do need enough Premise for this conclusion.

It is not so clear how humans follow from nature, and as a result it is not so clear how we humans can follow nature. We promote nature's grand design, and after that there may be some consideration of nature's laws in human affairs, a bit of dialectic about spirit in antithesis with matter, or some theory about economics or genes as driving history. Still, nature sets us no specific tasks. All the natural preface commands us to no particular performance. It is singularly difficult (indeed, most say it involves a naturalistic fallacy) to discover an *ought* for interhuman ethics out of what merely *is* the case in nature. Ought "children of the universe" to love each other? But what in nonmoral nature authorizes this imperative? There is little capacity in soft naturalism to deal with the distinctive features of cultures, with an idiographic stretch of history, with personal affairs. We are told how to behave toward nature but not toward one another. We are taught the meaning of nature as it lies behind and around us, but not the meaning of the human drama that now proceeds center stage.

Nature hardly seems up to the guidance of the child she has delivered. For some, that is cause for freedom and relief. Humans are, we have earlier said, self-defining animals. They do not need to consult nature, but are intellectually and morally free to do their own thing. But others would like for humans to be defined in their place. Otherwise, we cancel all promise of showing a systematic unity between human life and cosmic or earthen nature. The introduction to the story is, in the absence of such a relationship, discontinuous with the fulfillment. Some light is thrown on the context in which we live, on nature. But no light is thrown on the text of life, on historical eventfulness. We are given a womb and stage, but no role in the drama. Perhaps the normative structures of theism are sometimes overbearing, but those of soft naturalism do not give us enough support to orient human life. There is no intelligible interdependence of nature and history, no weaving of the two stories into a single narrative. We do not have humans-in-nature after all, but only nature-and-humans. It is one thing to be set free in the world, another to be set adrift in it. The difference lies in having a world model that is competent to instruct conduct. One wants an *ought* that can be more persuasively grounded in what *is*.

We have omitted "humanism" as a religious option, for humanists further subdivide into hard or soft naturalists, or into some more or less stable hybrid of the two. Humanism took its name in protest against theism, rejecting the supernatural, but in so doing it often elected to celebrate human worth in isolation from its natural matrix, not causally but valuationally. Human values are supposed to be independent of their natural roots, and in this sense humanists also have rejected the natural for its religious qualities, not less than the supernatural. They often have bought into the concept of nature found in hard naturalism. Or they may simply celebrate human affairs, while remaining agnostic about naturalistic metaphysics, in which case they operate in default of any explanatory theory. More recently, humanists have begun to find their all-central "humanism" rather awkward. The term is too narrow, anthropocentric, and provincial. Many have preferred to become religious naturalists in their background metaphysics, while retaining a central focus on human values. This results in a commendable joining and enjoining of awe before nature and values in human life, but there is not really present any increased explanatory power by way of unified theory.

4. EASTERN PERSPECTIVES

Science arose in the West, with logical connections to monotheism. God rationally orders a good creation, disenchants it, forbids polytheism, animism, and nature worship. God sets humans in dominion over nature and calls them into a historical covenant. Such beliefs can grow scientific, especially when combined with Platonic and Aristotelian philosophies, which value rational thought. The early scientists hoped to "think God's thoughts after him,"[26] and science and theism, even in warfare, have since managed to coexist in kinship. Even those who hold that science has now displaced religion may find that monotheism was an effective, necessary preparatory stage.

Whatever the ancestry and nurturing of science, monotheism comes to be challenged by its child. If belief in God first makes science possible, science, say some, later makes belief in God impossible. But science seems unable to deliver a replacement value set, and the naturalistic schools may do less than enough explanatory work. A plausible turn at this point is to ask for the resources of nontheistic religions, which is in effect to make a turn East, where religions have prospered with neither monotheism nor science, neither at least dominantly and in their Western forms. Can they help us to handle monotheism's unruly child? Have they models of nature and of history that offer insight into the nature of this nature after science?

One would need to be either foolish or very learned to say briefly what the final relationship between the several Eastern faiths and the multiple sciences can be. The Western dialogue that has been going on intensively for three centuries has hardly begun in the East. It is difficult enough to conclude anything in the West, much less to predict what may happen in the East. It is not easy to find spokesmen from the religious East who are thoroughly grounded in Western science and philosophy of science. But, at some risk of appearing shallow and blunt, and claiming our findings only to be provisional and schematized, we will nevertheless take a scouting trip eastward. *Our appraisal is not of the complete adequacy of these creeds, but only of their promises to make unified sense of nature and history.* Thus, we do not inquire whether they privately save the believer (we allow that they can), but whether imported models from India and China can better absorb or criticize the nature that science leaves us, especially where this troubles theism. Can they better interpret the historical eventfulness that troubles even the explanatory skills of science?

This procedure accords with our basic methodology from Chapter 1. When a theory is challenged by the observations coming in, we should not only consider lesser revisions that protect the main theory. We should explore a paradigm overthrow and cast around for rival theories that can better accommodate what we are now experiencing. We hope that we do not ourselves have one more theory grown arrogant (of which we have had too many in science, and which becomes no religion), but hope to take an honest look at options. Coached by our scientific methodology, we want here more of what characterizes good scientific debate than what typifies nowadays the easy tolerance of so much religious inquiry. Thus, we want to listen for challenges to Western paradigms, facing up to our own epistemic crisis, but also (what we would not do in the more genteel discipline of comparative religion) we want to be quick to attack the soft underbelly of a creed, rather than to praise its strengths.

We will look for falsifications more than for confirmations, at the same time that we ask whether the other paradigms can handle what are anomalies to theism or to secular naturalisms. To some that will seem unfair, uncharitable. But, short of arrogance, it will at least be good scientific methodology, and—we think—one profitably followed in serious religious inquiry. We want truth more than tolerance or civility. Need it be said that we assess theories, not persons?

Natural History as *Māyā,* Illusion, Superimposed on Brahman

By the Hindu account, the classical paradigm of nature (*prakṛti*) is *māyā,* illusion. The world is a kind of "play" or "sport" (*līlā*). But "play" is little to be taken in the dramatic sense, nor "sport" in the pleasurable sense. The trifling, mocking, deceptive senses are dominant. All the words designate a kind of appearance (*vivarta*) we fall for, fall into, a *superimposition* (*adhyāsa,* "image," "mirage") over something else. The common worlds of nature and history are the production of our *ignorance* (*avidyā,* "not-seeing"). *Saṃsāra,* the spinning world, is a rotating wheel spun by our ignorance, like a firebrand whirled so rapidly around that it appears to be a solid wheel. The ancient swastika (Figure 6.2) serves as a diagnostic symbol of this rotating world. Here and in what follows we may take as representative Sankara (ninth century A.D.), whose Advaita Vedanta has remained central even today in Indian thought: "*Saṃsāra* is only based on *avidyā* and exists only for the ignorant man who sees the world as it appears to him."[27] That is a description, but inseparably also an evaluation. Taking it by itself, we do not yet have much promise of a comprehensive model making sense of nature and history. Nevertheless, we might have an accurate model of what is going on.

Is there any unifying principle? Indeed so, for behind the *māyā* there is Brahman. Brahman is the Hindu nonpersonal and undifferentiated divine Ground, the ultimate explanatory reality. Brahman is neither object nor subject, more like silence than speech, more like light than something illuminated, more like space than something contained in it, more a void than something articulated and formed. Brahman is the divine ocean, a plenum, pure (objectless) consciousness, yet not a center of experience, not an agentive actor. Compared to the phenomenal world, Brahman is at once "all this" and "not this" (*neti-neti*). "In ascertaining the true nature of Brahman, men of wisdom should not think of it in terms of whole and part—unit and fraction—or cause and effect. . . . We must give up all such conceptions and know Brahman to be undifferentiated like the sky." "It is devoid of attributes, for it is one only without a second." This utter "homogeneity of Brahman" is pure being, consciousness, bliss (*sat-chit-ānanda*).[28] Brahman is behind all the gods, which are *saguṇa* Brahman, Brahman "as if" with name and form, but Brahman beneath is without attributes (*nirguṇa* Brahman).

Our inquiry now becomes: In what sense is nature descriptively explained as a derivation from Brahman, and what evaluation is to be made of it? If we posit Brahman (B), the manner in which natural and human history (N-H) follow by implication will be crucial. If B, then N-H? There are two tracks of explanation, difficult to reconcile. The first, which we have already suggested, is that the implication is a mistake (*avidyā*), but the second is that Brahman somehow manifests itself in this mistake. Nature (*prakṛti*) is emitted in the indifference and purposeless play

Figure 6.2 The swastika as *saṃsāra*

(*līlā*) of Brahman. "It thought, may I be many."[29] But this involves our corporate mistake quite as much as the divine production. "There is only one highest Lord ever unchanging . . . who by means of ignorance [*avidyā*] manifests himself in various ways." Brahman is "fictitiously connected with *māyā.*"[30] But it is difficult to see how a "fictitious connection" based on *avidyā* (not-seeing) can be an implication or an explanation at all, not in the scientific sense, where out of a theory one derives the phenomena.

The Hindu scriptures occasionally use a term for "create" (*pariṇāma*) that approximates the Judeo-Christian concept. But the Advaita Vedanta commentators steadily sublate objective, cause-and-effect models in favor of subjective, projectionist models in which the world is a superimposition (*adhyāsa,* mirage) over a Brahmanic substrate, an appearance (*vivarta*), like a fence post mistaken for a person. It is not genuine creation. For this reason, no effective route to Brahman lies in the study of the phenomenal veil. Rather, one sets aside the illusion and moves by meditation inward to the deeper self.

Certain elements here are congenial with scientific descriptions. All natural items use common tectonic materials, manifestations of a simple primordial energy. Apparently substantial matter dissolves into a fluxing wave set that at microscopic levels loses its picturability and becomes (in Hindu vocabulary) "formless and undifferentiated." If one does go back to the big bang, all "name and form" collapses in the incredible energies of the primordial state, when the universe was collapsed into nothingness, which was also the consummate greatness. Usually, however, Hinduism takes as given the ever-cycling epochs and needs no start-up creation.

Brahman is the spaceless, timeless matrix for the particle play. To use our crude model from Chapter 2, Brahman is the basketball over which the dents travel, warps called wave-particles, intersecting to form the world. Brahman is the infinite potential, the superposition of quantum states, from which there bubble up in random play the material states of our native levels. The entropic and negentropic tendencies in phenomena can be demythologized from Shiva, the destroyer, and Vishnu, the creator, aspects of Brahman, although if all is *māyā,* neither creation nor destruction need be taken as more than mistakes. Evolutionary development is a queer sport (*līlā*). We are caught up in ignorance of all this. Living things fight to preserve their organizational and informational states; they call forth this superposition state and not that one by genetic, neural, and volitional actions. We thus carry forward the illusion of multiple, separated realities, blinded thereby to the foundational unity. The mixture of order, chance, surprise, and skill in the games of creation does sound a little like "sport," a game, and good games need not be so serious; indeed, they require a kind of play illusion.

Māyā is the doctrine that appearances are deceiving. The world seems flat, but isn't; the table seems solid, but isn't; the sun seems to set, but doesn't; the tree seems green, but isn't. *Māyā* is the theory (as physicists would say) of observer-dependence. The scene on a television set is hardly "real"; it is just a show held up by a shower of electrons hitting the luminescent screen from behind. But then again, after the revelations of physics, the world of "real" phenomena (the real world that the television pictures) is itself hardly any more real. It too is just a show held up by a whir and buzz of electrons, protons, neutrons. And beneath all this sparkle of electronic particles? Brahman holds up all the phenomena.

But can one also say that it is really our ignorance that produces the television picture, or the electronic world? Certainly the latter was there eons before we came; it produced us, not we it. The "fine-tuned" anthropic constants of the big bang do not sound so much like *māyā* and *līlā,* illusion and sport, as do they suggest purposive creation. If all that is meant is that something more Ultimate supports the phenomena, monotheism knows that God continuously creates the world drama. The question is not whether there is a transphenomenal support, but whether *Brahman-avidyā-māyā* versus *monotheistic creation* is the most adequate model for it.

Is there really much logic in the derivation of nature and civilization from Brahman? All the terms—"illusion," "ignorance," "superimposition," "fictitious connection"—suggest a kind of senseless operation. These are in fact nonexplanatory terms, quite as much as they are explanatory. Nature is just appearance, sport-play, mirage. We are reminded of "nothing but" fallacies elsewhere, as that humans are nothing but epiphenomena, that the world is nothing but mathematics, or that life is nothing but matter-in-motion. Something irreducible gets left out. Not enough attention is paid to the emergent reality for what it intrinsically is. Too little is premised in the prior reality or the composing steps by way of logical explanation for the mysterious consequence. The Hindu account seems vexed by the events of nature, history, and personality, more anxious to explain the phenomena away than to explain them.

Natural History as *Śūnya,* Empty; Ultimate Emptiness, *Śūnyatā*

In Buddhism there is much intramural diversity, but in the main schools the master word *śūnya,* "empty," is applied across both phenomena and any noumenal substrate. Perhaps (as we next wonder) this is only *agnostic silence before ultimacy, describing nothing noumenal.* But *śūnya* seems undeniably meant to be *descriptive of phenomena,* nature and history. If it says nothing (*śūnya*) here, if nothing is asserted, the inquiry we pursue is foiled at once. But *śūnya* as applied to phenomena intends to assert that the world affairs are not absolute but relative, that they are fluid, impermanent, and changing (*anicca*). With that judgment anyone acquainted with nature after science, and any theist, will entirely agree. But in the classical Madhyamika schools, of whom we may take Nagarjuna (second century A.D., and still thought to be the greatest Buddhist philosopher) and his disciples as representative, there is more.

The *saṃsāra*-world is our corporate mirage. "This universe like a magic play comes from nowhere and goes to nowhere; being due to mere mental bewilderment, it does not stay anywhere."[31] "Common mankind, whose power of vision is obstructed by the darkness of ignorance, imputes to separate entities a reality which they do not possess, a reality which for the saint does not exist at all. . . . The separate entities of the phenomenal world have never originated and do not exist."[32] Superimposed over some ultimate nothingness, we have a phenomenal "mock show." Phenomena are "not real"; they are "insubstantial" and "empty." Again like the whirled firebrand that we mistake for a wheel, the world is a whir that superficially seems real.

This peculiar genius seems alternately almost insane and remarkably like conclusions on the frontiers of physics. There are no "things," not absolutely, only a series of moving changes that carry relative identity in an incurably successive world. The world-play is like a motion picture, an illusion played upon what is in truth a blank screen. We know well enough what matter is, in transience, for tables and chairs, but ultimately it has evaporated on us. It is some kind of spin and not much of anything; to seek its essence is like peeling an onion. We do not know what is there; there is nothing there.

All world building is caused by "clinging" in a matrix of dependent origination, that is, of fluxing causation. Living beings try to hang on to their holdings; by desire (*taṇhā,* "thirst") they try to maintain their separate existences. But they have no permanent reality; they are in fact empty and transient. The world is impermanent and unreal (*anicca*), but so also is the separate self unreal (*anatta*). In reality all is empty. Meanwhile, everything burns. This is offered as a description of what is going on, with some suggestions about what makes the organic parts of it cohere (clinging, thirst). It posits causation for the inorganic parts too (dependent origination), although we have, once again, heavy components of ignorance and mistake, rather like those that troubled the Hindu account. It is also the sort of description that seems inseparably to carry a negative evaluation.

Is there a comprehensive underlying explanation? Concerning any such principle, there is great insistence on negation and reserve. The master word *śūnya* is used here quite as vigorously as with phenomena, but it seems to do a different job. It is nondescriptive and reverts to silence about anything more. "No Reality was preached at all, nowhere and none by Buddha." "About the absolute, the saints remain silent."[33] The denials here, nevertheless, point to a formless and undifferentiated womb, called *śūnyatā,* Emptiness, symbolized by an empty, open circle (Figure 6.3), one that the enlightened can use just as freely to symbolize the world, when seen through. Staunch Buddhist interpreters hold that the Absolute (Brahman) is being denied, although others suspect something rather kin lurking in the emptiness, posited in the very negation. Other Buddhists will supply a few nondescript semipositive descriptions. The "ultimate reality" (*dharma-dhātu*) underneath is "suchness" (*tathatā*), "thatness" (*tattva*).[34] All things contain Buddha nature. Others say that *śūnya,* though it refers to nature and history, is not meant to apply at all to the ultimate. It is a transformative, not a descriptive term. It gets a job done (salvation), but it asserts nothing (makes no truth claims).

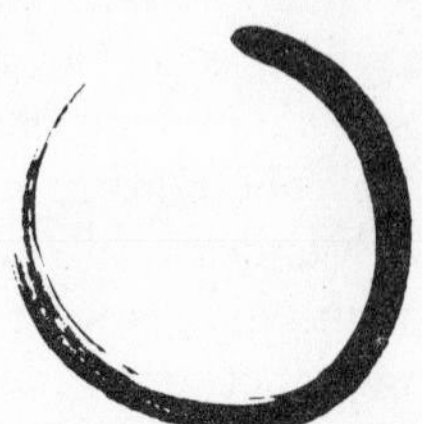

Figure 6.3 The *śūnyatā* symbol

That job, we can pause to notice, is finding *nirvāṇa.* The Buddhist end state, *nirvaṇā,* pictures a "going out," as with a candle flame. *Nirvāṇa* is the cessation of all desires, the stopping of the worldly whir. It is peace, the loss of separate, anxious individuality, the cooling of the ego, the final stillness, the great "quiescence of plurality."[35] It too is empty; one enters *śūnyatā.* Toward this all things move, a state something like the heat death of the universe, to adapt a scientific image. Buddhist schools may develop this differently, and often *nirvāṇa* is an opposite of *saṃsāra,* a heaven different from earth. But there is a paradoxical and powerful countertheme, featured in Madhyamika, by which the *śūnya* model equates the two. *Saṃsāra* is *nirvāṇa,* provided that it has been seen through to its emptiness.[36] After the fluxing whir (*saṃsāra*) has been seen for the empty (*śūnya*) mirage it is, one has *nirvāṇa* immediately in the here and now.

If nothing is asserted about any underlying principle, then, once again, we have nothing (*śūnya*) claimed, and our inquiry is stalled. If Buddhists wish to contribute to the debate in which we are now engaged, they must advance some juxtaposition between the noumenal "thatness" and the phenomenal diversity. They must get something out of their nothing; some concept of nature and civilization must issue from their ultimate emptiness.

Many themes here—the fluxing world, the lack of a substrate, the denial of permanence, the struggle for existence—are congenial to nature after science. Some will find the elements of agnosticism welcome and realistic. If, indeed, Buddhists claim anything about the ultimate, then this formless void (*śūnyatā*), from whose bosom all things come, suggests the primordial undifferentiated chaos out of which evolution has taken place. It has a maternal tone. Perhaps the vast universe squashed into the formless, minute fireball at the beginning is a sort of nothingness.

The separateness of diverse things, driven in living forms by their thirsts (biological, genetic, and psychological programs), is a local, provisional illusion, destined to eventual quiescence. Perhaps we can stretch the concept of "clinging" to cover even the gravitational and electromagnetic forces. Everything in one way or another "burns": energy in process, life in suffering. In the end, all things are but traveling concavities in a plasma; matter is just a space warp, a bubble. The *śūnyatā*-void, like the Newtonian space-time receptacle-void, may first seem passive and empty, but it has a way of proving active and formative, mysteriously so, a suffusive, universal ground like the relativistic void-ether-plasma of Einstein and quantum physics, even though in the end Buddhists wish to still this activity. "There nothing moves, neither hither nor thither."[37]

At the same time, many Buddhists deny that they have made any claims, and if they have, this explanation is largely accomplished by negation. With all forced into the *śūnya* paradigm, everything is set equal to emptiness, a kind of unity achieved by equation through zero. Things are all dissolved, not really brought into *unity* but rather into *nothingness.* The explanation, if there is one, seems to work something like this. If *śūnyatā* (S) underlies all, then do nature and human history (N-H) follow in course? If S, then N-H? Now if we set S = 0 (*śūnya*) and N-H = 0, the implication does follow, since zero implies zero. But it is not so clear at this point that *śūnya* is doing any explanatory work at all. There is no world derivation, no logical movement from nothing, or from "nothingness," into something. It is rather the ultimate reduction—to zero! It is scarcely a soft explanation; actually, it is an *empty* (*śūnya*) one. Emptiness is a sort of inchoate and nondescript model, a nonmodel, for the unconditioned ultimate, and also a paradigm, quite descriptively so, for the natural and historical worlds. The *śūnyatā* symbol sweeps over nature, history, and personality, and vanishes headed in the direction of the ultimate.

In addition to wondering whether nature, history, and personality receive adequate causal or logical derivation from some underlying ground, we also worry that none are properly valued. "There is no good to be attained by the knowledge of the narrative of the creation," warns Sankara. "Those whose ideal is the attainment of the highest good do not entertain any respect for creation in its diversity because it can lead to no purpose."[38] The bliss of Brahman brings the dissolution of the saints. "Having thus attained identity with the supreme Immortality, they discard individuality, like a lamp blown out or the space in a pot when broken."[39] When we turn to Madhyamika, the language is equally severe. The world metaphors move through mirage, magical illusion, dream, mock show, airy castle in the sky, a bubble on the water, a passing stir in the wind. Nagarjuna much dwells on the horrible unloveliness of everything in the world. The world is "a disease, a boil, a thorn, a misfortune."[40] "There is here in this world neither reality, nor absence of illusion. It is surreptitious reality, it is cancelled reality, it is a lie, a childish babble, an illusion!" But likewise in *nirvāṇa* there is "no existence, no ego, no living creature, no individual soul, no personality, and no Lord."[41] The saint there does not exist, "just as in space the track of a bird does not exist," "just as the track of a dream, an illusion, a mirage, an echo."[42]

The world models *māyā* and *śūnya*, while not serving very well as constructional paradigms, serve with overkill as valuational terms. They fail to explain nature, history, and personality; and, worse, they villify them. It is hard to keep *māyā* from meaning "crooked," "deceitful," "false." *Śūnya* means "empty," and it is hard to keep it from meaning "hollow" and "worthless." The formational and informational levels of evolutionary development, the historical narratives of the particular cultures, all the myriad individual personalities—everything is dissolved in these powerful yet formless and static solvents, Brahman and *śūnyatā*.

That, if you like, coincides with the "scientific" view that personality is epiphenomenal, not really a key to understanding the whole. This belief can seem easier than the theistic belief that personality is a telling category, revelatory of God. But it reduces persons and natural histories to something less than even hard naturalists would make of them. They are not even epiphenomenal tangents to nature, rather just ephemeral illusions. One lives by sitting apart from these transient world values that bear so little connection with what is ultimate. These accounts only seem to endorse process and change; they in fact recognize them so as to eliminate them. They find the world a sham and recommend individual disengagement from what we have found to be an engaging natural and historical story.

Everything is by-play over an untouched core. But the core is not ennobled by its distance from the trifling play; the play is not ennobled by its dissolution into the core. We have consulted only classical sources, and more modern reformers have tried, and will continue to try, to put a better face on these difficulties.[43] But when

Brahman and *śūnyatā* become positive enough to activate and endorse nature, history, and personality, in which they have been classically uninterested, the result may not be so different from the theism that they now oppose. It might be alleged (by soft naturalists, Hindus, Buddhists) that Christians and Jews have been too anthropic; but, if so, it must surely also be alleged (by soft naturalists, Christians, Jews) against Hindus and Buddhists that they have cared too little about nature as a developmental system, about historical narrative, and about the culmination of this drama in personality.

The Tao and Binary Nature

In China, a more world-affirming climate, we meet the Tao as the ultimate explanatory principle, the "Great Form." Nature is conceived as bipolar interpenetrating opposites, the *yang* and *yin* ever passing over into each other (Figure 6.4). "In Tao the only motion is returning."[44] The original theme was diurnal and seasonal, later deployed into an oscillatory model finding binary opposition the central secret of nature. The *yang* is active, warm, dry, bright, positive, masculine, associated with the sun, the day, air, mountains, summer, southern slopes, northern (less shadowed) shores of a river, with foods grown aboveground. The *yin* is passive, cool, wet, mysterious, negative, feminine, associated with the moon, valleys, soil, winter, northern slopes, southern (more shaded) shores, with foods grown underground. Every natural thing, every social enterprise, every personal career is a complex cycle of epicycles of the two. This way (Tao) is at once nature's inevitable operation and its goodness.

There are merits to this model.[45] Assuming that the *yang* and *yin* are scientific principles, or proto-scientific principles (which may be a mistake, since science is about causes, religion about meanings), how does their operation describe the forces of nature and history? At everyday levels, life is stimulated by diurnal and seasonal turnings; organisms and ecosystems are homeostatic systems; organisms reproduce themselves over and over again; ecosystems undergo succession. There is genetic pairing, life and death. The electronic nature of matter is a complement of positive and negative charges. Particles have their antiparticles, magnetism has its poles. Mathematical equations, found throughout science, balance positive and negative values on both sides of an equals sign. In the nature described by such equations, there is much conservation through changes.

Figure 6.4 The Tao symbol

Physics has its complementarities, although there is nothing oscillatory or binary about quantum theory in general, or excitation levels, or half-life decay, or relativity theory. Things begin and end, and begin again. There are the rotating galaxies, the rotating stars, the rotating and revolving planets. There are interconversions of energy into mass, and back into energy again, the wave-particle dualisms, the interpenetrations of space and time, the vibrating atoms, the orbiting electrons, the restless particles. Everything is a great cosmic dance. Even stars are born and die, and there may be, on an unimaginably astronomical scale, a pulsating universe that expands, contracts, expands, contracts. The reversing takes account of growth and decay, of the impersonal creative forces that persist interminably, the entropic *yin*, the negentropic *yang*. The ultimate is a natural law, without a lawgiver. The natural way is to be followed, not aggressively manipulated and controlled. Perhaps here science is *yang* and religion is *yin*.

But the fatal defect is that the model is irredeemably cyclic and nonhistorical. True, science is full of balanced equations, corresponding to balanced opposites, something like the *yang* and the *yin*. But equations tell a poor story. The world history seems less a fiction than the balanced mathematical equations that science abstracts from such limited parts of the history. In any event, history is as real as mathematics, and more in need of explanation.

We now know too much of the depth of historical change, and here we do not get any explanation for the evolutionary upslope or any hope for linear societal progress. The world is incurably successive, but not in cyclic reiteration. There is an extensive historical serial in which time proceeds as the carrier of innovative change. The world is a drama, a novel, more than a repetitive dance. The course of nature, though cyclic at close range, is in fact rather a linear spiral. There is no returning; the future is never like the past. There is local microsameness, but there is no identity in the large. There is rather irreversibility; rebirth is always renewal; re-formation is reformation. For all its wandering, nature's self-assembling is crucially a vector, but Taoism gives no account of the miraculous emergence, of the long-range assembling of primordial energy into the human personality. It makes no sense of the ascending formational and informational levels.

The Taoist cosmic model offers only an indiscriminate conflation of natural elements (for example, sexuality, seasons, moisture, politics) that have no essential connections with each other at all. Everything is forced oversimply into a provincial, rural model, one that prevents the enlarging scope, power, emergents, possibility, knowledge, and emancipation we confront in nature. It is not the oscillating and ever-returning balance that interests us, or that we want to follow. Rather, some disequilibrating imbalance impels an Earth pageant, and makes for storied history. Compared with this, a fluxing, ever-restabilizing homeostasis is at best a stagnant good.

History in the Eastern Perspectives

Brahman, *śūnyatā*, the Tao—the models all predict that nothing much will happen, only illusion, mock show, eternal recurrence. Nor can they retrodict what has in fact happened. Brahman and *śūnyatā* are too pacific and the Tao is too cyclic for the aggressive course of natural history. What counts for a good theory, by our criteria in Chapter 1, is its capacity to draw together and make sense out of the available material. But these theories rather slip into making nonsense out of nature. The problem here is not merely that they do not have any rigor as explanations. Worse, they do not seem to want it.

In terms of history, of course, the East has biographies (as with Gandhi and the Buddha), though it does not always rejoice in the historical detail of the phenomenal life. But the biographies seldom accumulate into any longer-storied narratives of history. One thing (as we shall later complain) that Judeo-Christian theism lacks, and that the hard and soft naturalisms lack, is any model that makes much sense of all the histories of non-Western peoples, and so theists would welcome such from Eastern sources. Would that they could supply their own models, parallel to the Western ones, that make sense of their regional and local histories.[46] Theists would like to find the divine presence in all of history, in every biography, but it is difficult to find it for others unless they can find it for themselves.

But by these theories, things spin and move nowhere. Sometimes, it seems that Eastern religions would like to protest that our efforts to take seriously nature and history are misguided, for these make no final sense, but have to be seen through, with a concentration on something underlying. They want a short circuit beneath, and an escape from, our problem. But if one believes what they believe, there is little point to further inquiry about the natural world, about history. One knows already what it is (*māyā,* illusion, *śūnya,* empty). One gets on with more important things, the quest for Brahman, *nirvāṇa.*

Even the Taoists, who do wish to get with the world flow, find it directionless. Perhaps the most favorable account that one can give is that the East fell repeatedly into what the ecologists call homeostasis, or what the theologians call the eternal presence, and while that is not a bad fate, it is like those evolutionary lines that fell into survival without advancement. They were not (to speak theistically) God's chosen lines, not the lines chosen to wrestle hardest (as Israel would say) with redemptive suffering, with the worth of historical tragedy and the regenerated Earth beyond. It seems, alas, that if a comprehensive account is to be found, we are thrust back to the West, back to be debate between science, naturalisms, and theisms.

The Eastern creeds seek *nonduality,* seen in the Hindu desire for *advaitan* (nondual) consummation in Brahman, in the Buddhist equation of *nirvāṇa* with *saṃsāra,* in the nirvanic quiescence of plurality, in the Taoist confluence of *yang* and *yin.* Now, while none of the Eastern spokesmen wishes to fault science in its place, this nonduality is an experiential state that leads to a final tension with the scientific outlook, which can be judged unfavorably to be rationalistic and analytical, to foster separation of the knower from the known, to feature name and form. In the end, science blocks the route to unitive knowledge, and even the search for unified theory is misconceived and misleads. What is rather to be sought is the nondual state. This is seen especially in Zen Buddhism, which incorporates much Taoism and is the most world-affirming of Eastern religions, sometimes delighting in phenomenal things. But it is also the most insistent that the route to *satori* (contemplative union) is nonrational and nonscientific.

With minor exceptions, Eastern religions are driven by a background belief in *karma* and *reincarnation.* These preface Hinduism and both Theravada and Mahayana Buddhism, and they widely permeate but are less essential in Taoism. (In Zen Buddhism they may become muted.) The traditional soteriological problem arises within the wheel of rebirths. One seeks eventual escape from this, though perhaps in the next rounds only more fortunate rebirths on Earth or in heavenly realms. The force of good or bad deeds (*karma*) carries over from life to life. One's status here is the result of previous *karma,* and present deeds lead to better or worse status upon rebirth. Many forms of animal life, down at least to insects, contain souls or selves that have been or may become human.

Such belief can hardly be said to be scientific. Weakly perhaps, the *karma* belief in moral causation, as the perfect conservation of good or evil, is an analogue of the laws of conservation of mass and energy. *Karma* involves a kind of spiritual genetics, a passing of information from life to life. Its forces can be hidden in the unconscious mind, or in seeming randomness or determinism in the genes. Out-of-body experiences had by dying persons may suggest reincarnation. But science cannot guarantee the conservation of good and evil, much less endorse rebirths. *Karma* and reincarnation must be regarded (at best) as extrascientific beliefs, and (in fact) as difficult to reconcile with science, which finds these categories superfluous. But in the East *karma* and reincarnation form the axiomatic scheme for the common operations of nature and history, and the two set for humans their religious assignment. If these beliefs collapse, much of the quest for Brahman or *śūnyatā* loses its urgency and even its possibility. Failing such beliefs, one lives only in this life and dies into extinction.

5. THE DIMENSION OF HISTORY

We have now reached the point, often foreshadowed earlier, where *history* must step out and demand its own, to strain the leashes of science and pull us further into religious inquiry.[47] It is sometimes argued that science, with its systematic objectivity and empirical care, has made history possible in the modern sense. If that is half the truth, the other half is that history makes science impossible as a full explanation of either nature or history. Reductionist accounts incline to make physics the ultimate science, to which biology, psychology, and social science are to be referred back for their methodology, ultimate laws, and constituents.

But the truth lies nearer to a compositionist account. Physics is really more abstractive than it is empirical; it leaves out almost all the actual goings on of higher interest. History, not physics, is the full-grown science, but is at once the sort of science that metamorphoses into something else. The events that are partially abstracted out for study by physics, biology, and social science are seen full round only in the narratives of history. This reckoning with history will take us into deep waters, as will a reckoning with suffering in the section to come. But it is in deep waters that every explanatory theory must be tested for the limits to its strength.

Some historians in recent decades have hoped to make history a social science, using statistical, quantifiable methods for the empirical verification of causal connections and even the finding of lawlike regularities in history. Such historians pull in the direction of the scientific method. Humanist historians continue to insist on the uniqueness of history and the necessity of the narrative form, or something approximating it. In this they are joined by a resurrection of interest in narrative in both literature and in theology. Robert P. Swierenga, a historian, comes to a double conclusion: "The discipline of history is in a state of virtual anarchy today." And: "Social scientific history is a companion, a complement, to traditional narrative history, rather than a substitute."[48]

Another historian, Edwin Van Kley, concludes, "History is to serve present society as its cultural memory, and . . . this can best be achieved by carefully reconstructing past epochs and events in their full variety and complexity, by artistically telling the stories, and by writing it all in ordinary, jargon-free language for the general reading public."[49]

Idiographic Richness in History

In a kaleidoscope each pattern is different but there really is nothing new, always and only shifting recombinations of an old repertory. History is kaleidoscopic, and more. In embryonic development, each stage is only an unfolding of the genetic program, recycled in each new generation. History is reproductive, and more. In natural history, the acorn becomes an oak, which becomes an acorn, which becomes an oak, but the latter oak is not the same as the former. It is a mutant, a new being. Summer turns to winter, winter to spring, but the next year is not the same as the last. It is a new chapter in the story. In cultural history there are Abraham, Isaac, Jacob, Joseph, fathers and sons all over again, but there are Moses and David, Isaiah, Jesus, Augustine, Aquinas, Luther. There are Babylon and Egypt, and Israel, Greece, Rome, the Holy Roman Empire, Britain, the Soviet Union. Things have proper names, and if in common usage we bother to give proper names only to persons, places, communities, events of special interest, we could in principle give proper names to all the individual events that succeed each other, to every star, mountain, dinosaur, or oak tree. The succession of these proper names forms the idiographic historical narratives.

There is enormous repetition going on here, with discernible laws. A law is a collection of recurrent cases, but how similar? Well, similar enough to come under the law! Similar enough that the margins of error, where our law fails to fit the real, nonrecurrent world, do not bother us! The law and the similarity codefine each other, and banish the differences. If "law" seems overstrong, we can speak more weakly of statistical "generalizations," so as to allow the easy ousting of poorly fitting cases. Or we can speak of developmental "trends" and allow each stage to be different from all others, understood despite that in its continuity and likeness. Laws, generalizations, trends enormously help us to understand proper-named events in their similarities. They help us get at the nature of, the Nature in, a thing. But they do not really touch their crucial differences.

Single events, even in the sciences, are unique, though often in trivial respects, and there we can abstract out properties that are similar enough to be brought under covering laws. Single events in history, differentiated and standing alone, may not be interestingly unique, but when they are integrated into complex narratives to become a biography or a storied history, they often become interesting. What is unique is not just the event but the event in its narrative. Properties can be abstracted from some unique events and brought under causal law, but such properties cannot be abstracted from narrative lines so as to provide complete insight into the intelligibility of those lines.

What we want is a complete account of the changes, but since each new phase is a bit unprecedented, we cannot have such an account in a covering law. The fittest survive, over and over again. But why does the Indian rhinoceros grow one horn, the African rhinoceros two? Why this time are mammals the fittest and not reptiles? Why this time is there an increase of cranial capacity and not of muscle? Why this time is it cooperation and not competition, why once was it power but next ethics that saved the day? Why this time was it the Israelites and not the Egyptians who made history? Sometimes we want to answer that the difference was only happenstance, but if we always answer so, the detail of development quite escapes the law that so loosely embraces it. Sometimes we want to answer that the differences are causal enough, but result from the impingement of unrelated causal lines, interactions too messy for us ever to write causal laws about (the plate tectonic movements that separated Madagascar from the African continent and preserved there, owing to reduced selection pressures, lemurs, which became extinct on the mainland).

Physics and chemistry achieve their explanations too cheaply, by leaving to the inexact physical sciences—geology, geomorphology, meteorology, climatology—all the vicissitudes of Earth's course. Even these latter sciences gain their laws by overlooking the proper names—the Shenandoah River, Mount Saint Helens, Hurricane Hazel, the Wisconsin glaciation in the Upper Pleistocene, Earth—that are the actual field of natural history. These names may be reduced to index points, instances in a geomorphic theory about peneplains, volcanism, weather patterns, or planetary formation, but the particular actual fact is richer than the general explanation, which discounts some of the reality. This discounted element, however, over a long enough course, demands entry into these sciences. In historical geology, there is a time flow of periods—Cambrian, Pennsylvanian, Triassic, Jurassic. Here, we first say, there is only orogenesis, erosion, sedimentation over and over again. But any who notice the fossil record will see that in biology at least there is more than mere recurrence. There is the singular, nonrepeated, and nonrepeatable evolutionary odyssey. There are cycles, but also spirals and vectors, axial lines. Confronted with the richness of Earth's course, the natural sciences do a little explaining and then lapse into randomizing, with little theory for the branching and ascent.

The anthropological sciences may describe the recurring evolution of societies, and this helps to make some preliminary sense of the hopeless array of cultures. Societies move from hunter-gatherer to tribal agricultural communities, then to chiefdoms and petty states, then to agricultural states, and finally industrial nations. But why are there long arrested stages, then again sudden bursts of development, declines, extinctions, reformations, transformations, and axial lines? Why does India take this route, China that route, and the Mediterranean—Israel, Greece, and Rome—still another? Confronted with the richness of historical material, the social sciences do a little explaining, but then stammer and grow vague, especially at mutation points where the future is novel to the past.

We are not just getting recombinations of what was there before, similar cases all over again, unfoldings from a predictable or causal process. We are getting unshared novelty with significance. It is again and again in the historical developments on Earth (the appearance of life, of sentience, of subjectivity, personality, Buddha, Jesus, England, the civil rights movement) as more complex and more surprising than the nonhistorical processes (planets circling the sun, sodium and chlorine combining to form salt, enzymes digesting food, hot weather producing irritable humans) that the scientific form of explanation seems inadequate, that the consequents seem something more than the causal results of antecedents.

Lawlike explanations, causes, or trends do not handle the appearance of new "information" well, and it is often in just the information—in genetic codings, learning structures, revelations, or the expanding of cultural libraries—that the later events differ crucially from the predecessors. Repetitions in newly informed contexts may bear revised meanings. Science tends to understand humans by suppressing their unique claims and adventures, by showing how they obey natural laws, how they too have been naturally selected, how they have a psychology or a sociology in common. But each human has also to be understood by situation in her idiographic history, and humans have to be understood collectively in their narrative histories. The latter is the context of meaning (where many causes also apply); the former, lawlike account can feature causes that may or may not complement meanings.

Whether the historical and theological disciplines can handle this distinctive material either is not yet evident, but at least they are prepared to notice it. They worry that any scientific account will fail by half, or more, because it values laws and causes, and disvalues the novelty. The historian, insists Daniel Boorstin, director of the Library of Congress and a senior historian of the United States, is "the high priest of uniqueness. If the historian has any function in the present welter of the social scientific world, it is to note the rich particularity of experience, to search for the piquant aroma of life. . . . The historian as a humanist is a votary of the unrepeatability of all experience, as well as of the universal significance of each human life."[50]

The Bible, for instance, is a book of surprises. A simpler theology will find there laws of God, a plan of salvation, commandments timelessly reapplied across each generation. But there is also Abraham's memorable call, Israel's unlikely Exodus, its unprecedented survival in Exile, Judaism's mutation into Christianity following the birth of a child in a manger, crucified on a cross. Unexpected events continue as this emergent faith conquers an empire, and is itself periodically reformed and preserved by its very mutational diversity, from New England Quakers to African Nazarites. The story is about the steadfast faithfulness of God, but this plot is found in ever-surprising places.

The surprises "make history," even though the laws also count. A black woman's feet hurt on a bus. She takes a notion not to budge, despite the driver's request for her to move to a rear seat. She is, predictably, arrested. But Martin Luther King, Jr., is moved to respond, the Montgomery protests are triggered, a civil rights movement is rallied, and the course of American history is different. One may say that something else would have triggered this, if she had not, for the time was ripe for such events. Individuals are just particles across which social waves travel. Perhaps. But perhaps events would also have been significantly different under a leader less committed to nonviolence. The individual particles do some directing of the waves they bear. Personal ventures do supply the mutants, which are selected only if the *Zeitgeist* is right, and yet the *Zeitgeist* can do no selecting if the mutants do not come forth. All the creativity is in the individual increments.

A cop on patrol looks again to notice a strip of paper tape over a lock, and guesses that someone has rigged a clandestine reentry. This launches events that topple a president and force a nation, again as in Montgomery, to soul-searching over its values. But without Watergate Richard Nixon might have stayed his full term. Without Three Mile Island, and a reconsidered nuclear policy, greater disasters might have ensued. Such accidents—the Chernobyl reactor explosion or the Challenger space shuttle tragedy—mix equipment flaws and human misreactions and they can change the courses of Soviet and American history. Individual events make a difference. Luck and contingency, genius and resolution shape the distinctive picture. The very notion of great persons suggests that some individuals have import, far advanced beyond others, but this can also be true of particular peoples. Take away Jesus or Muhammad, or Israel and England, and Earth's narratives would be something else in their axial streams.

If we consider the complete person, each is an idiomorph, in a class by himself. Each is, no doubt, in many shared classes (British, a jogger). But each occurs in a specific space-time place on a once-only world line generated by travel through a once-only environmental setting. To adapt language from physics, here relativity generates uniqueness. Each person has a distinctive genetic combination, operates in his own reference frame at a particular, never-repeated stage in a developing culture. Each coagulates the events of his career out of open potentialities, like no other person past, present, or future.

Jimmy Carter decided against the B-1 bomber in June 1977. In understanding why he did what he did, psychology and sociology, so far as they give us universals of personality and social forces, are not going to help much. No explanation can afford to ignore Carter's roots—the South, the church, the soil—or his leadership style, or his career stages—Annapolis, his turns as engineer, businessman, farmer, governor. Without biography, we can never really understand individual choices. But biography is never science, and is again and again the key to particular history, which is the only kind of history there is.

The Dramatic Character of History

Historians seldom invoke covering laws when they give explanations. They recount dramatic sequences. They do not, as might a botanist, place a thing in its genus; they rather offer a story of events that took place. It is occasionally illuminating to classify the event—a civil war, a rebellion, a rural-industrial disequilibrium. But the mere taxon is an impoverished explanation. Scientific statements, ultimately, are logically plural; but historical statements, ultimately, are logically singular. Scientific statements and historical statements may both contribute to a pattern, even a developmental pattern, but the one pattern wants sameness; the other wants drama. Here is one civil war, like twenty others, but unlike a half dozen that turned out otherwise. But what usually happens in civil wars is only one element of explanation in a unique chronicle. One wants the epic of the War between the States, the detail of the Dred Scott decision, of John Brown's raid on Harpers Ferry, of Lincoln's decision to send federal ships into the Charleston harbor. One is not so much satisfied by an explanation that shows how this war was like others, as by an explanation that can find a plot in the events—turning points in the history of slavery, citizenship, states' rights and the federal union, democracy, freedom.

Phases of nature do not have much history, at least in short-scope runs or grossly abstracted. This planet circles the sun as do they all, with counterbalanced centrifugal and gravitational forces. Even here, finding eight more cases like the first does not explain what it makes familiar, not until we see a higher-order pattern, a logic that comprehends these instances. Meanwhile, planet Earth, circling on, undergoes a steadily unfolding history, very different from that on Saturn or Jupiter, and here there cannot be simple deductions from a covering law. The interesting part now is not the repetitions but the dramatic narratives. A law, "The fittest survive," covers repetitions, but the larger drama has, in fact, taken place only once, whatever parallel evolutions or polyphyletic origins may be subsets within it.

There is no class in which Earth and its history can be put. Covering laws can be stretched out into genetic laws, trends, and we may come to see the shape of the curve. Forms with exoskeletons remain small; those with endoskeletons grow larger. Neural power trumps muscle power. New groups tend to arise from small, unspecialized ancestors. Forms within a group tend to grow larger. Societies in temperate climates outstrip those in the Arctic or the tropics, where environments are too harsh or not challenging enough. Societies at crossroads are more creative than those in isolation. But the accumulation of these explanations does not relate the full historical odyssey.

In narrative explanation, one does not want a covering law except in a subsidiary way. One wants a covering historical model. Events are not to be joined with others of like class so much as with their precedents and consequents, with what led up to them and what they led to. Scientific laws interpret things in terms of causes and predict effects, what went before and what came after; but historians interpret things along narrative lines that include what came before and after as the beginnings and endings of the stories in which the actors played their parts, actions the significance of which often becomes clear only decades or generations later. One gets the plot, a moving picture, a theatrical show. One wants causes, where these exist, but one also sees events as parts contributing to wholes. The narration of a crisscrossing series, together with related explanations of situations, conditions, analyses of character, success, failure, glory, tragedy, is en route to catching the meaningful patterns, the connections between which are not simply serial but gestaltic. Perhaps there is not some single correct description of every particular historical sequence; there are many stories interwoven, along with story fragments. Still, sometimes there are genetic developments beyond causal effects, unfoldings, upslope climbs leading toward some ends, great or small, culminations that may give way to new beginnings.

The Torah does not work like Newton's laws to define a class; it rather calls forth a chosen people, and under this historical covenant Israel is to be understood, past, present, and future. Jesus, with his proper name, is the Christian model, who does not so much include his disciples in a class of Jesus people as call them into a church community with its ecclesiastical history. The model here is the sort of thing that led persons to put all those sixty-six diverse books, each detailing its local stories, into one book and call it a Bible. The theme is universal, catholic, just because it is dramatic and historical, many stories, one Story.

But the Earth story too is a kind of calling of special pilgrims up in a special place. It is another Exodus, another land of promise, another heroic fight for survival, for freedom, for learning truth, the primordial exodus of spirit from matter. Perhaps the best that rigorous science can say is that we have random mutations selected over by the law that the fittest survive. But these achievements have also been so channeled as to form stories, and if science cannot supply narrative principles for the ascending journeys, then history, philosophy, or religion will. Natural history, properly conceived, will not be despised as taxonomic classification, museum work; rather, it is the most insightful work of all, that of lifting out nature's most crucial thrusts and placing them along narrative wholes. That converts mere chronicles into a saga.

One reason that evolution is a much richer and more welcome theory than was the former belief in fixity of species is that it makes possible vaster depths of story. Here the recent tendency among biologists to question slow, steady gradualism in evolutionary development, and to posit instead punctuated periods of explosive development, other periods of stagnation, and still other crises of life, can further enrich the story. Derek V. Ager, a paleontologist, concludes a reflection over the stratigraphic record: "The history of any one part of the earth, like the life of a soldier, consists of long periods of boredom and short periods of terror."[51] That is remarkably like cultural history and personal biography; and the military metaphor covers what are both evolutionary and revolutionary developments, all ingredients in narrative crusades. In that sense, even those who are not prepared to say that the natural order is a designed system may nevertheless be impressed with how nature, inclusive of natural selection, is a story-spinning system. It no doubt selects for survival, but its arrivals tell ongoing stories, those of contributions, conflicts, and crises in a "storied fight" from matter to life, life to mind, mind to culture, and culture to spirit, a fight through good and evil.

Persons are distinguished from machines by how they possess historical properties. Actually, even machines do not escape having their histories, as when an engine's metal fatigues during the events of its use. But in its machinelike properties a machine is lawlike and timeless; it has no narrative. When machines come to have developmental histories, we begin to think of something more, perhaps next of a clockwork system such that the laws of the system, together with initial historical conditions, determine everything subsequent. But natural history is not really that kind of developmental machine either. It is not a computer on program. In a historical system of the earthen kind, accumulating experiences are added into the character of the product. These further exploits are carried forward to become later-coming levels in explanation of events that ensue, and the stories over time make sense. There is adventure; and the adventure moves on, and some of the adventures cohere.

One does not want mere narratives, just stories, good tales, but narrative models, the analysis and synthesis of events under governing models, historico-critical models that catch the meanings in the stories. One wants a principled narrative, a systematic story, not merely lawlike repetition. Nor are we merely saying that *human experience in the world* is inevitably in story form (while the world itself might not be storied). We are claiming that *the world is story,* eventful sequences that were taking place before humans appeared and out of which humans have appeared, who continue consistently with their world environment, enlarging the story and alone able to enjoy the story form of the world.

Narrative-laden History and Theory-laden Science

Science does not dissolve history into nested sets of classes with lawlike operations. Rather, it is the other way around: history absorbs every bit of scientific work. No science is universal and timeless, but all sciences are historically conditioned and evolving. One can date a bit of science—the theory, to say nothing of the apparatus used—about as one can date a fossil, an arrow point, a work of art, a novel, an old wireless set, in its historical epoch and setting. The "survey of the literature," which characterizes the beginning of scientific papers, is really history, a recounting of previous developments in this particular story. We will need to understand Newton, Einstein, Darwin, Durkheim, Skinner, Freud, and their theories as historical stages through which human intellectual history has evolved.

Theories too have their stories. Like religion, science is part of our life story; it is still more exodus into truth and freedom. Each new theory is another adventure in travel and exploration. If, earlier on, science does much explaining for us, later on even science itself becomes part of what needs to be explained in a larger, historical creed. At this point those religions (among them Hinduism, Buddhism, Taoism, and sometimes monotheism) that seek to abstract out the universals (Brahman, *nirvāṇa,* the Tao, God) are frequently to be faulted not because they are nonscientific, but because in this hope they are too scientific to be adequate to historically occurring reality.

Even those who think they have timeless, abstract, universal, rational, objective views in science and religion, devoid of story elements—they too, as much as any of the others, have been caught up in some of the stories. Further, they may have to puzzle whether their story-less accounts fit well with the objective reality of the histories of religion and of science that they inhabit. Likewise, stories that are not history (such as the Genesis myths, or Milton's *Paradise Lost,* or Harriet Beecher Stowe's *Uncle Tom's Cabin*) can shape history. The course of Western history has been what it was under the meaningful influence of its myths, creeds, and novels.

"Narrative" is used here in a broad sense that includes storied history, characters and events moving through time and space, through challenge toward achievement, through conflict toward resolution. In storied history, as distinct from fiction, the plot is detected and evidenced in the events, not merely "made up" and inserted by the storyteller, although the narrator's vision determines what plots she can find and it may, on the cutting edges of history, reform the plots she finds.[52] Here once again, as has been true in science, what we catch is a function of the nets with which we fish; the data in the best of the sciences are theory-laden. The events in the best of storied histories are caught up into the plot because the constructed plot enables the interpreter to catch them, while it is likewise true that events newly taking place or discovered can falsify a purported historical story. They can tear our nets and force us to mend them or make others. In this respect social science history, with its data and statistical analyses, its quantifications of voting records or land holdings or immigration records, is welcome for its capacity to test, corroborate, or falsify hypothesized movements in the stories.

Just as the facts of science are theory-laden, so the facts of history are story-laden. What the history was, or is, or will be depends on the stories we hold—as certainly as our stories can be tested against and reformed by events. One has to make up a story to catch and interpret the history, just as one has to make up a theory to catch and interpret the empirical facts. Recounted events are story-laden, but this makes them no less true than the theory-laden data of science, where theories are "made up" too. Differential calculus is as "made up" as the Exodus story, but both may capture world events. *Plot* for narrative history does what *theory* does for science. Both have to be justified in experience. Both may require one to ignore or argue away some of the data, or to live with anomalies. Both have to stand against rivals. A center of gravity is as much and as little fiction as is Jesus at the center of history, but the latter involves a plot, the former only a law.[53]

The world, it will be complained in reply, does not always come in the form of well-made stories, with beginnings, middles, endings, with plot, resolution, closure. It is often a mixed bag of episodes and incidents. Narrative fits the world no better than law, because both have to leave out anomalies to corroborate the model. One has to fictionalize to impose narrative on the real world, rather as one has to abstract and idealize to impose law on the world. That may be so, and there is here no claim that the world in all its minor eventfulness composes well-made stories, not even short stories, much less novels. Ends can often be new beginnings, and fragments of stories are often in collision. Stories come in nested sets, and there are superposed story possibilities only some of which coagulate in the actual course of history. Some events may not be intelligible or interesting or significant; some are like mutations and worthless trial ideas that, though parts of groping stories, get selected against. In the background of the storylike signals there is meaningless noise, events without coherence or point, just as there are noise and randomness in the background of theories and causal laws.

But there are also stories bubbling up through the noise, and often what narrative can do, made up though it is, is catch the story patterns, and with this the meanings, just as law and theory can unscramble the causal patterns. What natural selection does, for instance, beyond preserving the survivors who are better-adapted, is by trial and error to spin advancing and diversifying stories. The laws provide the algorithms out of which the stories are told. The noise is the openness superimposed over which stories can sometimes develop, and, over time, always develop sooner or later. There is always some level at which any event can be caught up in a story, and here events that are too messy for us to write causal laws about are often not too chaotic to be lifted up into a story.

History need not always be narrative; it may be analytic, comparative, topical, even statistical. It may record chronicles and data, or describe in cross section a slice of the past. Historians may make maps or paint portraits of the past. But the world is not a map, and lives are not portraits; the dynamism is missing. History is never finished until one searches out how many stories there are and whether they have meanings. History is not solely narrative, but it is necessarily eventually so. One does not have all the explanation until one has the stories, although subroutines of explanation may not require story.

The world is in that sense as irreducibly narrative as it is causal. One has to formulate plots to detect the stories, no less than one has to construct theories to find causes systematically. The plots mix facts in the world with artifacts of the narrator, no doubt, but this is just as true of the numbers of science, which sift for facts in the world using artifacts of the theorist. Both enterprises can err and fail, they can caricature, but they can also sometimes be right, as right as right can be, whether in science, history, or religion. The logic is: if S, then O. If this were the story (S), then this would be the observed evidence (O). Only now we are in less hurry to think that there is only one right story, analogous to one finished theory; there may be such richness that multiple stories—at diverse levels, complementary, parallel, and competing—will be required and welcomed. Both science and history hope to tell about the world in a successful way, and they can really succeed only when they carve nature and history at the joints where they have articulated themselves.[54]

Life is thus a narrative text. We see this foremost in the idiographic histories of particular cultures, but this paradigm for explanation, so far from being epiphenomenal and reduced away when one descends into physics, chemistry, and natural history, rather overpowers the simpler scientific kinds of explanation used there. Earth, indeed Nature herself, is an idiographic text to be interpreted. What is undeniably there is not just matter, energy, law, information, consciousness. Above all, it is *story*.

Meanings in History

To make connections in the narratives of history is eventually to seek meanings. It is sometimes said that history studies events with "insides" to them—what persons thought and intended—while science studies events that have only "outsides," causal connectedness without inwardness. Such a contrast fits only some science, for the social and psychological sciences study subjects, too, and even zoologists and ethologists study animals that are centers of experience. The difference is better put by saying that natural science is satisfied with causal hookups, as social science may sometimes be, and that even when social science investigates meaning systems in its subjects, it is satisfied with finding out how ideas or creeds function in a community. But historians want to find out more than this: they want to know what were the meanings in history. This moves them from meaning systems functionally in a society to the meaning of a singular stretch of history—for example, the War between the States, as this contributes to the ongoing American story.

Historians preeminently notice what sociologists, anthropologists, and psychologists also notice, that they cannot always, indeed cannot often, use the word "cause" as natural scientists use it of nonhuman nature. For their *causes* now dominantly include *reasons* in the agents. Their *facts* include *acts.* When historians use "since," "because," and "therefore," they are usually following an intentional stream along a checkered course, tracing (at least at first) the meanings these events had to the actors. These intentions can, if you like, be called "causes," since they control behavior. But what are the causes of these causes? In dominant part, the "causes" are meaning systems in the agents, interspersed with decision points, as a creed is tracked out, inlaid onto, or modified as a result of singular adventures. The word "cause" in history has meanings and deliberations woven through it, superimposed on whatever naturalistic causes remain present. The hookups in the narratives are not simply causal in a scientific sense and are not to be understood by reference to like causes. They are meaningful at least so far as there were actors underlying a singular stretch of events.

This leads some to deny that nature has any history, on grounds that historical events must have "insides" of the type that form a storied awareness. R. G. Collingwood, a philosopher of history, wrote, "Nature has no history."[55] The passage of matter and energy through the world can be merely causal, but the passage of consciousness through the world will be reflectively narrative, and this will require the category of meaning. This bifurcates the world into nonhistorical and historical sectors. The processes of nature are causal sequences of mere events, but those of history are thought processes.

But the truth is more complex than this. Stories may have meanings, while mere laws are hardly complex enough to have meanings. Still, when the causal chains begin to produce stories, as happened on the primitive Earth (and had in some sense been happening from the big bang onward), one needs more than *causal* or lawlike explanations to understand the emerging drama. One needs to know the *meanings* of the storied events, a category more religious than scientific. After life begins, there are events informed by information coding. Though without awareness (since before the threshold of consciousness), there is a historical tradition, the communication of narratives in genetic form. Thus, a single living cell, unlike a merely physical crystal molecule, has a billion years of history in it, which have made it what it is. It has storied information, a "tradition," as inorganic molecules do not, although one might say that even atoms are fossils of stellar events. Tigers are historical natural kinds in ways that oxygen is not. If physics can understand things in great abstraction and with no attention to history, certainly biology cannot.

T. A. Goudge, a philosopher of biology, insists, "Narrative explanations enter into evolutionary theory at points where singular events of major importance for the history of life are being discussed. . . . Narrative explanations are constructed without mentioning any general laws. . . . Whenever a narrative explanation of an event in evolution is called for, the event is not an instance of a kind, but is a singular occurrence, something which has happened just once and which cannot recur. . . . Historical explanations form . . . an essential part of evolutionary theory."[56]

The passage of life, as well as consciousness, through the world has been narrative. It is not true that humans are historical beings, while animals are not. Tigers are historical beings, but they do not know it; humans are historical beings, and (some of them) know it. Even the geoplanetary incubating steps, such as the hydrologic cycles and generation of the atmosphere, are prolegomena here. En route to evolution's becoming conscious of itself, events had their significances as memorable preparatory stages in the dramatic achievements of life. They were remarkable, worth noticing, even though there were no persons around to notice.

Likewise, even in human history the meanings of narratives may be unknown to the actors, and seen by others only later, in retrospect. We cannot study past events on their own terms, because they were our terms. They were greater than they knew. They had impact. This is true of the trilobites, the lemurs, the Neanderthals, the Israelites, the Hussites, the Puritans, true of the first writing of the code for hemoglobin, or of the Deuteronomic code, or the Magna Carta. There were issues, crises, and ends not always in view. They were meaningful for their time, so as to be meaningful beyond their time.

But this influence is not that of causal law; rather, they set patterns of meaning that persisted and proved seminal. They have to be understood in terms of categories that were not available at the time of the events, for the events themselves are the first stirrings in the generation of these categories, through which they are afterward interpreted. We will need the flashbacks and foreflashes that are so common in narrative accounts, because they show the logic of the story. Events make more sense in retrospect or in prospect. We will need to know the interconnections with complementary or rival narratives. We will need evolutionary and revolutionary categories not available or not fully available to the actors.

History is not restricted to discerning what its actors thought (although it is written falsely if this is misjudged), and history does not fade away where there are no thinking actors. We do not write true history until we include everything that went into the significance of the story, and trace the significant connections. Very much went on over the heads of the dinosaurs, and over the heads of the dignitaries at the signing of the Magna Carta—even though history also turns on what goes on in the heads of its prophets, princes, and presidents. They sometimes aim higher than they reach.

Perhaps there is no single "great meaning" of history, or perhaps we may not be able to discern this until after a long time. Perhaps the world that nurtures plurality will refuse to be simply summarized. But meanwhile there may be multiple meanings in history, both natural and cultural, nested sets of meanings that we are challenged to detect; and we are likely at least to hope for some unified theory about the whole. We can ask *when* history has meaning long before we ask about *the* meaning of history. There can be piecemeal episodes that have intrinsic value without necessary contribution further. The thrills of the hunt, the song, the dance, the poem, the rewards of springtime and harvest, of parenting and maturing, love and friendship, craftsmanship and leisure, the joys of laughter, the pangs of tragedy, the day spent at play, the life at a task—these are recurrent meanings in the routines of each generation. But they do not take place isolated from a distinctive narrative cultural flow. They are woven into and spiral out from and around it.

Indeed, without the orientation that culture supplies these degenerate into wearisome anomie. There will be worthwhile personal values, local careers that do not play any constitutive part on the grander scales, though even such noncontributors are likely to be defined by their traditions. The mere transmitters of a heritage will share in its meanings, as well as its reformers. History contains trials and errors, mutations, cults, movements, most of which are nonconserved. It drifts in part, but not in all parts, sometimes not in regional trends, and perhaps not in the scheme of things entire. History can be a trail of mistakes as much as of successes, but persons can learn from the former as much as from the latter. There can be meanings for the better and for the worse. Vital parts of the historical experience can become cruxes and keys for longer reaches, and one can be religious about meanings in history even while keeping in abeyance the question of universal history.

All cultures have considerable meaning, but this does not require that they all be equally meaningful, any more than individual lives are equally meaningful. Nor is it necessary that the meaning of a culture, or career, be everlasting. Things can have point (both inherent meaningfulness and instrumental significance) without permanence. In both social life and biological life, meaningful forms greatly outlast the individual. Even the extinction of a species or the collapse of a culture is typically followed by other forms of equal or higher value, using the achievements (hemoglobin, photosynthesis, the wheel, writing, the Magna Carta) of the supplanted forms. It is myopic to say that Neanderthal lives had no point, or that bird or beetle lives are without meaning (although the latter two may be without experienced meaning). These lives have enriched the world by being what they were and are, intrinsically and instrumentally.

It is a mistake to say that the three million hunter-gatherer years were bad years, and that only the last two hundred years, since the Industrial Revolution, have been good ones. The hunter-gatherers generally enjoyed good health, we are told, and rarely worked more than eighteen hours a week! They ate a wider variety of foods than we do, were less warlike, and were more independent of remote powers and markets. They had fewer unsatisfied desires.[57] The preliterates were not pointless because they were not modern, just as we do not pity ourselves because we do not live five hundred years hence. After all, some of the most historic achievements were made in those (so-called) prehistoric years: brains, hands, language, culture, fire building, clothing, toolmaking, agriculture, conscience!

At the same time, there has been progress, and we welcome it, with its mixed blessings. There is something noble and irreversible about the forward moving that has taken place. While one can enjoy local meanings without religion or cosmic theory, one does need a world view for systematic storied meaningfulness. Disconnected, anecdotal meaning, while possible, is less meaningful than connected, integrated meaning. Too grand a theory, like too large-scale a map, leaves out too much detail, and the supertheory can be provincial just because it does not attend enough to all the provinces of history. But if there is no inquiry about supertheory, the full story can remain untold.

It is important not to be too Western or too anthropocentric, and one wants to leave space for independent dramas, intrinsic values noncontributory to the whole, and for wildness and otherness, since even anthropocentrism is a form of provincialism. Here some will say that any imperial theory of the whole forces too much condensation, too much Procrustean cutting of data to theory, too much overlooking of diversity and plurality. But the individual always lives in an ecosystem and cultural system, and, modulated over all local events, loosely but also definitively, there are certain irreversible and linearly dramatic themes. These do in central motifs come to a head in humans, who dominate the globe, and some of these motifs in recent centuries have flowed through the West, as seen in how traumatically Western science has revised the human understanding of our place in nature.

History is in one sense fixed and unalterable; the past was what it was, and no one can revise it. But history is differently seen with the present formulating of novel approaches. The level of involvement here deepens, although we have already begun to recognize this in considering the social sciences. The historian cannot supply paradigms of meaning (which may complement or contest those in his subjects) without having these abut his own commitments. A scientist, in contrast, might supply paradigms for causation without such challenge. The kind of narratives the historian writes looks to draw morals, if such there are, from what one studies. There may be pure history, which avoids application, but any pure history that finds meanings, advanced beyond the scientist's causes, is going to be tempted to cash in on such meanings found. Meanings have a logical leaning toward relevance. One asks not merely "What next?" as in a chronicle. One asks "So what?" as in a plot. One wants not merely an account, but to know what in it counts.

Historians are on the front line of the history they write. The doing of history is the making of history, and their transformations of the past are themselves historical events. One has to write the records, but also to tackle them, and to live on in the outcome of that struggle. That is a matter of the uses to which historical conclusions will be put, but it is also the recompounding of the story.

The hypothetico-deductive method, as this may characterize science, must pass over into a historico-critical method and find meanings in unique, dramatic narratives. The sorts of cultures we build will depend in part on how we write natural history. Our modern technology, for instance, has followed the era of Newtonian mechanism and is importantly its product. So the story is carried forward in the light of what we believe (on the basis of science, interpretive history, and religion) is going on. Our cultures even determine, in the Earth story at least, what the course of natural history will henceforth be.

Historicism and Directionality

It does not take much history or anthropology to disabuse social scientists of the naïve belief that they have discovered much of the permanent logic of the human mind, of the way all cultures must work. Any fundamentals that we do get, cutting across history from Neanderthals to New Yorkers, are likely to prove too general to illuminate these two very different historical contexts, or the others in between. Part of the problem lies in contingencies, but more lies in how humans are self-defining animals, not from scratch in each generation, but at some choice points along their route. Part of the problem lies in the growth of knowledge and value, with shifting standards of rationality and worth. When we look behind, judgments that once seemed coherent or just may no longer seem so. We cannot fully envision a century ahead what it will mean to be logical in the light of what has then taken place or come to be known.

We cannot foresee what will count as a finished theory, a satisfactory explanation, a just society, a worthwhile life. No one could predict or specify what evolutionary theory, relativity theory, non-Euclidean geometry, or genetics had to be like before those theories came along. We did not know even the form of the theory before its arrival. No one could say in advance what the religious response to such theories had to be. No prophet can predict just who will be the twenty-first-century prophets and what they will find unjust and ungodly. What counts as good science, good religion, or even good logic before evidence alters over time. Radical conceptual innovation cannot be predicted, whether in science, ethics, or religion.

Does it follow then that historical development is not predictable or retrodictable, and, in turn, that one cannot have a science of history, since the formal principles are not and never have been set—not firmly enough to control the course of events? In every natural science, notably in evolutionary theory and quantum mechanics, we do not know with a full set of hard causes why this event and not that occurred. Nevertheless, we know statistical trends, and have limited explanations of what is going on, for example in the growth of the brain case or in the half-life curves of radioactive decay. But when the plot thickens to include persons with their storied awareness and decision-making capacities, can we say anything at all about historical trends, past or future?

One might think that with statistical laws, economic forces, natural selection pressures, and the like, we cannot make specific local predictions (when Sam will die), but that we can make long-range, overall predictions (what the death rate in Los Angeles will be). Trends average out over fluctuations. The individuals rattle around in the statistics but make no difference in the outcome. But in another sense it is just the long-range predictions that are hardest to make, because we do not and cannot know what the emergents will be. We do not know what policy decisions concerning carcinogenic substances will be made, or what medical breakthroughs will come. So one would be foolish to predict the death rate in Los Angeles a century hence, though one can easily predict where the moon will be then. Astronomy does not study a mutational, informational, learning, or policy-deciding system.

Perhaps celestial mechanics is only one sort of science, and evolutionary science a more typical kind? It explains after the fact. But even evolutionary science is powerless to show us why much of what has happened had to happen. Some events, perhaps, are not unexpected. But we are forbidden to put much stock in natural trends, and cautioned that the upslope phylogenetic sequence is only random, without explanation. In this most would-be-historical of the natural sciences, we are not given any genuine predictive or retrodictive capacity. Perhaps social science can be a more competent predictor and explainer of what will happen, knowing as it does about the functions that undergird history, about the laws of society? But we have seen even social science falter over history, and likewise when it takes a forward look. There is no way to predict what use can be made of a "defect" in a crystal lattice, not from theories of crystals, theories of society, or trends in history. But we are now seeing social revolutions taking place and global military strategies revised, owing to computers, springing as these did from the discovery of transistors, depending as these do on crystal defects.

Every historical prediction is made from an open set of premises, although no historical event occurs without issuing from its antecedents. There are limits to what can happen, probabilities to what will happen. One might hope for a historical science that will understand enough of the connections between things that have been going on so that it can guess well about the future. In the subroutines of history, this can perhaps be by discerning like kinds and by inductive generalizations about trends. But in the overall and once-for-all directions this will need that sort of predictive sense that comes from insight into how a plot is unfolding. Such a belief that history has its discernible trends is called, usually pejoratively, historicism, or, more favorably, a belief in directions in history.

Historicism is a hope of science quite as often of religion, seen in Comte's law of the three stages, predicting that religion would disappear, or in the Christian hope of a coming Kingdom. Historicism can be reductionist and naturalistic, but also holistic and cultural. Broad social forces overpower individuals; historical trends are inevitable and irreversible, as, say, the growth of technology or increased diversification of labor. Here historicism may be objected to not only as being false, but, were it true, oppressive, a juggernaut suppressing any liberation of those who believe in it. But with other kinds of historicism the coercion is logically persuasive. We may not object to the overruling of individuals, persons, and movements, ourselves and our denominations included, where individuals and their local groups are in the wrong, or are ignorant of the more worthy ends they serve.

None of us would object if we could find in the evolutionary sequence an upslope pressure toward sentience and intelligence, or if in the sequences of civilization we could detect a benevolent yeast working for more rationality, love, freedom, and recognition of human rights. Most will not object to increased understanding of, or control over, the environment. G. W. F. Hegel held that humans move toward rational goals, often in spite of themselves, and that they become increasingly free as history develops.[58] Arnold Toynbee found that civilizations rise and fall, even in their declines with overall an increased awareness of God.[59] Alfred North Whitehead could say, "Religion . . . is the one element in human experience which persistently shows an upward trend. It fades and then recurs. But when it renews its force, it recurs with an added richness and purity of content."[60] Such theories may be false but need not be oppressive.

About all we can say is that the shadow of the past often outlines the future. Some lines stagnate, some die out, but others develop, and there is a growth of knowledge, technology, industrial power. There is increased urbanization, increased connectedness between regional societies, and, in some nations, the growth of democracy. We can predict that when Europeans encounter Bushmen, the Bushmen will (over a few generations) more likely be converted to the European views than will the Europeans be converted to the Bushmen's views. There is the growth of universal faiths in replacement of provincial ones. There can be loose thrusts in history, even if there is no determinism, and despite the surprises. There are brooms that sweep in the same direction, even if there are also revolutions and new directions.

There really are no repeat performances, but there are one-way tendencies in the natural and historical phenomena. There is in culture something analogous to the informational climb in genetics, and this too is not everywhere in the global system, but in leading currents. These trends can remain even when cultures wander and wane. Seen in broad-enough scale, some upslope themes seem to be there; indeed, these headings are among the most impressive of the phenomena. But just what sort of "progress" they involve is a difficult element to get into a theory about nature and history.

Past what trends can be demonstrated chronologically (urbanization, growth of universal religions), the larger logic of national, cultural, or moral lures in history (sentience, rationality, freedom, knowledge, justice, spirit, the Christ) will take an act of faith to see. It will take a different faith even to call historical things nonmeaning, epiphenomena, empty, a mock show, an illusion, a superbly aimless random walk. Historical appreciation on this scale is not an act of scientific knowledge, and yet it can be rational even in faith, so long as the creed seeks correction in the events, is wary of how it uses theory-laden facts, and is at least as open as the historical system it evaluates. The most we can have is a *soft historicism* that detects globally developing themes in nested sets of local histories, with the sets in their kinships cumulatively forging world dramas. If we had more than this, if we had certainty of implication, the story would be compromised, since narrative logically requires surprise and contingency amidst constancy and continuity, all en route to promised conclusions. Likewise, the transience so often lamented in religions East and West proves in fact to be logically required for story.

A feature of scientific, historical, and theological thinking alike is that one needs to have it both ways on the question of inevitability. There are aspects of knowability and aspects of unknowability, and among the former some can be known in outline, some in detail. Among the latter some future events are, in whole or in part, there to be known but veiled from us because of our ignorance, and other future events are contingent, some random, some to be decided. One needs to be able to say, out of one's model, something of why the past was and what the future will be like, but also to have humans free for rational self-definition. Depending on the character of the inevitable power, if the model closes too much, it does oppress. But if it becomes too open, it has no unifying or explanatory power.

We cannot get more than the grasp of faith in a long-range model of history, and even in faith this cannot be made very coherent. But this need not be taken as an unexpected or dismaying defeat. Since we are dealing with emergents, and groping for what is going on over our heads, it is reasonable to presume that we humans do not have enough cognitive power fully to understand the whole dramas in which we are taking part, all the subplots, their cross references and relations to any whole, to distinguish between what must be and what may be. But the riddles of history can, on this reading, be taken as hints of a mysterious transcendence. These thrustings and headings, these freedoms and responsibilities, just because they involve us but are greater than we can fully fathom, are a sign of God. We need lessons from history to face a future that is unlike the past. We have to deny knowledge to make room for faith and to be forced to labor for the kingdom to come.

Thus, the nature that has been so powerfully revealed by science is still a nature after science that escapes its clutches. Nature casts us into an open and unbounded historical system that is past the powers of physics and chemistry to deal with, past the powers of biology and psychology as well. Nature casts us into cultural systems in which we are becoming responsible persons, called onstage to write the next act, a task beyond the judgments of social science, since the choice is between good and evil. Once again, but at ever-higher levels—past the necessities and contingencies of astrophysics and microphysics, past the necessities and contingencies in the evolutionary story, past those in personal and cultural life—now again in history it is neither the necessities nor the contingencies alone that impress us, but the mixing of the two by tandem turns to further the narratives and catch the upstrokes, telling this remarkable, mysterious story that is earthen-human history.

6. SUFFERING

Suffering has evolved with life; it too is among the emergents. The physical world cannot feel suffering; the cultural world is replete with it. On the evolutionary upslope, depicted earlier as an advance of formational and informational levels, there is a steadily increasing capacity to suffer. Indeed, the story could be titled, perversely, *The Evolution of Suffering.* Each seeming advance—from plants to animals, from instinct to learning, from sentience to self-awareness, from nature to culture, and often within cultures as well (as with our increasing capacities for warfare)—steps up the pain. In the prescientific world, suffering might have been caused half by others, half by nature, but in the scientific era this has become 90 percent by others, 10 percent by nature. In the stricter sense, pain comes only later in evolutionary development, after subjectivity appears; but even the objective forms of life, though devoid of inwardness, undergo duress. Written deep into the nature of the planetary drama, struggle deepens through time into suffering. We eat our bread with tears.

Suffering in Nature and History

This problem is not a scientific one, although the descriptions of science are relevant in framing the question. Suffering is not a feature of mere causal relations; there is no suffering in astronomy or geology. It appears in bioscience, though even here it exceeds causality, when the fight for life deepens into sentience. Irritability is universally present in life; suffering in some sense seems copresent with neural structures. There are endorphins in earthworms, which indicates both that they suffer and that they are naturally provided with pain buffers. On the other hand, in humans the relationship between bodily wounding or deprivation and pain is very complex, involving cognitive factors such as cultural conditioning and psychological evaluation of the situation. It is difficult to extrapolate to animal levels and make judgments about the extent of their suffering. A safe generalization is that pain becomes less intense as we go down the phylogenetic spectrum, and is often not as acute in the nonhuman as in the human worlds.

But when all these descriptions are made, science still only describes the facts, including any feelings (with which it has minimal descriptive power), and has no resources with which to evaluate them. The question metamorphoses into one of the meaning of problematic experiences. Even where a nonparticipant observes suffering, as with our earlier worries over the pelican chick or the anemic African child, the question is not simply one of what is going on, but of the value or disvalue of the events. To judge suffering to be bad (and hence to refute belief in God, or to confirm the Buddhist emptiness of the *saṃsāra* world) is a religious-order judgment. "Tragic" is not a scientific word. The question of suffering in natural history, in the idiographic historical narratives, or in personal *Existenz* escapes the competences of science. Yet it is one of the central issues we face.

The problem is so traumatic that many despair of much capacity to put the issue into philosophical or religious models. We cannot solve the problem of pain with words, for arguments abstract too far away from that immediate experience which is the only reliable context of considered judgment. The need is not to think better, but to suffer through, and learn whether it is worth it. Evil is absurd, the surd in life, not merely neutral or empty, but irrational disvalue. To explain evil rationally is a contradiction in terms; this would be a religious equivalent to giving a scientific explanation for random events. Some events do not have causes, not at the point of their randomness. Some events do not have meanings, not at the point of their evilness.

But neither the appeal to existential ordeal nor the appeal to impenetrable anomaly is wholly correct. Pain is subjective and we experience it personally, but pain still has its origins and consequences that are objective and public, about which we can critically think. Some paradigms are able to make more sense of, to absorb more suffering than others.

Notice too that the problem of suffering, a classical obstacle to belief in God, is not solved or dissolved by eliminating God in favor of some naturalism or Oriental nontheism. It stays on our hands as a given in natural history, a given in personal life. Any theory that makes sense of things must have its substitute for a theodicy, must recommend some conduct in view of its analysis here, or admit to failure. When God goes away, evil remains, and sometimes grows the more urgent and bleak. Notice how, for instance, to ask why there is suffering is already to suppose that perhaps there should not be unfair suffering, and thus to suppose that there is, or ought to be, fairness in the universe.

We can discern some essential logical connections between the heroic and the harsh elements in life. An organism can have needs, which is not possible in inert physical nature. A planet moves through an environment, but only an organism can need its world, a feature simultaneously of its prolife program and of the requirement that it overtake materials and energy. But if the environment can be a good to it, that brings also the possibility of deprivation as a harm. To be alive is to have problems. Things can go wrong just because they can also go right. In an open, developmental, ecological system, no other way is possible. All this first takes place at insentient levels, where there is bodily duress, as when a plant needs water.

Sentience brings the capacity to move about deliberately in the world, and also to get hurt by it. We might have sense organs—sight or hearing—without any capacity to be pained by them. But sentience is not invented to permit mere observation of the world. It rather evolves to awaken some concern for it. Sentience coevolves with a capacity to separate the helps from the hurts in the world. Even in animal life, sentience with its counterpart, suffering, is an incipient form of love and freedom. A neural animal can love something in its world and is free to seek this, a capacity greatly advanced over anything known in immobile, insentient plants. It has the power to move through and experientially to evaluate the environment. The appearance of sentience is the appearance of caring, when the organism is united with or torn from its loves. One can have little narrative plot in sentient life unless one has good and evil in the world relationships. The earthen story is not merely of goings on, but of "going concerns." The step up that brings more drama brings suffering.

Further, pain is an energizing force. Suffering not only goes back-to-back with caring sentience, it drives life toward pleasurable fulfillment. Not only does the good presuppose concomitant evil, but the evil is enlisted in the service of the good. We come up in the world against suffering, but we could not come up in the world any other way. This truth is both paradoxical and partial, but nevertheless it penetrates into the essence of pain, in both short- and long-range perspectives. Individually, one wants to be rid of pain, and yet pain's threat is self-organizing. It forces alarm, action, rest, withdrawal. It immobilizes for healing. The experiences of need, want, calamity, and fulfillment have driven the natural and cultural evolution of the ability to think. Such benefits are the biological and psychological purpose of pain, even though there is an overshooting of this in cases where pain is of no benefit to, and even crazes, particular sufferers. More generally, where pain fits into evolutionary theory, it must have, on statistical average, high survival value, with this selected for, and with a selecting against counterproductive pain.

Early and provident fear moves half the world. Suffering, far more than theory, principle, or faith, moves us to action. We should not posit the half-truth for the whole; we are drawn by affections quite as much as pushed by fears. These work in tandem reinforcement; one passes over into the other and is often its obverse. In this sense, pain is a prolife force. Not all suffering is thrust upon us from without; much of it comes from internal collapse, as with the pains of failing life in age or cancer. Even here, the body typically does things that make sense in fighting the collapse, postponing the end, although death is inevitable. The death of individuals is superseded by what this makes possible, new exploratory forms, mutant beings, which will be selected for their better adaptedness to the problems that beset their progenitors.

The world is a theater that, if full of possibilities provided at the level of physics by the nested sets of quantum states, is also full of nested sets of goods and evils at the level of history. The particles in physics undergo their annihilations, reformations, and identity flips. So also these goods and evils in history are made over one into another, sometimes relative to the reference frames of the sufferers (to adapt terms from physics), and yet there is also a sort of conservation of good over time. Or, in a more adequate historical model, there is a storied development of goods winning out over evils, and evils returning to test the powers of goodness afresh.

A generalization from evolution and history, safely predictable for the future, is that all advances come in contexts of problem solving, with a central problem in sentient life the prospect of hurt. In the evolution of caring, the organism is quickened to its needs. The body can better defend itself by evolving a neural alarm system. In psychological development, a person's values are defended through conflicts both by stimulus-response reactions and by self-actualization. There can be no will without a testing of will. At this level, there can be no compassion without pain. Culture is a foil to the hostility of nature, though it is also a product of evolutionary inventiveness and requires ecological support. Within culture, the creative advances come when humans, facing difficulty, are roused to some unprecedented effort. Arnold Toynbee expressed this in the "challenge-and-response" formula, finding it characterizing the emergence of every great world culture.[61] The major advances in civilization are processes that have often wrecked the societies in which they occurred.

The evolution of suffering is also to be seen as an exodus out of suffering into lands of promise. Pain is a disequilibrating force that elevates life, not always, but archetypically. We do not really have available to us any coherent alternative models by which, in a painless world, there might have come to pass anything like these dramas of nature and history that have happened, events that in their central thrusts we greatly treasure. There was naïveté in the divine-blueprint model that was so upset by Darwin's discovery of nature red in tooth and claw. (This is still true, whatever one makes of the "fine-tuned" universe of which cosmologists speak.) It was a bad religious model, really, as well as a nonscientific one, for it knew nothing of the constructive uses of suffering. It knew nothing of the wisdom of conflict. There are sorts of creation that cannot occur without death, and these include the highest created goods. Death can be meaningfully put into the biological processes as a necessary counterpart to the advancing of life.

Life needs death, if there is to be more life. Anything that would give the individual organism immortality would destroy the evolution of species. The individual life comes to a stop, but the evolutionary sequence? Whether it will ever stop we do not know. It seems to thrive on the tragic accidents that slay all the successive individuals.

A Cruciform Naturalism

In the biblical model in either testament, to be chosen by God is not to be protected from suffering. It is a call to suffer and to be delivered as one passes through it. The election is for *struggling* with and for God, seen in the very etymology of the name Israel, "a limping people."[62] The divine son takes up and is broken on a cross, "a man of sorrows and acquainted with grief."[63] The element we seek at the moment is not the monotheism in this, but rather the note of redemptive suffering as a model that makes sense of nature and history. So far from making the world absurd, suffering is a key to the whole, not intrinsically, not as an end in itself, but as a transformative principle, transvalued into its opposite. The capacity to suffer through to joy is a supreme emergent and an essence of Christianity. Yet the whole evolutionary upslope is a lesser calling of this kind, in which renewed life comes by blasting the old. Life is gathered up in the midst of its throes, a blessed tragedy, lived in grace through a besetting storm.

The enigmatic symbol of this is the cross (Figure 6.5), a symbol we adopt here not yet for God, not for some extrahistorical miracle, but as a parable of nature and history. This symbol is an earthen sign, and it too has its limits. One needs also the sign of the Logos, of intelligibility and order. In nature there is first simply formation, and afterward information. Only still later does nature become cruciform. But the story does develop so, at least in the rich plot on this Earth. The cross here is not nature's only sign, but it is a pivotal one. It would also be a mistake to say that

Figure 6.5 The cross as a natural symbol

life is nothing but a cross, for life is gift and good news too. Still, all its joys have been bought with a price. The drama is Logos and Story, Cross and Glory.

Moral evil in history, as this amplifies the spontaneous evils of nature, deeply compounds the story. We suffer disease and accident by onslaught of nature, but we suffer more by crimes and carelessness at the hands of persons. Here again, half of history is moved as brother slays brother, though it is likewise true that the other half, often the obverse, is moved by the loves that we have for each other. By some accounts, this adds absurdity with a vengeance. Amoral nature produces moral persons, forces upon them much suffering, equips them with a conscience, and then systematically tempts them to be immoral each against the other, thereby multiplying suffering the more. But by logic alone, the possibility of morality contains the possibility of immorality; and by the logistics of life, we cannot help each other in a world where we cannot hurt each other. We cannot have responsibilities in a world without caring. This education and evolution of moral caring inevitably introduces guilt into our storied awareness. This leads on to themes of forgiveness and reconciliation, likewise gathered into the symbol of the cross. Here, supremely, one suffers through to joy. "Blessed are those who mourn, for they shall be comforted."[64]

It is false to think (as earlier theologians did) that chronologically suffering entered the world after sin and on account of it. There was suffering in the biological world for long epochs before the human arrival, and in human history ample suffering continues that has no origin in sin. Following the emergence of conscience, suffering doubtless produced sin, and later on sins also produce suffering. Pain is, we can say, a sign of something gone wrong, but pain operated in the biological order for eons before humans came. When they too come, and go wrong, the pain is intensified, as sin produces suffering at new levels. In this sense, to sin is to betray oneself and others. When moral responsibility does come, this does not change the sign of natural history, or produce any fall. It rather intensifies a theme already crucially there, enriching this motif because it adds moral self-awareness. After this, history begins to turn on concepts of right and wrong, justice and guilt, obligation and retribution. But the way of history too, like that of nature, only more so, is a via dolorosa. In that sense, the aura of the cross is cast backward across the whole global story, and it forever outlines the future.

"I believe in Christ in every man who dies to contribute to a life beyond his life," confessed Loren Eiseley.[65] Adolf Harnack observes that it is the "self-sacrificing deeds, and not only self-sacrificing deeds, but the surrender of life itself, that forms the turning-point in every great advance in history. . . . Did Luther in the monastery strive only for himself?—was it not for us all that he inwardly bled?"[66] But that theme, willingly or unwillingly, is everywhere in the plot; it is the alpha and omega, prefigured in nature and essential to history. All the creatures are forever being sacrificed to contribute to lives beyond their own, like the lamb slain from the foundation of the world. Blessedness is success on the far side of sorrow.

We can now understand why the Genesis creation was pronounced "very good," not yet perfect, but adequate to the story that in these creatures unfolds, where even the imperfections are a key to future goodness. What happens to the humans—their lapse from innocence into the struggle for life redeemed from evils—also (and long before the human arrival) has happened to every living thing. Every organism is plunged from naïveté into a struggle in which goodness, as it is known at that organic level, is given only as it is fought for. Every life is chastened and christened, straitened and baptized in struggle. Everywhere there is vicarious suffering. The global Earth is a land of promise, and yet one that has to be died for. All world progress and directional history is ultimately brought under the shadow of a cross. The story is a passion play long before it reaches the Christ. Since the beginning, the myriad creatures have been giving up their lives as a ransom for many. In that sense, Jesus is not the exception to the natural order, but a chief exemplification of it.

Can this paradigm for nature and history be defeated by evidence? Yes and no. Simple counterexamples, even chains of them, will not suffice. Each factual case requires also a value judgment. Nevertheless, since the cross is taken faithfully to map both earthen nature and history, a sufficiently revised data set will lead to a reformed interpretation. One will have to encounter a degree and order of suffering that can never be justified by any subsequent good. Such a judgment will involve not merely past and present facts but a prediction that the suffering cannot later be redeemed, a prediction made even though nature and history are open systems. These counterexamples, which may be many, cannot do their work singly but will have to erode the dominant pattern that suffering is productive. Hence, this narrative-model both arises out of experience and answers to it, although not in a simple or direct way.

Both hard and soft naturalisms have characteristically been offended by suffering, which has often driven naturalists to deny theism. It may be possible to be atheist or agnostic and still retain the cross as a symbol of this paradoxical, transformative power in nature and history. One would have a cruciform naturalism. But hard naturalism, with its axiom that subjective events are epiphenomenal, will find it difficult to think of redemptive suffering as a key to the whole. Soft naturalism will find nature thereby to be the more mysterious, hardly the better understood, since here Nature, which (unlike God) has no center of experience, has to be understood under the model of transformative suffering, a phenomenon found in the parts but not known in the whole. Once again, the emergents are nobler in their capacities than the mother matrix from which they spring.

Neither Buddhists nor Hindus have ignored suffering, especially not the Buddhists, who find suffering, *duḥkha,* to be the first noble truth, and thirst, *taṇhā,* to move the world. Suffering prompts Hindus as well to call the world process a *māyā* illusion, rather than a genuine divine creation. But neither of these models attempts seriously enough within the context of nature and history the transformative uses of suffering to reach something higher, to yield social relevance, or any bringing of what others have called the Kingdom of God on Earth. Life is suffering, but life is suffering through to something higher. Life is unsatisfactory, as it is also satisfying, for the dissatisfactions drive the creative process discovering new satisfactions. It is not the extinction of desires and thirst that we wish, but rather their elevation and, at length, their sanctification.

These two Eastern creeds seek peace in their ultimate, but the *saṃsāra* world is not redeemed by suffering; to the contrary, nature and history, as illusion and mock show, are to be extinguished and stilled in the quiescence of *nirvāṇa* and Brahman. The cessation of suffering is the cessation of the world. Neither faith expects redemptive suffering to bring the salvific upbuilding of the world. Buddhists and Hindus may object (with some plausibility) that nature and history do not form the only categories of religious life; mystical absorption may be another, fundamental one. Still, nature and history do form crucial categories, and, escape to the Ultimate though one may, one does not have any comprehensive, ultimate explanation that reconciles science with religion, which is our concern here, until one has dealt constructively with both nature and history.

The grass, the flower of the field, is clothed with beauty today and gone tomorrow, cast into the fire. The sparrow is busy about her nest, sings, and falls. Jesus knew these things, and noticed in the same breath that trouble enough comes with each new day.[67] Tribulations come as surely as does the Kingdom. The hard, straitened way leads to life. But day by day we press forward in trials, in the will that this pageant continue. We believe that we could not have come this far and would not have the strength to struggle on were it not for some power greater than ourselves at work in nature and history, providing for the drama. Earth is a providing ground. Some providential power (and can it be merely a naturalistic one?) guarantees that the story continues across all its actors. In this perspective, regenerative suffering makes history. Tragic beauty is the law of the narrative. But we do not pretend in this model to have solved the problem of suffering, which remains a *crux interpretum.* Perhaps, just as pain itself comes in contexts of problem solving and leads to solutions beyond, so also the problem of pain as a religious assignment is an advanced phase in this problem solving, and our cognitive dissonance before it is an opportunity for growth.

In communication systems, for those not keyed in, a really complex signal is difficult to distinguish from noise. Computer signals seem only whines, or the color television picture collapses into a blur. When we listen for signals from space, hoping to eavesdrop on extraterrestrial cultures, we hardly know how to construct our instruments. The evolutionary sequence, in that miraculous power by which life persists over the vortex of chaos, is a really complex informational signal, and thus it is not surprising that much in it, or even the whole, seems like noise to many interpreters. In the cruciform model, the evils both in spontaneous nature and in history, symbolized as death, are transformed and reinforce a larger pattern, symbolized under the themes of resurrected life.

To recall the gestalt of Jesus used in our opening chapter, on the logic of inquiry, there are surrounding blotches that frame the portrait. These first are received as a kind of static that destroys legibility, but they are later tolerated by the governing gestalt, at least in outline and shadow. To discover the good news in the noise takes an act of faith; here a solution of the conflict is inseparable from a participant's resolution in it. One believes that life will be provided for through suffering, and bettered by it. But this is not something one has scientific reasons to predict, though it follows from what has happened in historical trends. Still, many are willing to bet on this result, and to invest life in it.

Such a faith might be only a self-fulfilling prophecy, but also this kind of courageous faith might tune us in to the complex signal. For this faith, *nature*, eventuating as it does in our *history*, is a signal from God, a signal of intelligence out there on the grandest scale of all. But it is difficult to decode. There may be various keys to unlock its multiplex meanings, but the puzzles to be unlocked are the storied meanings in nature and history. One needs not merely electronic apparatus, but an inquiring philosophical and saintly mind, one that can absorb science and welcome its revelations, reflectively to ask about the meanings of the signal, past all causes found. Alternately put, the kind of decoder needed is that exquisite sort of electronic apparatus called the human, the person in all modes of operation and sensitivity, and especially in that input which the spiritual life registers as the sense of the presence of God. To these theistic convictions we now turn.

NOTES

1. Albert Einstein, "On the Method of Theoretical Physics," *The World as I See It* (New York: Covici-Friede Publishers, 1934), pp. 30–40, citation on p. 36.
2. Edwin F. Taylor and John A. Wheeler, *Spacetime Physics* (San Francisco: W. H. Freeman and Co., 1966), p. 5.
3. Victor K. Weisskopf, "Of Atoms, Mountains, and Stars: A Study in Qualitative Physics," *Science* 187 (1975): 605–12, citation on p. 605, quoting Paul Ehrenfest.
4. Norbert Wiener, *Cybernetics* (New York: John Wiley and Sons, 1948), p. 155.
5. *The Letters of John Keats,* vol. 2 (Cambridge, Mass.: Harvard University Press, 1958), p. 102.
6. J. B. S. Haldane, *The Causes of Evolution* (Ithaca: Cornell University Press, 1966), pp. 167–69.
7. Fred Hoyle, *The Nature of the Universe*, rev. ed. (New York: New American Library, 1960), pp. 117–18.
8. Albert Einstein, *Out of My Later Years,* rev. reprint ed. (Westport, Conn.: Greenwood Press, 1970), p. 61.
9. Victor F. Weisskopf, *Knowledge and Wonder,* rev. ed. (Garden City, N.Y.: Doubleday and Company, 1966), p. 101.
10. These theses can be found variously in George Santayana, Herbert Feigl, Sterling Lamprecht, Roy W. Sellars, John H. Randall, Jr., Ernest Nagel, Morris Cohen, Bertrand Russell (earlier more than later), Sidney Hook, Karl Marx, Sigmund Freud, and generally among positivists and tough-minded philosophers of science. But recent proponents seldom state the view so baldly, and we may not find the full set owing to agnostic provisos and antireductionist modifications.

11. Ernest Nagel, *Logic without Metaphysics* (Glencoe, Ill.: Free Press, 1956), p. 7.
12. Herbert Feigl, "The Scientific Outlook: Naturalism and Humanism," in Herbert Feigl and May Brodbeck, eds., *Readings in the Philosophy of Science* (New York: Appleton-Century-Crofts, 1953), pp. 8–18.
13. Bertrand Russell, *Religion and Science* (Oxford: Oxford University Press, 1961, 1978), p. 243.
14. Bertrand Russell, "A Free Man's Worship," *Mysticism and Logic* (Garden City: Doubleday, Anchor Books, 1957), pp. 44–55, citation on p. 52.
15. Ernest Nagel and James R. Newman, *Gödel's Proof* (New York: New York University Press, 1958).
16. C. W. Misner, "Cosmology and Theology," in Wolfgang Yourgrau and Allen D. Breck, eds., *Cosmology, History, and Theology* (New York: Plenum Press, 1977), pp. 75–100, citation on p. 96.
17. These theses can be found variously in Henry David Thoreau, Ralph Waldo Emerson, John Muir (and generally in Romantic naturalists), Julian Huxley, Henri Bergson, sometimes in John Dewey (and many evolutionary naturalists), and more recently in Gregory Bateson, Paul Shepard, Loren Eiseley, Kenneth Denbigh, Joseph Wood Krutch, Lewis Thomas, and many of the current generation of environmentalists.
18. Anthony F. C. Wallace, *Religion: An Anthropological View* (New York: Random House, 1966), p. 38.
19. David Lipset, *Gregory Bateson: The Legacy of a Scientist* (Englewood Cliffs, N.J.: Prentice-Hall, 1980), pp. 301–302.
20. Gregory Bateson, *Steps to an Ecology of Mind* (New York: Ballantine Books, 1972), p. 434.
21. Sanborn C. Brown, "Contributions of Science to the Unitarian Universalist Tradition: A Physicist's View of Religious Belief," *Zygon* 14 (1979): 41–52, citation on p. 49.
22. Chapter 3, p. 132.
23. R. W. Sperry, "Science and the Problem of Values," *Zygon* 9 (1974): 7–21, citation on p. 19. Sperry thinks, however, that by the scientific method he can reliably arrive at value judgments.
24. Alfred North Whitehead, *Modes of Thought* (New York: Macmillan, 1938), p. 185.
25. Eric J. Chaisson, "The Scenario of Cosmic Evolution," *Harvard Magazine* 80, no. 2 (November–December 1977): 21–33, citation on p. 33; Mike Corwin, "From Chaos to Consciousness," *Astronomy* 11, no. 2 (February 1983): 15–22, citation on p. 15.
26. A phrase going back to Kepler, who was driven to discover natural laws because "God wanted us to perceive them when he created us in His image in order that we may take part in His own thoughts." Letter to Mästlin, April 9–10, 1599, cited in Gerald Holton, "Johannes Kepler's Universe: Its Physics and Metaphysics," *American Journal of Physics* 24 (1956): 340–351, citation on p. 350.
27. Sankara, *The Bhagavad-Gita,* trans. A. Mahadeva Sastri (Madras: V. R. Sastrulu and Sons, 1972), 13–2. Occasional minor editing and repunctuation occur in the citations from Oriental translations that follow.
28. Sankara, *Commentary on the Brihadaranyaka Upanishad,* trans. Swami Madhavananda (Calcutta: Advaita Ashrama, 1965), 2-1-20, 3-8-8, 5-1-1.
29. Sankara, *The Vedanta Sutras of Badaranya,* 2 vols., trans. George Thibaut (New York: Dover Publications, 1962), 2-1-33, 1-1-5.
30. Ibid., 1-3-19, reading "ignorance" for "nescience," 2-2-2.
31. "The Ratnavali of Nagarjuna," 2-13, trans. Giuseppe Tucci, *Journal of the Royal Asiatic Society,* 1936, pp. 237–252, citation on p. 242. For a recent effort to relate Theravada Buddhism to science, see R. G. de S. Wettimuny, *Buddhism and Its Relation to Religion and Science* (Colombo, Sri Lanka: M. D. Gunasena and Co., 1962).

32. Nagarjuna, *Karikas,* with Candrakirti's *Commentary,* in Th. Stcherbatsky, *The Conception of Buddhist Nirvāṇa* (The Hague: Mouton, 1965). The citation is from the *Commentary,* pp. 138–39. *Śūnya* is a key word in both works.
33. Ibid., *Karikas* 25.24, p. 78; *Commentary*, p. 138.
34. See K. Venkata Ramanan, *Nagarjuna's Philosophy* (Rutland, Vt.: Charles E. Tuttle Co., 1966), for *śūnyatā, tathatā, tattva,* etc.
35. Nagarjuna, *Karikas,* 25.24, p. 78.
36. Ibid., *Karikas*, 25.19, p. 77.
37. Ibid., *Karikas* dedication, p. 69.
38. Sankara, *Commentary on Aitareya Upanishad,* 2-1, and *Commentary on the Mandukya Upanishad* (including Gaudapada's *Karikas*), at the *Karikas* 1-7, both trans. R. P. Singh, *The Vedanta of Sankara* (Jaipur: Bharat Publishing House, 1949), pp. 277–78.
39. Sankara, *Commentary on the Mundaka Upanishad,* 3-2-6, in *Eight Upanishads,* trans. Swami Gambhirananda (Calcutta: Advaita Ashrama, 1973).
40. For the metaphors, see *The Large Sutra on Perfect Wisdom,* trans. Edward Conze (Berkeley: University of California Press, 1975), pp. 38, 193; citations on p. 204. For Nagarjuna's commentary, see *Le Traité de la Grande Vertu de Sagesse de Nāgārjuna,* vols. 1–4, trans. Étienne Lamotte (Louvain: Institute Orientaliste, 1949–1976), pp. 357ff., 1311ff., 1431ff.
41. Candrakirti's *Commentary* in Stcherbatsky, *Conception of Buddhist Nirvāṇa,* pp. 125, 127.
42. *The Large Sutra,* p. 118.
43. For an effort to interpret *māyā* more positively, see L. Thomas O'Neil, *Māyā in Śankara: Measuring the Immeasurable* (Delhi: Motilal Banarsidass, 1980). For efforts to put evolution into the Hindu scheme, see Aurobindo Ghose, *Evolution* (Pondicherry, India: Sri Aurobindo Ashram, 1950), and Rama Shanker Srivastava, *Sri Aurobindo and the Theories of Evolution* (Varanasi, India: Chowkhamba Sanskrit Series Office, 1968).
44. *Tao Te Ching,* stanzas 35, 40, trans. Arthur Waley, *The Way and its Power* (London: George Allen and Unwin, 1934, 1965).
45. For favorable interpretation, see Huston Smith, "Tao Now," in Ian G. Barbour, *Earth Might Be Fair* (Englewood Cliffs, N.J.: Prentice-Hall, 1972), pp. 62–81; and Fritjof Capra, *The Tao of Physics* (New York: Bantam Books, 1975).
46. For an effort to interpret history from within Advaita Vedanta, see the essays in the *Indian Philosophical Annual,* vol. 16 (Madras, 1983–84), a special number on *Philosophy of History: Indian Perspectives,* ed. R. Balasubramanian.
47. For a fine discussion of history as differing from science, discussing also religious themes in interpreting universal history, see William H. Dray, *Philosophy of History* (Englewood Cliffs, N.J.: Prentice-Hall, 1964).
48. Robert P. Swierenga, "Social Science History: An Appreciative Critique," in C. T. McIntire and Ronald A. Wells, eds., *History and Historical Understanding* (Grand Rapids, Mich.: William B. Eerdmans Publishing Co., 1984), pp. 93–135, citations on pp. 98, 100.
49. Edwin Van Kley, "History as a Social Science: A Christian's Response," in George Marsden and Frank Roberts, eds., *A Christian View of History?* (Grand Rapids, Mich.: William B. Eerdmans Publishing Co., 1975), pp. 89–97, citation on p. 96.

For a fuller introduction to these issues, see Gordon Graham, *Historical Explanation Reconsidered* (Aberdeen: Aberdeen University Press, 1983); W. J. T. Mitchell, ed., *On Narrative* (Chicago: University of Chicago Press, 1981); Louis O. Mink, "Narrative Form as Cognitive Instrument," in Robert H. Canary and Henry Kozicki, eds., *The Writing of History* (Madison: University of Wisconsin Press, 1978), pp. 129–49; Lawrence Stone, "The Revival of Narrative: Reflections on a New Old History," *Past and Present,* no. 85,

(November 1979), pp. 3–24; W. B. Gallie, *Philosophy and the Historical Understanding* (New York: Schocken Books, 1964); Morton White, *Foundations of Historical Knowledge* (New York: Harper and Row, 1965); Arthur C. Danto, *Analytical Philosophy of History* (Cambridge: Cambridge University Press, 1965); Maurice Mandelbaum, "A Note on History as Narrative," *History and Theory* 6 (1967): 413–19, and Richard G. Ely, Rolf Gruner, and William H. Dray, "Mandelbaum on Historical Narrative: A Discussion," *History and Theory* 8 (1969): 275–94; A. R. Louch, "History as Narrative," *History and Theory* 8 (1969): 54–70; Haskell Fain, "History as Science," *History and Theory* 9 (1970): 154–73; J. H. Hexter, *Doing History* (Bloomington, Ind.: Indiana University Press, 1971).

50. Daniel J. Boorstin, *America and the Image of Europe, Reflections on American Thought* (New York: Meridian Books, 1960), p. 66.
51. Derek V. Ager, *The Nature of the Stratigraphical Record* (New York: John Wiley and Sons, 1973), p. 100.
52. See Chapter 7, p. 338–39.
53. A historical novel, to be distinguished from historical narrative, can be true to the history of a period while not true history, as *A Tale of Two Cities* genuinely reflects the French Revolution, not in spite of but by using its fictions. Still, such a story must catch real history; it cannot be pure fiction.
54. "The second principle [of analysis, after the unity of particulars in a single idea] is that of division into species according to the natural formation, where the joint is, not breaking any part as a bad carver might." Plato, *Phaedrus* 265e, trans. B. Jowett, *The Dialogues of Plato,* vol. 1 (New York: Random House, 1937), p. 269. See also the *Statesman,* 262b.
55. R. G. Collingwood, *The Idea of History* (New York: Oxford University Press, 1956), p. 114, endorsing G. W. F. Hegel. Compare G. G. Simpson, "Historical Science," in Claude C. Albritton, Jr., ed., *The Fabric of Geology* (Reading, Mass.: Addison-Wesley Publishing Co., 1963), pp. 24–48.
56. T. A. Goudge, *The Ascent of Life* (Toronto: University of Toronto Press, 1961), pp. 71, 75, 77, 79.
57. Davydd J. Greenwood and William A. Stini, *Nature, Culture, and Human History* (New York: Harper and Row, 1977), pp. 428, 436, 450. Notice also that year comparisons are deceptive in terms of head counts. There may have been five million hunter-gatherers alive at any one time; there are four billion persons alive today, as many as in perhaps a thousand generations of the hunter-gatherer years.
58. G. W. F. Hegel, *Lectures on the Philosophy of History,* trans. J. Sibree (London: George Bell and Sons, 1888), pp. 17–21.
59. Arnold Toynbee, *A Study of History,* vol. 7 (New York: Oxford University Press, 1954), pp. 420–44.
60. Alfred North Whitehead, *Science and the Modern World* (1925) (New York: Free Press, 1967), p. 192.
61. Toynbee, *A Study of History,* vol. 1, second edition (1935), pp. 271ff.
62. Gen. 32:22–32.
63. Isa. 53:3.
64. Matt. 5:4.
65. Loren Eiseley, in an address accepting the Lecomte du Noüy Award, condensed as "Our Path Leads Upward," *Reader's Digest*, March 1962, 43–46, citation on p. 46.
66. Adolf Harnack, *What Is Christianity?* (New York: Harper and Brothers, 1957), pp. 158–59.
67. Matt. 6:25–34.

Chapter 7

Nature, History, and God

Both *Nature* and *God* take plurals poorly, only provisionally or penultimately, with a certain logical and experiential drive toward the singular. In earlier times, persons have often believed in multiple *gods,* of course; but "gods" shifts to the lower case and changes its meaning. Further, such beliefs have tended to be replaced by monist or monotheistic creeds. The spirits are but fractured omnipresence. There can be but one God, at most. Likewise, there are *natures* in local things—the weasel has one nature, the oak another. Granite is hard, while water flows. By this we refer to the distinguishing characteristics of phenomenal entities. But we can recognize and enjoy these natures and still find, as science does, that lawlike operations govern the whole. There are fundamental constituents, origins, kinships, patterns, connected levels. Then we come to think more comprehensively of systemic *Nature,* which is expressed in the diverse natures of particular things.

Just as polytheism in religion gives way to monotheism, so science makes connections, more and more, and pluralism shifts toward more unified theories of nature, a universal nature omnipresent but delimited in particular things. Both science and religion are driven toward collective terms. The many are referred to the one.

We do not check these impulses by admitting that we are assigning to both fields more than they have yet delivered—an integrated model of nature and history, or one of God as the warrant for nature and history. The sciences are plural, with their multiple paradigms. The differing models of persons and societies in the human sciences mesh poorly with one another, and these in turn have not yet been fully correlated with models used to describe biological or physicochemical nature. The religions are many, with their multiple creeds, and even within theism there are several leading denominations. Believers and scientists both live with a certain hope that we can gradually envision the unity of things more clearly, but both know that we travel hopefully and slowly arrive.

We are always on a frontier where what is known mingles with what is believed and hoped about things incompletely known, wrongly known, and unknown. This drives the ongoing quest. We should hardly have predicted that our intellects would know nature as well as they already do, but, given where we now stand, we may believe that we will know nature yet better still. But to hold that nature is corporately singular, a Universe, or that there is one God grounding this systemic unity, claims that are believed amidst the diverse phenomena and the competing, only partially convergent theories and creeds, will at present and in the foreseeable future require acts of faith.

1. NATURE AND SUPERNATURE

How far does the one term, "Nature" (with its history), compete with the other, "God"? Does the latter notion, *God,* complement what we believe, or know, or fail to know about the former concept, *Nature?* Our answer will involve, in this section, approaching the concept of the supernatural by an ever more comprehensive look at the natural. In three sections to follow, we will look at leading options within theism, examining each for its relative capacity to overarch or accommodate the place of the natural. These positions we call scientific-existentialist theism, process theism, and transscientific theism. Afterward, we close with what is more a commencement than a conclusion, an invitation to continue religious inquiry, past scientific inquiry, by doing the truth.

Emergence, the Natural, and the Supernatural

To believe in God is (by most accounts) to posit the *supernatural* beyond the natural.[1] But no theory can move toward a complementary union of God and nature so long as there is an unresolved dualism between the natural and the supernatural. Further, no contemporary creed can convincingly claim belief in God unless it first passes through, rather than merely bypasses, what we now know about the natural. This need not mean that the categories of science are error-free or complete, much less absolute or canonical, for these categories have no more claim to infallible finality than do the claims of religion. Nor are they less theory-laden. But we must nevertheless reckon with them.

It is often thought that scientific conceptions of the natural make belief in the supernatural impossible, superfluous, or superstitious. Some secular equivalents of religion may be viable in the future, possibly some naturalistic or socialistic faiths, but supernaturalism will disappear. "Belief in supernatural powers is doomed to die out, all over the world, as a result of the increasing adequacy and diffusion of scientific knowledge. . . . The question of whether such a denouement will be good or bad for humanity is irrelevant to the prediction; the process is inevitable. . . . Put in this way, the evolutionary future of religion is extinction."[2] So we meet again, this time from an anthropologist, Anthony F. C. Wallace, the now-familiar prediction that theism will vanish.

But perhaps the matter is not so simple. While scientific and supernaturalistic explanations are sometimes rivals, as we have seen, must they be so inevitably? To reach an answer we will enrich the category of the natural and try to leave it open-ended enough to permit its own transcending. We need to soften the rigid, mutual exclusiveness of the categories of the natural and the supernatural.

Once there was no consciousness, no learning, no life, no Earth, no solar system, neither compounds nor elements, but only the simplicity of primordial energy-particles. If we let our thoughts run backward in time and run downward over ontological structures-processes (reversing the ladder of Figure 6.1) to imagine *nature* as it primevally was, any account we could then give would need to be stripped of all subsequent emergents. But such an account, though seemingly accurate for that period in time and level of development, would, as we now know, be manifestly incomplete. With the evolution of each later stage in the world story, the tectonic potential of nature actualizes into something *higher*—into stars and their elements, into planets and their compounds, and, on this planet, into life, mind, society, world history. That notion of "higher" is the critical dimension in the "super" of *supernatural.*

Each of the emergent steps is "super" to the precedents, that is, supervenes on and surpasses the principles and processes earlier evident.[3] When life appears, the organic *trans*forms the inorganic. Properties are *super*imposed on materials that before bore no such qualities. Needs appear, as do hunger, struggle, and disease. There arises the cybernetic steering by a life core, keyed in DNA. From the point of view of the categories of physics and chemistry, these phenomena are *super*natural, that is, *super*physical. They transcend previous ontological levels. This is not to deny biochemistry, only to insist on irreducibility. The phenomenon of information transfer through time by its instantiation over successive material sets, essential to life, is absent from the causal sequences of astronomy and nuclear physics.

The conscious search for meanings, central to personality, is absent from plant life. At changes of state, we may find blurred quantitative-qualitative transitions; there are twilight zones. But there are also genuine passages into novel phenomena. *In that sense, we do not say that naturally inexplicable things never occur; they occur in every emergent increment that breaks previous records of attainment and power.* They would come as a surprise to any science based on previously known nature. When they come, it may be possible partially to develop a science of the new phenomena; but they also come, above all, as developments in a story with increasingly rich historical dimensions.

When subjective inwardness appears, based on neural structures, it too is *super*natural to its precedents in objective life, for all earlier somatic life was devoid of any centers of experience. With the coming of mind and culture, there is again a dramatic shift that is *super*natural, that is, *super*biological. The human capacities for language, for abstract symbol manipulation, for toolmaking, for cultural transmission, for doing science and religion, for historical self-awareness would, if viewed by monkeys, *per impossible,* be considered as supernatural to any orders of nature evidenced within themselves, despite the anatomical likenesses of their hands and brains with ours. Culture and the literate mind are *super*natural to spontaneous organic nature. Employing the richest categories of biological science, we can only stammer before the phenomena with which sociology and anthropology have to deal. A sacred scripture, such as the Bible, or a scientific instrument, such as a spectroscope, is nonsense to chimpanzees, because these entirely transcend their capacities. They are nonsense even to biologists who are restricted to using only biological categories.

From the viewpoint of those placed at any particular stage, or forced to operate with the categories of nature available there, the higher steps will be supernatural, that is, transcending and irreducible to existing manifestations of the natural. All the lower steps, however, will with rather more plausibility be regarded as natural, having progressively come to seem merely natural with their introduction across the stages of natural history (Figure 7.1). These phenomena are able to be understood (at least partially) by the probing mind. But we may also be forgetting when we term what lies behind "merely natural" how amazing is what has already managed to happen, and how incompletely our natural science categories explain the subsequently emerging developments in the story thus far.

Further, events have to be understood not just in their particular, plural natures, not in their classes, nor even in their causal connectedness or their lawlike operations, but in the parts they play in a drama. Sometimes a thing needs to be understood not merely immanently, in terms of what it now is in its own-being, but in terms of what it is becoming, as a link in a story. But the higher principles that it foretells on the story line are not yet evident, and are indeed, in our sense, supernatural, not immanent in that thing nor anywhere yet evident in the natural system, but ulterior in as-yet-uncreated, never-yet-natural states. In that sense, every emergence presents a kind of emergency in its challenging of theories and laws competent for the previous levels. Every lower science proves a limited case within some higher theory, and for adequate explanation we increasingly must pass through science into history.

Supernature and Supercharged Nature

There is a mixture here of epistemic advance with ontological advance. Some may object that events all along have been purely and simply natural, but when humans in their earlier ignorance first studied them, they mistakenly regarded as supernatural those events that we now know to have been natural (lightning, birth, the creation of life, mind). With epistemic advances, the enlarging category of the natural will eat up the category of the supernatural. But more is involved. The category of the natural is elevated as it enlarges, so that the seeming victory is Pyrrhic; the "super" reappears, even though digested, in a now-spiritualized "Nature." Nature proves richer, more fertile, brooding, mysterious, sacred than was recognized before. "Nature" conquers by rising to just those levels that were before defended by the term "supernatural," and none of the high ground is lost. A spirited history, a history of spirit, supervenes on matter-energy. The generative power is, after all, the lure of Spirit, and whether we call this *Supernature* or *Supercharged Nature* is little more than a semantic difference.

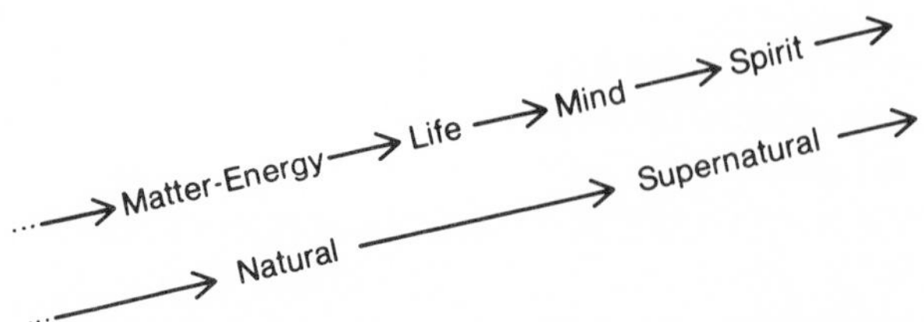

Figure 7.1 Emergent natural states: a storied drama

We must add a further complication. We humans can with considerable success look downward, backward (so to speak) at the matter and life over which we have advanced, making relative sense of these in causal and scientific terms. But the level of mind, at which we stand, and the level of spirit, now incubating and toward which we pass, lie on the frontiers. They may well surpass our capacities for self-referential explanation. We can expect our human-made sciences to stumble over them. Things are going on in our heads that are over our heads, just as things were going on in earlier performers that were over their heads. The emergent steps currently in progress are, and will remain, super-to-the-natural, supernatural from the vantage point at which we stand. Actually, this grants too much to our capacity to fathom matter, energy, life, and information naturalistically, for even these phenomena outdo our capacities for analysis. But especially as mind and spirit emerge out of matter and life, our analysis stalls at the cutting edge on which we live historically. Here the category of the supernatural, past the natural, becomes more urgent and inevitable. It is this power in nature to move over our heads toward increased spiritedness that we call supernatural. This inexhaustible open-endedness is greater than we now know, or can foreseeably know.

At this turn of thought, we want naturalistic explanations that are open toward being subsumed under a *super* account. The upper-level accounts cast their light back across what might in short-scope perspective have seemed complete naturalistic accounts. They cast shadows over them. The earlier events begin to figure as subplots within a larger story. Afterward, the scientific explanations do not look so compelling, exclusive, or nonnegotiable as they earlier did.

It is not, for instance, the impressive consensus within evolutionary theory about the upslope directionality of the life process that impresses us. It is the chaos of indecision and nontheory about life's critical turnings. This softness drives the religious interpreter to use additional premises not found in science. It is not the satisfactory capacity of psychology or sociology to explain human nature and to generate meanings for personal or social life that extinguishes our religious impulses. Rather, their limping causal explanations and their stuttering incapacity to supply enough meaning to locate the ego in its environment prompt the theologian to invoke principles past all natural science and social science. It is not the capacity of physics, biology, neurophysiology, psychology, sociology, anthropology, or any other science to explain how subjective consciousness, with its center of experience, arises out of objective nature, devoid of felt experience, that satisfies us. It is the utter mystery that remains in their silence as to how this happens.

To believe in the *supernatural* is to believe that there are forces at work that transcend the physical, the biological, the sociocultural. These spiritual forces sway the future because they are already breaking through and infusing what is now going on. To term these *supernatural* forces, transcending *natural* ones, is not to make an absolute bifurcation between the secular, natural, and the sacred, spiritual, realms. It is not to posit forces antinatural, unnatural, or foreign. It is only to speak from our present vantage point and believe in a fourth dimension (spirit) when three dimensions (matter, life, mind) are already incontestably evident and the fourth is secretly and impressively also at work.

Almost anything can happen in a world in which what we see around us has actually managed to happen. The story is already incredible, progressively more so at every emergent level. Nature is indisputably there, and what are we to make of it? Both good induction and good historical explanation lead us to believe in surprises still to come and powers already at work greater than we know. For all the unifying theories of science, nature as a historical system has never yet proved simpler or less mysterious than we thought; the universe has always had more storied achievements taking place in it than we knew. To suspect the work of spiritual forces is not, in this view, to be naïve but rather to be realistic.

When Jesus likens the kingdom of heaven to the seed growing secretly, using (as he often does) the natural order as a parable of the supernatural, he does not make a merely heuristic use of fortuitous, disconnected analogues. He draws an ontological bond between nature and spirit. The power organically manifest in the farmer's field is continuous with the power spiritually manifest in the kingdom he announces. To put the matter into contemporary idiom: the spontaneously evolving Earth, bringing forth its seed and harvest, secretly hides the Spirit of God. The biological miracle is a preface to the later-emerging, still more marvelous, realm of spirituality. No doubt there is a natural autonomy in the biological processes. "The earth produces of itself [Greek: *automatically*]."[4] But there is a fuller account. The Spirit first gives organic life, and afterward still more abundant life. In the communion that Jesus institutes, the bread and wine remain what they naturally are, but they also become sacraments in the fuller context of his kingdom, as this is launched in passionate tragedy. The Spirit adds meanings continuous with, but influxing over, the former values.

Superintending Levels in the Earth Story

Each of the *super*vening levels piggybacks on the precedents, but so far from being reducible to the precedents, each takes the earlier levels under its *super*intendence. Organisms are material and energetic, but they inform matter and energy according to their novel programs. Mind is based on the brain and hand but comes to direct these. How many diverse mentalities and careers are possible choices for one individual, options for one brain and pair of hands! Culture is based on psychological and organic life, but thousands of historical cultures can be superimposed on the one kind of psychological-anatomical structure that is common to all human nature. The control of the successive material-energetic states is not so much "from below" as it is "from above." To believe in the supernatural is to believe in downward causation on this historical scale, that the upper levels are controlling the lower, even while there bubble up from below the materials and energy, the nested sets of possibilities, on which the upper-level drama emerges and proceeds.

We have in this more comprehensive theory to add arrows reverse to those that we first met chronologically (Figure 7.2). The macrohistory draws the microhistories after it, although the microhistories (genetic mutations, individual careers and choices) emit the novelties that perfuse the macrohistory. There is something now on the historical scale that seems reminiscent of the randomness and interaction in physics in relation to living organisms, although physics omits history. Something acts as a sieve to catch creatively the fortuitous histories; something acts as an

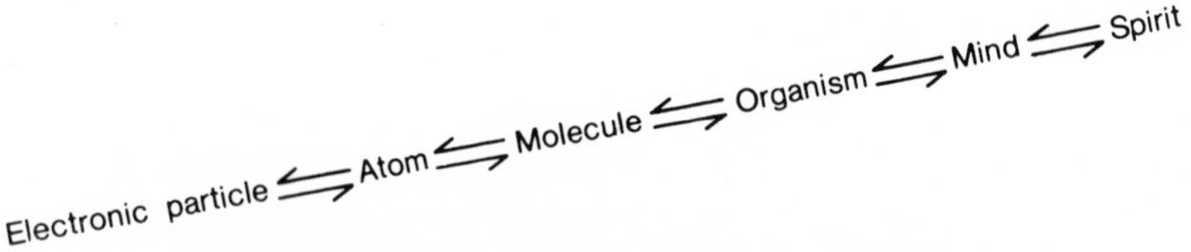

Electronic particle ⇄ Atom ⇄ Molecule ⇄ Organism ⇄ Mind ⇄ Spirit

Figure 7.2 Superintending levels in the Earth story

interaction apparatus to call forth this world course and not that one from among the nested sets of historical possibilities. Each later level is taking up into itself the mechanisms and processes that preceded it. Each has to be understood in terms of something higher than itself. Just this sense that human affairs, emplaced in the natural history, are being implicated into something higher is a sign of the divine presence. We will expect here a sense of rationality mixed with that of mystery. Faith is an openness to the next higher level, to things not seen.

To believe in the supernatural is to take the epiphenomena seriously, despite the fact that we have as yet no scientific theory that gives much of the unity we seek in the emergence of these successive phenomenal stages. *Life* does appear, and afterward *mind*, but are these (as hard naturalism maintains) nothing but epiphenomena, nonrevelatory and adventitious episodes that provide no key to the nature of nature? If, as theists hold, *spirit* is likewise detectable, is not this noblest of "epiphenomena" more revelatory still, to be enjoyed as a critically emergent category, not explained away as a fluke? This exceeds science. It exceeds even the natural, as the "natural" can be currently referenced. But it does not exceed experience. To the contrary, it is faithful to experience, more so than are simplistic reductions of everything to matter and energy.

We have within ourselves marvelous evidence that mind haunts and transcends (however much it is grounded upon) matter in a spatiotemporal matrix. Consciousness is something more than a material, spatiotemporal affair. At least where we ourselves incarnate matter, outwardness is only half its face. There is an inner face, seen surely in the quasi-empirical, psychosocial dimensions of mind, seen, we may as well believe, in the dimensions of spirit. Emergents lie so startlingly around, behind, and within us that it is too conservative not to believe in one more. So much self-transcending in nature has already taken place, so much surmounts earlier, lesser natural modes, that it is no great stretch of thought to believe that the superseding of nature is greater than we know. A further, supernatural power would not be any more or less miraculous than what has already taken place under so-called natural powers.

To believe in the supernatural is to insist on keeping the concept of the natural open-ended, to refuse to close the system. It is to listen for supersignals. It is to take an aerial view. This is a high-order antireductionism, which is no more prepared to reduce the spiritual to the natural than it is to reduce the psychological to the biological, the biological to the physical. Aphoristically put, to believe in the holy is the ultimate holism. It insists on the truth, but nothing less than the whole truth, the holy truth about the forces working for expression in our world and in ourselves.

Some will complain that in this account we are forgetting how the later stages

—mind, life, spirit—did not, in point of fact, exist at the earlier times (although nature even then had the potential for their subsequent production). Supernaturalists do not want to say merely that spirits, like minds, later emerge. In theistic creeds at least, one wants to believe that Spirit is omnipresent and eternal, always there, regardless of chronological unfolding and structural development. The movement is not only *to* spirit; it is *from* Spirit.

While this is true, and reveals some inadequacy in our approach, we have only indicated how the category of *spirit* is credibly to be reached "from below," not what more can be done with it upon reflection "from above" once it is obtained. The deploying of this category back across the whole, moving from emergent spirits to a superintending Spirit, requires doubling back over the whole from the later end. Physicists, who come from the former end, will think that matter-energy, primitive nature, has always been there, present from the start and persistent in all the transformations that form natural history, the promoting substrate for everything. However much this may be true, theologians will think to the end, think back from this end, and think that in view of the momentous outcome that the world stuff reaches in ourselves, a fuller account will need enough explanatory power en route to bring the precedents up to their narrative outcomes. It is this latter power that is elemental, not the particles that first appear in the phenomena. And, as we have seen, even the physicists who think back to the beginning, and who think about those anthropic constants with which our world is so fortunately constructed, have also to puzzle over whether the end is not somehow controlling the beginning.

We can say, in short-scope perspective, that life is immanent in the organism, in the DNA. But this is only half the truth. Life requires transcending relations. It takes place in an environment and is nourished and shaped by it. The ecology is as vital as the biochemistry. Further, over time there is a transcending of individual organisms and ecologies, with their local powers, by dramatic evolutionary developments. More and more life appears, as the system elicits advanced life, and the individual organism becomes a story link. The intelligibility of the process is only partly immanent in the organism; it ultimately transcends the organism and is resident in the system.

The Divine Spirit in Historical Nature

But to put this in a theological perspective, even the system is animated by the Spirit of God, a Spirit-field, known to prescientific writers as the Divine Wind (Greek: *pneuma,* wind, spirit), which transcends and makes itself increasingly evident in the storied developments. Spirit animates the whole. The Spirit is, in Judeo-Christian conviction, the giver of life. Here theists actualize or reify what naturalists are willing only to call disposition or potential, but even naturalists must somehow manage to think that some such capacity was always there, even in the absence of its manifestation, at least enough for what has managed to happen. In this, theologians are simply positing enough premise for their conclusion. They too want a unified theory, but they prefer to explain the less in terms of the more, not the more in terms of the less.

Theologians explain the beginning of the drama in terms of the end, not the end in terms of the beginning. Causes are what they are from the start; a story is what it is at its end, more than at the beginning. Theologians have an a posteriori and not an a priori position! They adopt here the advice of Jesus that we can best know a thing by its fruits. The whole material-energetic performance stands under the narration of Spirit. On the one hand, they are taking emergence seriously, but on the other, they believe that the emergent phenomenon of "spiritual forces" cannot be dumbfounding, but must have its explanation in a Spirit implicated over evolution. "The end preexists in the means."[5] God lies at some order of magnitude and level beyond the superposition of quantum states, beyond the trans-space-time ether-foam, beyond the anthropic arrangements; God is the supernature out of which nature congeals. In this sense, God precedes that which follows—matter, life, mind, spirit—being revealed progressively in, because omnipresent in, the *superb* evolutionary sequence in which more *supernature* emerges within nature.

By now, alert to these storied developments, we have to be careful about thinking that there is no interruption of the routine natural or historical orders, since there is no such thing, ultimately, as a routine natural or historical order. Every historical and natural sequence is unique. It is not so much that the laws of nature are never interrupted as that they never more than partially explain the idiographic narrative, which is always being interrupted by chance, or mutation, emergence, creativity, decision, resolution, or surprise. The particular is always something else than natural laws or historical trends can fully specify, something significantly autonomous from its precedents and determinants. If natural law, historical trend, causal sequence, mere randomness, or their combinations will not catch this element that puts meanings into the adventuresome plot, is there anything else that can? The theistic answer is that of Spirit brooding over history.

God is present in the natural-historical world somewhat as the person is in his body, and as the organism is present in mass-energy, not to violate it but to superintend its processes, although there will also be emergent dimensions in this superintending of which we have little inkling. We may remember, too, when using the former analogy, that we do not understand human action in the body very well. Indeed, at the crucial points of the interaction of the subjective mind on its material body we understand little indeed. The extension of the model of the human in his body to God in the world is the sort of deploying that is bound to distort a model. God is not local, as are individuals-in-the-world. Nevertheless, the model can suggest how God is incarnate in history, narrating it, as the person uses his body to narrate a career.

That would be more supernatural than anything that the present sciences can describe. But it would be a form of the supernatural with precedent, sacrament, and analogy in the experienced reality of personal, earthen life. This view would want even for God a passage through the world in narrative form. The universe, Earth is God in the story mode. The historical form of such explanation will be more adequate to nature than the best of merely lawlike scientific explanations. It can reach the level of meanings beyond causes. It reaches for the sense of the Presence of God, the Divine Thou inhabiting the It world.

God and nature become ends of a progressive spectrum, one spectrum viewed from alternative ends, although this claim is made relatively, from our reference frame in this universe, and is not an absolute identification of God and nearby nature. In this spectrum there are quantum jumps, both microscopically and macroscopically. There are emergents and natural selections that lead to a constructive upslope. This development continues across moral and spiritual unfoldings to orient successive cultures, despite their wandering tragedies. To believe in the supernatural is to believe in the Kingdom of God in our midst. It is to believe, using a poetic metaphor, in that lofty land where the great mists lie, but from which also the great rivers spring. Nature's most startling mystery is this river of life and spirit that flows from on high.

2. SCIENTIFIC-EXISTENTIALIST THEISM

We turn next to options within theism. As with the naturalisms and Eastern religions of the preceding chapter, we are giving only outlines and not documented historical specimens of faith. But we do try to portray essential, paradigmatic attitudes toward science and religion, toward meanings and causes, nature, history, and God. We are arranging these, moreover, in terms of reaction patterns to science, and not asking about theism in categories that might arise in comparative religion.

We will be both describing and criticizing each as we go along, reaching what evaluative conclusions we can, though we want to say of theism in its creeds about God, as we have said so frequently of science in its theories of nature, human nature, and culture, that we have at best a fallible knowledge of open systems. We "see in a mirror dimly," not "face to face."[6] In such an epistemic condition, one must make an appreciative survey of the strengths of each creed. But one wants, as we have so readily done with the sciences, to attack the soft underbelly of each creed, an attack that will be the likeliest route to further truth. "Iron sharpens iron, and one man sharpens another."[7] But we do not close with criticism; rather, with an invitation to live the truth in a historical, critical drama.

What we will call *scientific-existentialist theism* must bear a hyphenated name because it tends toward a dualism or, more favorably, a complementarity between nature and God, between the objective and the subjective dimensions of reality. The "scientific" half of its name cedes entirely to science the outward realm of nature and of cultural history as this is a consequence of causal forces. But the "existentialist" half of its name defends intensely the privileged inner core of spirit, the province of religion with its detection of meanings. With the other theisms, it is convinced of a realm beyond the natural, beyond science. It believes in the supernatural, but does not try to locate this within nature or history. Rather, we look for it within the existential self. By this acceptance and delimitation of the two spheres, scientific-existentialist theism becomes at once the most "scientific" and the most "spiritual" of the creeds. But the cost of this achievement is unclarity about the intersection of the two halves of the creed, or, in other words, about the unity of the story of God, nature, and history.[8]

God beyond Nature

The conception of God's acting in nature is essentially deist. God created the world, but assigned to it an autonomous integrity. God is not now immediately present in nature, nor does God violate it. Nature is a self-contained order, neither overseen by God's immanent hand nor perforated by miraculous interruptions. From the viewpoint of the ongoing natural order, there is nothing supernatural, and we may expect science to achieve ever more rigorous and complete explanations of events in the natural world. These will be causal explanations, but whether or not this will prove a fully determinist view is irrelevant to the main claim. Perhaps nature includes some random or statistical elements; perhaps it includes free decisions by persons.

But we do not try to reserve any gaps for God. The world goes its spontaneous way, chartered by God, sustained by God, but not now directly guided by God. Science will discover a thoroughgoing rational orderliness, inclusive of any indeterminisms, random mutations, nested sets of quantum states, historical possibilities, and the like. God will not be needed as a hypothesis in the scientific explanations of natural events. Rudolf Bultmann maintains that God stands outside of what we call nature and history; the objectivity that can be studied by scientists is "a closed continuum of effects in which individual events are connected by the succession of cause and effect. . . . This closedness means that the continuum of historical happenings cannot be rent by the interference of supernatural, transcendent powers."[9] *"God stands beyond all the great powers of nature and history.* . . . Here, in *this* realm, God is not to be found! . . . He is beyond them. He is their source; for from him are all things."[10]

We fall at once into bad science and bad theology to think otherwise, although this has been the most recurrent flaw in the interrelations between science and religion. As a result of maturing over four centuries of dialogue, we are now growing out of those conceptual confusions in which religion tries to make scientific claims, preaching something to science about God in nature, while science tries to make religious claims, progressively thinking to chase God out of life. Norman Perrin writes, "The world of nature and of history *is* a closed world in which God cannot directly be known. . . . Nothing is more pathetic than generations of theologians finding God in a realm which the natural scientists of their day have not managed to explore, only to find that the next generation does explore it. . . . The very idea of God as an effective cause at the level of the natural world is simply and basically incompatible with a true concept of God." It is "theologically obscene" and scientific "nonsense" to think of God as acting in nature.[11]

Such an older view is *myth.* Myth is the portrayal of the divine activity as if God were objectively out there in natural events, while in fact the divine activity lies in the inner, subjective life. Believers have projected inner events outward, unaware of what they were doing, as Freud helps us to see. So believers speak of God as acting to do this or that—creating life, forming Adam and Eve, moving winds and waters at the Exodus, sending lightning to consume Elijah's sacrifice, or rolling back the gravestone and resurrecting Jesus—while all these are in fact parables of inner, spiritual events. These existential events are not illusions, but genuine experiences, contrary to Freud. But the belief that God is superintending or interrupting natural processes is an illusion. Science drives the myth out of religion, finding causes where events were before alleged to come to pass by God's immediate command. The realm of nature must be *demythologized,* with benefits alike for science and for religion.

Traditional theism (even in what we will call its transscientific form) is faulty because it has not yet adequately cleared its house of quasi-causal claims. It keeps protesting that scientific explanations are causally incomplete; it wants to hide God behind quantum states, or genetic mutations, or brain synapses, or decision options and unique events, as an upper-level overseer of lower-level causal processes. It wants a ghost in the natural machine. All this is to hang on to lingering myth and to remain naïve before science. Science may next year or next generation complete its explanations to leave no space for God. Lookers who seek God's tracks in the space-time, material matrix are looking in the wrong place.

The emergent states do *super*sede each other, as we have noted in the passage from matter to life, life to mind, and to spirit within mind. But there is nothing *super*natural in this, not considered objectively from within the resolving powers of physics, biology, psychology, sociology, anthropology, as these describe chapters in the Earth story. It is only the next highest step that we want to call supernatural, and there look for God. The already accomplished steps, lying behind and around us, can be studied effectively with the categories of natural and social science, or at least with the categories of the historical, which, though they may exceed science, do not require the supernatural. Rather, we should look to what is going on in our heads that is over our heads for the Divine Presence.

God in the Existential Self

By the existentialist account, God directly acts only in the self. Since God is Spirit, God may be expected most appropriately to speak to spirit. We are to look for the tracks of God within our spiritual life—in our anxiety, despair, guilt, decision, repentance, and faith. Consider, for instance, twin themes prominent in the Judeo-Christian faith: love and freedom. Neither is to be looked for in rocks or trees, nor really in animals. These are emergents at the human level. Any precedents we find in mice or chimpanzees will hardly illuminate what humans come to mean by these words, surely not as religious virtues—agape love or responsible moral freedom. The latter are states of mind—indeed, states of spirit. If God is anywhere to be found, it will be here and, for all intents and purposes, here only.

Recalling our discussion of nature and supernature, we must take the epiphenomena even more seriously than before. Consider again (Figure 7.3) the evolutionary sequence, sketched earlier in Figures 7.1 and 7.2. The conclusion to be reached from the emergent states and superintending levels is not that God is everywhere to be detected, alike in levels below and above. Consider, by analogy, how it would be a mistake to try to understand life by studying inert matter, an

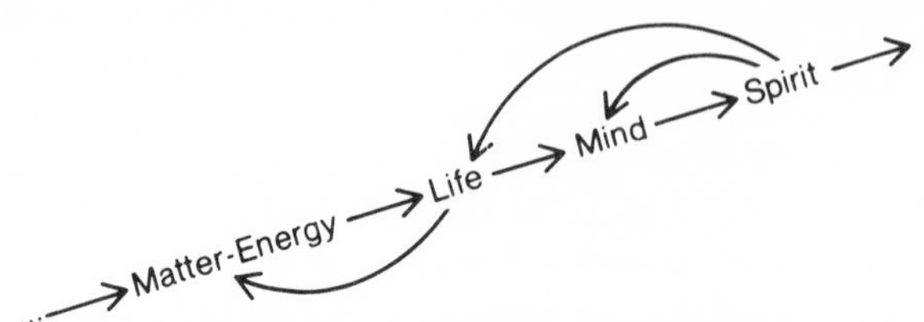

Figure 7.3 The detection of Spirit

attempt indicated by the arching reversed arrow at the lower left. Living qualities are not yet present there. The biologist would be looking in the wrong place if she expected from geology or mineralogy insight into information transfer, reproduction, irritation, or learning. The zoologist will not ask the botanist about sentience.

But so too if we look for spirit where it is not yet present, in biology or neurophysiology (much less chemistry or physics), as suggested by the reversed arrows at the upper right. Spirit emerges with inwardness, but not yet even at the lower levels of sentient awareness or with the merely pragmatic mind. It appears with the sort of higher reflection that comes at the level we call *Existenz,* with a self-conscious being in quest of meanings. After all, religion is about meanings, not causes, and *meanings* are sought only by *persons.* At prepersonal levels, meanings have not yet emerged in conscious awareness. It is thus bad theology and bad science to look for God as one among the causes of events in the natural world. God will be found only at the existential level. One needs an adequate detector. Rocks and plants are not adequate detectors of God, nor is the study of them an adequate place of detection. Persons are adequate detectors of God, at least in their spiritual modes. In the ordeals and joys of personal existence God can and does enter.

Only bad science thinks that it can say anything about good and evil, or the forgiveness of sins, or thinks to guide moral decisions, or to orient the person in love and freedom. Here only religion can serve. In this dimension of life, Christian theism offers saving power and grace. This gift is maintained over the centuries by a cultural education and passes from faith to faith. The gospel is preached and persons are moved in faith, freed for righteous love. The strong Spirit of God is detected here, not in bubble chambers or biological mutations. The Divine Spirit moves the human spirit, not rocks or rain, genes or quantum states. We do not want a *Cause,* for which monotheism has too long looked among the other causes in nature and history. We do not want a *Process,* a model process theists favor. We want a *Presence,* and such a Presence can be found only speaking to self-conscious subjects. God is not an *It* operating on other *its,* but a *Thou* addressing a *thou.*

God in History

Existentialist theists answer the question of God's acting in history both *yes* and *no.* Consistently with their denial of God's immediate presence in nature, God is not present in history, a theme we have already begun and now make more explicit. Norman Perrin, agreeing with Bultmann, is emphatic: "God cannot be the effective cause of an event within history; only a man or a people's faith in God can be that. . . . There never has been and there never will be an event within history (that is, world history) of which God has been or will be the effective cause."[12] Though contrary to what classical theism has believed (and to what transscientific theism continues to believe), this follows from withdrawing God from the sphere of nature. God is not a causal agent in the external events of history, any more than in nature. We now have to say that the mighty acts of God at the Exodus, the Return from Exile, or the Resurrection are myths, if objectively conceived.

But the revision is not as radical as may first seem, because in such events the factors always include human actors, and here God has been intensely present. God does not act externally in history, but he acts internally within persons, and persons make history. At the Exodus, God does not move the wind and waters. The winds of the Spirit are within Moses and his Israelite people. The Hebrews are empowered by the divine love for freedom, released indeed in the Exodus. But they are not saved by meteorological or tidal miracles, but by their faith, which enables them to struggle through, to give meaning to their journey toward the promised land. Science frees us from believing that God drove back the waters and similar primitive myths of a people who did not understand what was going on in their heads.

But science does not touch the belief, which theology still affirms with joy, that God makes persons free. God moves persons to cast off their bondages and calls them to an obedience of faith. God inspired the prophets and the scribes who preserved Israel in Babylon and by whose self-fulfilling faith Israel returned from exile. Jesus is not bodily raised from the dead; but a spiritual presence that he elicited before his death, by the strength of his personal faith, now reappears and continues in his disciples, who have Jesus experiences, realizing God's saving power within. The Spirit of God resurrects them, gives them new life in Christian faith. God is thus reinserted into history, that is, recognized as being present where God always and only was, mediately in history from the immediate divine presence in the personal life. In this way we have to understand God's presence in history in the Bible, in the American Revolution, or the Civil War, or the civil rights movement.

A Narrowed Story? A God-forsaken World?

In criticizing scientific-existentialist theism, we do not so much want to deny what it affirms (that God is eminently revealed in the personal life) as to affirm what it denies (that God acts in the natural order). Belief in God is possible only for humans (or perhaps extraterrestrial humanoids); it arises and is maintained within the personal life. But the personal life is not a private subjectivity, not a lonesome self defended in a God-forsaken world. Rather, the self is set relationally within an environing nature and history. Humans seek the meaning of a self helped or hurt in a natural ecology, incubated in an evolutionary ecosystem. We must deploy convictions that arise at the point of existential personality out across the world and back over history. Any God found within the self has also to be found out there, or else is not God of the whole. It would be odd of God to touch human life so intensely and be absent in the natural world that is our foundation and foil, womb and partner, odd if God cared so much about human *Existenz* and so little for all else. It seems illogical to have nature self-propelled across so long and remarkable a trajectory only to have God reappear in the drama late in time and immediately in subjectivity, bypassing the mediation of matter and energy, in terms of which everything earlier has been narrated. Deist-existentialist theism has a do-nothing God for twenty billion years, before subject selves appear. Even yet God is marginal to these objective forces that are of such enormous consequence in our lives.

Is it not equally credible to let God infuse the whole world-body? We humans are active in the world; our spirits work their wills upon matter-energy. Why cannot God do the same? If not, God is our inferior. We humans must live within this earthen evolutionary ecosystem, but we did not create such a context for our story. The "earthen vessel" we inhabit is of God. Even scientific-existentialist theists must say this sooner or later, and why not both sooner and later, and all along the way? In the midst of what science says about nature, we still can and ought to expect some embodied expression of the will of God. We may not have a God in charge instantly and omnipotently at every point; we may not have a God who violates the creation with outrageous miracles. But we do want a God who charges and supercharges the creation with divine purposes, a nature exceeding mere mechanical causes. The natural world is suffused with deity, and we ought to be bold enough to say this, and mean it, in the face of science.

Anything less is too limited an account of the presence of God. The world history is not the affair of existential beings only. All nature has an evolutionary history, from stars to planets, from dinosaurs to primates, and only in the last few minutes of cosmic time has cultural history been supermodulated onto natural history. Existentialist theism does not put God in very much history, only in a few of the latest episodes, and only when the actors have the right faith. A God who enters history so recently and reticently cannot be a God of all history, only of fractional parts of it. Here faith has not merely been demythologized; it has also been denaturalized and dehistoricized. We are not given much plot in the drama, little Presence in the narrative. Scientific-existentialist theism fails to give us a united world view. Before the criticisms of science, it has withdrawn into a kind of dualism, a narrowed story. God lives within, but there is only a radically secular world without.

The problem of the packing of God's presence into the lower earthen levels of being (rocks, plants, animals, ecosystems) is not solved by incarnating God only within subjects (persons) and eliminating that presence from objects. Indeed, it is almost as impossible to pack God into the human subject as to pack God into an object, for God vastly outgoes what we know as personal, localized subjectivity. The place to pack God is into the whole story, not just into the subjectivity of the latest of its actors, despite the fact that a self-conscious awareness of God arises only in the latest actors. God is to be discerned as the chief Actor, not as a kind of tutor or inner voice confronting the human players only. God is to be looked for "in, with, and under" nature, history, and selfhood.

Causes and Meanings: The Complementary Languages

The hyphen in scientific-existentialist theism indicates a kind of intended complementarity of object and subject, science and religion, but if this is not handled with care it congeals into a dualism that isolates the one half of the story from the other. We can explore this complementarity by noticing how we sometimes simultaneously use two languages to describe single events. The motion of my arm can be described in terms of muscle contractions, involving A-bands and I-bands, thick and thin filaments interdigitating and sliding past each other, regulated by calcium ions, nerve impulses, the firing of brain synapses using neurotransmitters, the movement powered by ATP hydrolysis, and the like. This would trace an unbroken causal chain, including perhaps a statistical averaging of random events at microscopic levels. But no amount of neurophysiological *causal* analysis could accumulate to produce the *meaning* of the arm motion, nor could it defeat my knowledge claims here.

I wave to greet a friend. This is talk on a different level. It expresses the intention of a subject, an entirely separate dimension from the empirical muscle movements and nerve impulses. One language gives causes, the other reasons. One is an objective account, the other a subjective account, although not merely of a private experience, but of a subject behaving in the world. The two languages parallel each other as complementary and noncompeting accounts of the same event, two windows into what is going on. Each account can be, in one sense, complete and without gaps or puzzlements. They are utterly disparate ways of knowing and coexist in irrelevance to each other, like independent television channels that never contact each other. Yet each account complements the one that it parallels, giving us further understanding.

Scientific and religious accounts are like this example.[13] Science is causal, tracing events in natural history or somatic and social processes in human affairs. But religion speaks a different language, about the meanings of these events. Such meanings are always personally and communally held; they have owners. Let us analyze more carefully those theological assertions that appear to be talking about God's activity in nature as though it were independent of the speaker. "I believe in God the Father Almighty, maker of heaven and earth." Theists who say such things are not really making claims about God as a causal agent in nature and so are not making scientific-style claims. They use such statements to report and recommend the meanings they themselves have found in life, experiences of its goodness, of awe and respect felt for the world, of the sacredness of life. Bultmann says, "Faith in creation . . . is not a theory about the past. It does not have its meaning by relating what took place at some earlier time, . . . but rather speaks precisely about man's *present* situation. It tells him how he is to understand himself *now.*"[14] The Genesis creation accounts, despite their superficial appearance, are not myths about beginnings, but rather are myths about dependence.

These are really reports from extrascientific territory, existential reports that are made in faith as a result of religious experiences. These reflect and project what is going on within believers as God helps them to find meanings in their lives. These are subjects' accounts of meanings found and confessed, not objective accounts of God's acting in nature. When we recognize this, scientific descriptions of natural processes can come and go, be missing, complete, incomplete, steadily revised, while the main religious claims remain untouched. Similarly, the knowledge that I greet a friend remains untouched by developments in anatomy and physiology.

Scientists can do what they please and theologians are majestically unaffected by it, since they are talking different languages—rather as the lawyers can do what they please and the poets are majestically unaffected by legal innovations, or vice versa. But is it this simple? Theologians do have to write a theology of nature on top of what the scientists have found out; they have to make the best of what the scientists are saying about the processes of cosmology, evolution, genetics, psychology, and the like. Or they have to challenge the scientific accounts by defending experience that reveals logical or empirical inadequacies, or by testing the complementarity of science and religion.

In criticism of the complementary language accounts, we need not deny what is affirmed. Causal explanations and meaning explanations do operate on alternate levels. But, contrary to the isolation supposed by radical two-language accounts, meaning explanations do interfuse with causal explanations; they do not merely run in isolated parallel to them. Arising within the personal life, as meaning explanations do, they are not assertions only about personal life in the world. They are also about God's meaningful activity in the world prior to and independent of God's activities within persons. In an attempt to narrow the range of the divine activity, restricting it to personal selfhood, the existentialist account shrinks the scope of religious claims. The double-language view sees only half their function. Believers are also trying to claim that God works in the world. So whatever causal explanations they can give or accept will have to be of such kinds as are also open to their superintending by an agent of meaning.

In the example of arm motion, only certain kinds of causal accounts can be reconciled with the intentional account. There has to be not merely a parallelism but a complementarity between the accounts. It is not so clear, for instance, that causal accounts can be complete and closed, especially in the cerebral traces, certainly not rigorously determined. There must be some spaces that can, from the other side, be seen as decision points, where the intentions of the "I" are expressed and modulated onto the causal sequence. There must be some gating at the synapses that is open to options, and hence an incomplete scientific account of the changes. In any case, the two languages are not describing two territories as insulated from each other as might seem. When we use them to describe one world we cannot avoid questions of interaction on their frontiers.

In the example of Creator-creation creeds, there is asserted personal dependence on God, and such dependence involves ongoing, continuing creation, but also there is asserted an original, start-up creation. While a great variety of causal accounts are compatible with the monotheist meaning account, not all such accounts are. An account that claims final adequacy for naturalistic explanations, or one that claims that life and humans are adventitious or accidental products of the causal process, will be incompatible with theism.

The fault with our analogy between double languages about arm movements and double languages about natural and historical processes is that in the one case we have a privileged access to immediate knowing. I wave to a friend, and I need consult no biochemistry to prove or disprove it. Science cannot touch my claim. But when I say that God creates the natural world or gives meaning to history, I have no such all-commanding privileged access, participant though I am. In the latter case, I am making broader claims where before I was making claims about just my personal experiences. Revisions in the descriptions of natural and historical processes can and often do force revisions of my claims about meanings there, as has steadily happened in the sciences originated by Newton, Darwin, Freud, and Durkheim.

Scientific explanations are not all and equally agreeable to meaning explanations; some are more congenial, some less so. What theists hope for is not merely the experience of meaningfulness had in disregard of the operations of the natural world, even if this experience is given by aid of the divine Spirit. They want to discover an appropriate meaningfulness by thinking God's thoughts after him, by detecting providence where it has been inlaid all along into the world, not merely as a gift erupting in the late-coming believer's *Existenz.* Scientific findings of intelligibility and order, for instance, go reasonably well with the theist's discoveries of meanings there, although the deistic God is rather remote from the creation. Darwin's descriptions of a junglelike nature have tended to dislodge belief. We cannot assume that causal explanations have no impact on meaning explanations, or vice versa. We do have to worry about conflict and incoherence between models, and complementarity cannot be assumed; it has to be wrestled with. The boundaries between languages and those between models are permeable, and the history of dialogue and warfare between science and theology is a result. We have to seek peaceful coexistence, not mere parallelism.

Meanings emerge in the personal life, but they do so in an evolutionary matrix where what were earliest causal processes are transformed into what in humans are meaningful processes. Do we find only a late, epiphenomenal parallelism of causes and meanings, and no productive evolution from the one to the other? Is evolution perhaps a prefixed, perhaps a more or less random unfolding of a spontaneous secular order? A unified account will need to discover more divine activity all along the route. Those genetic propositional sets, those programs defended in plant and animal lives, the pleasures and pains of sentient life, learned animal behaviors—all are valued precedents for the meanings that are found still more gloriously in the personal life.

Meanings need to attach to the systemic, storied whole, in such way that God does not just enter the drama in private lives or in faith communities, but is always there intelligibly and effectively. Believers cannot convincingly discover meanings in a world from which God is objectively remote. It will not do to posit some once-for-all setup at some aboriginal beginning, God's archaic creation of a nature that is thereafter an autonomous process. That is still to believe in a special creation and not in the continuing creation of God. The world needs God all through its course, not merely at the start as First Cause, or at this recent stage as Giver of Meaning in human lives. The causal and organic mechanisms operating out there need to be closely coupled with divine Presence. God is the Spirit brooding within the whole natural process, not merely the Spirit within our spirits. "Only if God is revealed in the rising of the sun in the sky, can He be revealed in the rising of a son of man from the dead."[15]

Existentialist theists try to be so scientific, but in some ways they fear science quite as much as they embrace it, and have therefore withdrawn theology from the province of the natural order. But this is premature and naïve about what science has already accomplished or gives reasonable promise of accomplishing. Impressive though scientific explanations sometimes are, they are very incomplete overall; and the better they get, the more they have a soft side, one open to God. This open texture is found not only on the frontiers of science, but again and again at its conceptual cores. The most recent "myth" that needs to fall is that of science as omnicompetent and omnipresent explainer of natural events. That is scientific nonsense. Perhaps God will not be needed in scientific explanations as such, but scientific explanations can never give a rigorous, complete account of the idiographic historical narratives in nature. Against Perrin's complaint about "pathetic" theologians who try with "obscene theology" and "scientific nonsense" to find God among the causes, it is "pathetic" to find theologians who have lost their nerve, withdrawing all claims about God's activity in nature. That, if anything, is "theologically obscene." Natural history too needs to be sacred history.

3. PROCESS THEISM

The second option we examine has incorporated the most from the categories of natural science into its concept of God. In that sense, it is the most scientifically sensitive of the theistic schools, radically extending science into metaphysics. Neither does it hesitate to make radical revisions in theism, nor does it mind superseding science where science impoverishes experience. In this account everything gathers around the word "process," symbolizing the dynamism that science depicts in the world story, incorporated now in the concept of God.[16] The resulting view supposes a powerful but not omnipotent God who transcends, but is in immanent interaction with, a processive nature that is neither originally nor entirely the divine creation. Nature and God both are limited and affected by each other. Process theism has (in the West) largely replaced monisms, which have gone out of style with the increased revelations of a processive, energetic, historical nature. Monism fixed on too much Absolute, and did not allow enough Divine Process.

Nature Is Organic Process

The twin themes that are paradigmatic for any concept of nature are change and organism. Alfred North Whitehead writes, "Nature is a process of expansive development, . . . a structure of evolving processes."[17] We do not anywhere know nature except as it is *becoming*. Nature is not (as earlier theisms supposed) the passive product of the divine will, not a crafted material, but is itself active, energetic, creative, with its own integrity, if also containing a dimension of the divine will. Further, although we know organic nature only on Earth, the relatively simple structures of preorganic nature have to be interpreted in the light of the evolutionary ladder on which they are the lowest steps. That ladder is organismic, not mechanical. Whitehead continues, "The whole point of the modern doctrine of evolution is the evolution of complex organisms from antecedent states of less complex organisms. The doctrine thus cries aloud for a conception of organism as fundamental for nature."[18] There is ever a flux of things, never permanence and substance, but in the flux there are relatively enduring patterns that spontaneously appear and ramify.

These loci of development have a gathering tendency. Lesser units are aggregated, pulled into orbit, bound, becoming wholes as composites of parts. At lower structural ranges, there are what Whitehead calls packages, pacts, "societies." At advanced levels, there appear organisms. Reality is not only dynamic and particulate, but social and creative. Our most mature view of nature is one of generation, beyond causation. There is a pressure to fill up many ontological niches, to make something actual for every possible slot. The prelife events are, one can say, merely physical and not biological; nevertheless, there is something generative about them that links them on toward organism, as will be seen over the long-range cumulations of the spontaneity and creativity that are always latently present, however dimly executed, in the simplest of phenomena.

In ourselves, the complex corporate events that constitute our bodies have a mental side. At the lower levels, rocks are composites of minerals, which are aggregates of atoms, and have a dominant physical side. But there is an evolutionary construction from the one to the other. There is emergence here, but also we need an overlapping of the physical and the psychical dimensions of reality. If we consider life at the level of, say, an oyster, there is striving. A life program is defended, although there is no conscious cognition. By extension, we can think of "inorganic" events as having a kind of apprehension, although it may be low-grade.

Only in this way can we explain what does in fact happen to inorganic materials, albeit rarely so, when they assemble into living forms. Over time, they do get somewhere, and so we assume that this attenuated pole is always nascently there, even when obscure and weak. It is spirit immanent in matter, a harbinger of things to come. Process theists do not object to being called panpsychists, believing in a psychical dimension to all nature. It is as logical and economical to think of spirit as being diffused and inchoate in matter as it is to posit a supernatural Spirit who stands outside and inserts it, or a deist God who is the remote Architect of a secular world-mechanism.

By contrast, a traditional monotheist will want to say that Spirit is transcendent to matter. When we say that atoms or rocks have the potential for being restructured into humans, we do not have to mean that the human properties, such as the psychical, are already there in rocks and atoms in some scaled-down way. One can with equal plausibility hold to a genuine potential and to the creative act that takes place when there is novel emergence so that things appear (mental qualities) that simply were not there before. The traditional monotheist, of course, will want an explanation for this creativity, which lies in the Spirit of God, the giver of life and mind. But such a monotheist does not have to put the explanation in a diffused and inchoate, rudimentary spirituality present all along in matter.

God Is the Ground of Order

Clearly, there is some assemblage of natural properties that provides for order. Laws hold, energies persist, conservation is maintained, structures are produced, organisms reproduce, and there is much logic in things. While some take this as a brute given, it bears further reflecting upon, and we may consider God to be the ground of order. The process God does not create the world *ex nihilo,* contrary to traditional theism. Nature ever coexists with God. But God supplies the order that is mixed with disorder in nature. In this sense, science is the first witness for God, since it finds order permeating nature. At a foundational level, this order is *causal,* sometimes mathematical, sometimes nonmetrically logical, and in other cases the order is regular but with operations that so far escape our logic.

But this order is also *meaningful.* Natural things come to carry meanings that intensify with the accumulation of complex event structures. "Order" is thus a term that begins in science but crosses over into religion. It bridges both disciplines.

Considering God as the ground of order, we can find God more evidently omnipresent, not only in the higher living creatures, but in lower structures, such as atoms and minerals. Although the organic and spiritual dimensions are inchoate and attenuated at low levels, the omnipresent order still attests the divine presence. "One of the attributes of God," says the biologist Edmund W. Sinnott, is "the Principle of Organization."[19]

God Is the Ground of Novelty

The order is not static. Nature continually displays fresh introductions on its cutting edges. There is both continuity of process and the emergence of genuine novelty. The future is more than the past, linked with it, but an adventure elsewhere. The natural system has its trends, but also its openness. Development is not the mere unfolding of inevitable clockwork. There are crisis points, options, opportunities, exploits, which give to nature its idiographic, narrative features. These make history possible, including the human history that overtops natural history. In the evolutionary movement from matter through life and mind (though science perhaps only approaches spirit), science is a second witness of God—just because science, with its featuring of causal regularity, poorly handles the innovative face of nature. The missing or disputed element in any evolutionary model is what Polanyi called "the orderly innovating principle."[20] Yet every attempt at explanation stalls without it. In reductionism, things "fall apart" into components. But what we want to know is why things do not fall apart but come together more and more.

We do not need to know about parts in their analysis so much as about the principle of their synthesis. We can expect, penultimately, better scientific models that gain increasing insight in naturalistic terms about how this synthesis took place. Yet such accounts, however relatively satisfying, will have an open side, invisible to science, that permits interpreting the orderly innovating principle as a sign of God. Like "order," "novelty" is a word that crisscrosses the languages of both science and religion. Causal explanations get toeholds, no more, on why this novel route and not that one is taken, or why there is novelty at all, why wholes are more than the sum of their parts. Novelty hides behind such terms as "random," "statistical," and "epiphenomenal." Religious explanations use this novelty to insert adventure, self-creation, local integrity, decision, and moral responsibility into the world story. God loves surprises, as God loves freedom and spontaneity. God both permits and insists on these, and in such way as gradually to innovate more and more of them, consistent with the constant divine aims. These are signs of the divine creativity interwoven with divine order. "Apart from the intervention of God, there could be nothing new in the world, and no order in the world."[21]

Thus, God is the source of the transformative principle that soft naturalism had supposed to lie inherent in nature itself. The process theist does not have to ascribe to God the existence of nature, nor its mere changing, nor all the options taken, nor all decisions made, nor nature's chaotic elements, nor its decay. God can be used selectively to explain some but not all of the phenomena, namely, the persistence of order and the recurrent introductions of novel developments. God is a relative explanation, not an absolute one, a partial explanation, not a complete one.

God Is Creative Persuasion

Process theism wishes to provide an overall aiming for the stories of nature and history, and at the same time to leave much up to local spontaneity. God presents possibilities in excess of actualities, together with a heading, but then draws back to suffer the entity its own increments of freedom. The routings taken are not all inlaid into the anthropic constants of the big bang, not all thrust up from superposed quantum states below, but they are actualized in part owing to elections that the organic entity superposes on the quantum states. There is provision for cocreation and continuous creation. God is the ground of creativity "from below," and yet aloft, aboveground, creativity is removed from God enough to be assigned locally to the creatures, who actualize themselves. They do their own thing, always in God but not always of God. There is the kind of parenting that puts local integrities on their own and yet educates them as they go. This influence is not mandatory or deterministic, but enticing, prompting. Process theism wishes to have it both ways on the question of historical inevitability. There are aspects of knowability and aspects of unknowability.[22] The shadow of the past outlines the future, but more. The shape of the future has lured things past "from above"; divine wisdom has outlined their routes, at once creatively yet also leaving their idiographic tracks up to them. Process theism supposes "soft" rather than "hard" directionality in history, but this is commended as a strength in its creed, and in God.

To some extent this conclusion is reached out of science, and to some extent this continues from, and reacts to, the religious heritage, especially Christianity. Science depicts plural natures, distinctive entities, nested sets of communities with crisscrossed historical lines, populated with private individuals, not marionettes. Biological science has its organic selves. Behavioral science needs to place an organism (O) between the stimulus (S) and the response (R):

$$S \rightarrow \boxed{O} \rightarrow R.$$

Humanistic psychology finds egos maintaining their centeredness, self-actualizing (SA) persons, while sociology and anthropology embed them in cultures that transcend the individual and yet are distinctive because of elements of individual self-definition. Our picture is of an open nature permitting multiple histories.

Nature's journey is not by evident linear progress but by the gradual natural selecting of zigzag proposals, trials conserving past successes intermixed with a spontaneous groping for more. Across the spiraling civilizations from the Neanderthals to modern humans, cultures are permitted their local integrities, and yet there is a cumulation of know-how and perennial lures such as freedom and justice. God as creative persuasion explains this in nature and culture, alike in the ends attained and in the open, patient texture in which these events take place. There is a constant, patient God, hiddenly and noncoercively present.

Here, though, theism has often erred in supposing a divine tyrant who oppressively predestines all, ruthlessly overriding his creatures, vitiating their industry and responsibility. In the Bible there is too much of the Oriental sultan remaining from the contexts in which monotheism was first engendered and not enough of the Divine Author who creates for us meaningful roles in a historical narrative. Nevertheless, the better biblical picture is of a persuasive God who wrestles with a wandering people, coaxes them to a land of promise, and who sends a Son, not to bring down fire from heaven, but to set loose the appeal of sacrificial love in the midst of a world of scattered aims and confused responsibilities. The twin divine aims from classical theism, love and freedom, both defy coercion. God can promote such virtues only as an influence operating on creatures who go their own way, subject to persuasion, not coercion.

God Is the Conserver of Values

A striking feature of the world is entropy, decay, tragedy, death. We do not need to assign these to God. But even more startling is how the life process climbs onward through shifting environments for almost everlasting millennia. In ecosystems, some lives are built on others. Nutrients, energies, and information all flow around and up through the system. In cultures, persons pass away, but their traditions pass on. Aims continue in the midst of a flux that might seem only to crumble them. Waves are transmitted and reincarnated across a succession of particles. Events in evolutionary and cultural history (as with the emergence of hemoglobin or the signing of the Magna Carta) have importances, which is to say that they are "imported" over to succeeding lines. There are losses, false starts, and dead ends; but these are replaced and recovered by new breakthroughs, with the same endings achieved by different routes. The most adequate explanation of all this is to regard God as the husband of values. Such a faith flies over the drifting vicissitudes of natural history, but it stays aloft because it perceives how natural selection preserves and innovates value, not mere survival. Process theism detects how the cultural heritages in their main currents are meaning systems in which the noblest aims are selected for transmission over the centuries.

Broadly, one can affirm with Whitehead, "There is no loss," despite the wreckage and transience of the centuries. God "is the poet of the world, with tender patience leading it by his vision of truth, beauty, goodness." That is to speak poetically, and the prose behind it is the world narrative in which values are distilled off and conserved over historical time. This permits an account of God as the "fellow-sufferer," before the cruciform character of nature and history.[23] A divine urging empowers the life forms to suffer through to something higher, achieved as this must be by the sacrifice of individuals (as with the pelican chick or the anemic African child) in behalf of the communities in which they participate.

We do not here say, as might classical theism, that God foreordains all these details, or even that God wills the fabric within which God works. But we do say that God is present to guarantee that any value that can be will be conserved out of the suffering. Charles Hartshorne assures us, "God is . . . a sympathetic spectator who in some real sense shares in the sufferings he beholds. He is neither simply neutral to these sufferings nor does he sadistically will them for beings outside himself. He takes them into his own life and derives whatever value possible from them, but without ever wanting them to occur."[24]

The Becoming God

In common life we take things that can grow to be superior to things that are static. Thus, a person, who can learn and be creative, enjoys a richer level of being than an inert rock. But how much of this applies to God, the Rock of Ages, the same yesterday, today, and forever? Process theism complains that classical theism erred in exaggerating invariant permanence in God. Nature becomes, and why not God? Process occurs in God, and process in nature reflects this. Whitehead concludes, "A process must be inherent in God's nature, whereby his infinity is acquiring realization."[25] God is no exception to the historical and evolutionary flow that is the chief feature of the world; God is, rather, its chief exemplification.[26] God is enriched relationally with the world; each contributes to the other. God and the world are engaged in a sort of feedback loop. God too comes into reality. The divine consciousness perpetually receives additions from the world. God is capable of additional values, and pursues them in Earth's history. "The Creator too is in process of being created, not simply self-created or simply created by the creatures, but the two together."[27] God, adds Hartshorne, is both the "supreme source" and the "supreme result" of the world process.[28]

But although process theism posits becoming in God, against immobility in older theism, it has found this difficult to clarify without dimming the divine omnipotence, perfection, and even reality that theism has long cherished. The changelessness of God is God's constant purpose abiding over the fluid millennia. But should we think of something in God that matches the emergence of richer levels of being (life, mind) out of the simpler things (stars, atoms)? Whitehead says that God primordially is "actually deficient, and unconscious," and only consequent to experience in the physical world does God become "fully actual, and conscious."[29] We are first tempted to think of God's evolving where earlier there was nascence, or of God's waking up to new levels of sentience or awareness, as did the sequence of creatures over time. It might seem as though God gains more power with more becoming. Later on, God is more successfully persuasive, though not more coercive, since the latter is unbecoming to the divine will. Does God gain more information, and does this make God formerly not omniscient?

But when we subtract all notions of temporal becoming over time, which characterize the creatures but not God, we have to say that God always is in interaction with physical nature, which always exists whether in this epoch or earlier or later ones, and thus God is always both primordially unconscious and consequently conscious. God's becoming is without going. But then just what categories in the developing natural history does God so supremely exemplify? Or does development in God not really involve anything like the emergence of matter and energy from the primordial plasma, the emergence of life from matter, or of the subjective within the objective, the complicated from the simple, even though these goings on originate with God and are reflected back to enrich God?

It is difficult to portray God as becoming more and more yet not be chronological about this. The cardinal feature of natural history is coming into being over time, even though relativity theory has taught us to wonder whether nature is ultimately spatiotemporal. Process theists have accounts of God's becoming along the lines of the actualizing of potential, accounts of nontemporal originative and consummatory phases in God. They manage to portray God as primordial and transhistorical becoming that is always fully actual. Perhaps they do not need a developing God who reciprocates the development in the creatures. But the notion of a changing God, who reflects the evolution in nature and history, has often suggested some evolution in the divine actuality. God too is in part created, and continuously so, a "result" as well as a "source."

Religious Adequacy of the Process God

This in fact opens up a series of critical questions about process theism, which gather about the religious adequacy of the process paradigm. Process is not Presence, not obviously so. Is this Process something to which one can pray? Is it nonpersonal, more like negentropic gravity, a lure rather than a Person? Can I address God as a thou? Is this what Jesus called "Abba," Father? Is the divine process so different, really, from the transformative principle of a nontheistic soft naturalism, which we found vaguely reasonable so long as it was kept reasonably vague, but which at closer look fell short of offering an adequate explanation of, or guidance for, the personal lives we are called to lead in history? In their concept of nature, process theists want to extrapolate back from personal experiences of becoming and find anticipations of psychical experience not only in organic but even in abiotic nature. They want to interpret evolutionary nature from the conclusions reached in ourselves, and not mechanistically or energetically in terms of physical particles. This is a "philosophy of organism."[30] It is also true that they want to attenuate the intensity of these psychical experiences as one moves rearward and downward on the phylogenetic scale.

But when we turn to God, who is supposed in some way to exemplify or parallel, as well as to lure, these developments in history, have we a Spirit who is as personal as, or more so than, we ourselves? Personality, not organism, or even consciousness, is the highest category we know, and anything less than this in God makes God our inferior. God as creative persuasion certainly sounds like a Thou. We are assured of a divine "subjective aim."[31] We are told that God is always conscious as a result of interaction with physical nature, as well as primordially unconscious. Is that enough guarantee of Presence? Scientific-existentialist theism did well to insist on the presence of God as Spirit to spirit, Thou to thou. It passionately defended a Divine Subject, although it withdrew God's immediate action from the objective natural world. Process theism has what scientific-existentialist theism lacked, an account of the divine activity in the natural world. But it is less assuring about the Presence over, above, throughout the process.

William Temple laments, "If only Professor Whitehead would for creativity say Father, for 'primordial nature of God' say Eternal Word, and for 'consequent nature of God' say Holy Spirit, he would perhaps be able to show ground for his gratifying conclusions. But he cannot use those terms, precisely because each of them imports the notion of Personality as distinct from Organism."[32] Transscientific theism, to which we soon turn, will insist on giving God a proper name, not as some one being among others, but with the conviction that the Universal behind the universe, the

One behind the many, is a Presence with a proper name: Yahweh. But it is hard to give God as process a proper name.

We can also agree that God acts as creative persuasion where this is appropriate, eschewing compulsion. We may further want God as "influence" on the natural process, so as to let the creatures retain their own autonomy and integrity, not tightly predestined by the divine will. But there seems no reason to think God absent from the necessities, the compulsions that also control the world and that often overcome us. Science has abundantly, though not exclusively, found compulsions in nature, and we need not banish God from this arena. Else God cannot be significantly found in the lower structural levels, in stars or rocks, where persuasion is not a relevant category, not unless "persuasion" comes to mean something radically different from what it first seems to mean, a minuscule influence over what are overwhelmingly statistico-causal processes. God works in the imperatives of the causal order, crafting forms that are cut in passive dependence, as well as in the education of independent, active selves. Here the notion of God as ground of order, if it is to assure meaningful, novel plot in the story, permits and even requires an omnipotent God at least sometimes closely coupled to the physical processes. Likewise at the human levels, the irresistible goings on over our heads do not always violate and oppress us. They can bless and free persons.

Perhaps the better model for nature and culture is not process but, to revert to a richer category, that of *storied history.* Development, order, novelty, perpetual perishing and the conservation of value, becoming more and more—these are really the ingredients of narrative. In some sense, that is what the Jewish and Christian monotheisms have affirmed all along. Genesis is the story of creation. The Old Testament is the story of a covenant people; the New Testament is a passion narrative writ large. History is God's story, and we can think of storied development even within God in the creative sense that God spins this marvelous story of the universe and its projects, of Earth and its peoples, of the Earth-Exodus into love and freedom, of the ongoing divine Advent here. One can pray to a *storytelling God,* who empowers the actors in the play for suffering through to something higher, rather more convincingly than to a lure in a process.

4. TRANSSCIENTIFIC THEISM

The third option, which we call transscientific theism, is classical theism become modern. It is the most robust of all the competing explanations, the most adventurous beyond (some will say oblivious to) science, and therefore, while it does not object to naturalism in its place, it is the least sympathetic to naturalistic canons. To some it will appear tender-minded and rationally soft, prone to overbelief, but its adherents try rather to be tough critics of weaker faiths, to be the least afraid of demanding more explanatory work than science can do.

God Is the One Who Loves in Freedom

All the classical attributes, or perfections, of God can be brought within the twin themes of love and freedom. We can array them as follows:

God's attributes of love:

God is perfectly gracious
and holy.
God is perfectly merciful
and righteous.
God is perfectly patient
and wise.

God's attributes of freedom:

God is perfectly one
and omnipresent.
God is perfectly constant
and omnipotent.
God is perfectly eternal
and glorious.[33]

The symmetries are complementary, as when the divine righteousness is the shadow of mercy, or holiness is the obverse of grace, or when constancy (faithfulness to purposes) yields omnipotence. We cannot enlarge these characteristics here, but only suggest how there is offered a unitary and coherent model of God. The Bible's name for this God is Yahweh.[34] In the divine tetragrammaton, YHWH, there is no positing of an abstract universal, an absentee God, or a process, but rejoicing in a proper name that signifies this Presence. Yahweh is the great "I am there," the continuing Presence who makes for love and freedom.

God makes God's presence known, yet also (to borrow a term from physics) this has to be "detected." Physics does not see the microparticles of its models, nor does theology see the God of its confessions. Both are inferred from the tracks they leave. To physics is assigned the causal tracking of the primitive object-processes in space-time. Theology tracks the richness of the divine Subject, who oversees the storied world history. Just as there are special phenomena to which physics turns for its revelations, there are crucial events to which theology turns expecting to detect the One who generates love and freedom. But these will not be manifest in the categories of physics; we can build no bubble chambers to register love and freedom. Such phenomena show up only at the higher organizational levels, primarily in events of the historical and personal life. They will involve transscientific categories, supersignals.

God Is There in Nature

Still, the earliest and ever-continuing dimension of the story is natural history. Nature is a sacrament of the divine presence, and remains so after the best descriptions of science have been received. Nothing known in science prevents the divine superintending of natural processes. To the contrary, whether we take the primordial big bang, microphysical processes and quantum states, or genetic sets and mutational potentials, science finds an open-ended nature that is a fitting field for the divine providence. Evolutionary developments and ecosystems have their intrinsic worth, but these have in global history been steered toward love and freedom, even though science cannot conclude so of its own resources. Nature's richest program is in this sense a godly one, the production of persons who can love in freedom.

The enormous amounts of time and space involved are no hindrance to those who are introduced to relativity theory and its placing of time and distance in our reference frame, or who are unafraid of infinity. Other worlds, if such there are, might make us less anthropocentric, perhaps less Christocentric in traditional terms, but they need not make us less theistic. Natural selection processes, though sometimes awkward, are no final embarrassment when understood as a continuing creation in which every emerging level is intrinsically good, if also tributary to the construction of something higher. The secondary causes are haunted by God. Since we have from science no model that can even explain how, much less why, the leading movements of the dance of life have taken place, it is entirely reasonable, and provides a more catholic explanation, to protest in the name of a divine orchestration.

Yahweh is present in the recurrent orders of nature, at seedtime and harvest, but these are still-frame shots, cross sections in a dynamic moving picture. Seen at longer range, as we know from paleontology, these are spiraling cycles, evolutionary ecosystems. They are but subroutines, and the executive program is narrative story. We describe historically (though with theory-laden models) what the biblical writers could only describe mythically in Genesis 1–3. But even they tried to narrate a creation story as though it were a sequence of historical events, in the limited reference frame of a divine Architect and a six-day, ready-made creation.

The story is much richer and more complex. We can be glad for the history that evolution has introduced into what was before thought a special creation. One can expect that the Yahweh who loves freedom will put much spontaneity into the creatures, showing up as randomness and self-actualizing, if it is also true that randomness sometimes veils the superintending divine presence, true that sometimes self-actualizing can image God. "The Spirit of God was moving over the face of the waters. . . . God said, 'Let the waters bring forth swarms of living creatures. . . . Let the earth bring forth living creatures according to their kinds!' "[35] We now know more of the complex mixture of authority and autonomy in that brooding command that Earth "bring forth" automatically.

Some say that nature is just physics and its epiphenomena, nothing more. But for the theist, the matter is not so simple, because matter is not so simple. Other conceivable universes could have gotten nowhere, and the one we do have has certain astronomical, microphysical, anthropic constants that provide for exciting happenings. Our universe contains the right physics to allow it to become aware of itself in us. On the global Earth this did happen through the magic of biology, by natural events whose ever more informed upslope climb seems rather to elude the explanatory categories of biological science. The setup dictated by chemical theory is remarkably propitious to life. Water, "the strangest molecule in all chemistry," seems almost to have been selected for its role in nursing and supporting life, concludes the evolutionary biochemist George Wald, whom we earlier heard claim, "This universe breeds life inevitably."[36] The contingencies and natural selections of the planetary Earth remarkably pump up life, and release it into culture.

Some will say that any universe where observers come to worry about these things will have to be of this kind. There can be no observers in other universes, where the physical and biological processes fail to produce them. But the fact remains that the only universe we know is of this miraculous kind. By the theistic account, this "observation" finds its best theoretical explanation as a sacrament of God. Those fortuitous physical constants undergirding and that transformative principle so constantly lurking over the biological contingencies are, seen again, the constancy of God. The whole earthen adventure is really a divine Advent.

It does not follow that nature provides a complete revelation of God, since love and freedom are only nascent there. For more impressive evidence we need to turn to more recent conclusions in our world drama, to the historical eventfulness of personal and cultural life. Indeed, only those who understand the complex fabric of love and freedom in the latter can rightly detect these threads running earlier through nature.

God Is There in History

We do not detect Yahweh by theoretical abstractions like those used to find the laws of gravity or relativity. The God with a proper name will be a God of story, where idiographic drama is as valued as is universal law. In Israel's struggle with God, it came to this creed early, and has since convinced much of the Western world. Yahweh is the covenanting God who effects an Exodus, a release into freedom, who holds forth a Promised Land, who struggles steadily to educate this people into righteous love, choosing them as a light to the nations. God is detected as the prophets "speak for" (*prophēteuō*) God, seeing the divine hand in the Conquest, in a land flowing with milk and honey, in a Davidic monarchy, in a post-Solomonic rebellion, in the Exile, the Return. In their history, the Israelites find a lure toward, a learning of, the divine Presence.[37]

Likewise in the judgments of subsequent history, especially for those empowered by such a faith, God is making peoples free again in the American Revolution, or the Civil War, or the civil rights movement, or liberation theology. On scales larger and smaller there is exodus, judgment, conquest, exile, return. Whatever can be said about the economic and material forces that drive history, these are perennially overlaid with thrustings and reformings toward more freedom, justice, dignity, love, events of tragedy quite as often as of fortune. This model finds a certain law (*Torah*, "teaching") permeating all, and yet detects the divine Presence, concrete and dramatic, if also hidden, in the singular narratives of human events. History is His story. The world is God in story form.

One cannot detect the historical God full-scale in each local event, any more than one can detect natural selection in each individual life, or relativity on every clock. Advancement is not discernible in each new species of evolutionary mutation. But the statistical trends are there in the leading lines. The effects show up on broad enough scales. We will not be surprised if there are mutation points, revelatory crises, emergents, particular inaugurations of themes that gradually grow universal. Earth is selected from among the planets for the launching of life in the solar system; Africa is selected for the launching of humanity on Earth. Israel is selected for the launching of monotheism in history.[38] More recently, the West is selected for the launching of science. Repeatedly yet surprisingly, there is (s)election for more story.

Sometimes the discoveries are transmitted by genes, sometimes by faith; sometimes the information spreads by interbreeding, sometimes by interbelieving. Nor does the story preclude crashes (the collapse of Israel and Judah, of Rome) that reset directions (like the great dyings at the ends of the Permian and Cretaceous periods). All history can be meaningful, as all life is sacred. But significant events (revelatory breakthroughs, prophecy, the Christ event) are not homogeneously dispersed in history, but (like breakthrough mutations in evolution) are rare and sometimes randomly scattered, sometimes featured in axial lines.

To be sure, the divine Presence can be enjoyed in nonhistorical activities, such as poetry or art, music and the mystic flight, in legend, myth, parable, proverb, in the contemplation of cyclic nature or the self-actualizing of a local ego. But these activities must be woven into and out of the meaningful flow of the historical process, into and out of cultural developments. The mystic vision is set in a cultural story (however novel to or redirective of it), just as the scientific theory is set in a cultural story (however novel to or redirective of it). None of these activities escape history, but they spiral around it as dated activities in the careers of historically situated persons, not as timeless disconnects from Earth's affairs.

Events are sometimes the visible consequences of empirical causes, more or less, but they are also the visible consequences of an invisible will toward narrative love and freedom. There is a sort of agency in the activity, an executive power that gives parts and wholes a creative upthrust and makes for life, diversity, culture, storied achievements. There are causes and precedents; ordered regularity is essential. But there is steadily more out of less: critical turnings, charismatic events, surprises, becoming and new being, information discovered, freedom learned, love enjoyed, peak experiences, and suffering through to something higher. The best explanation for these is that Yahweh is there in history.

God Is There in Jesus Christ

In Christian theism, Jesus is the living parable of God. Of all historical figures, he has the most aftereffect. Again we have a singular historical launching that turns the plot, this time a proper-named individual who casts his powerful shadow across subsequent centuries. Perfectly imaging God in human life, this divine Son loves in freedom, reconciling persons to God. To adapt metaphors from physics, through a kind of complementarity of the divine and human presences Jesus is locally present in Palestine, a particle there (as it were) and yet a manifestation of the Christ-Logos that is also a wave and ongoing presence over historical time, remanifest at subsequent probable and improbable locations within the lives of disciples after him.

Against those who say that suffering is too ugly ever to be godly, Christianity takes a supremely loving and free life and follows it to a hideous end, but to detect God's power in this sufferer who cannot be stopped. At the cross Christians are put in communion with this normative power for redemptive suffering. They are joined with this person who so dramatizes that he releases these divine energies at newly emergent levels, resulting this once only across world history in a pivotal resurrection faith. What first seems a hellish anomaly defeating the claim that God is omnipresent, a scandalous story of a God-forsaken and crucified Galilean peasant, is thus retold as the primary evidence for the presence of God. The Creator is present, perfecting his creation through suffering. "For it was fitting that he, for whom and by whom all things exist, in bringing many sons to glory, should make the pioneer of their salvation perfect through suffering."[39] Jesus on the cross is God in pain. The primal energy that with the coming of DNA turned into information, energy that with the emergence of suffering turned into pain, now, at a crucial innovation in the plot, turns into righteousness with such passion that it ever thereafter makes history.

Jesus becomes an observational and experiential base against which any world theory is to be tested. He helps us understand the world, and our accounts of the world help us to understand him. An acceptable theory must be able to account for him historically and for this long-lingering impact on believers. Only partly and subsequently, then, do Christians have a theory that colors and prefixes what they can believe about Jesus. They first and dominantly have a historical personage, a datum, his world line, and ongoing Jesus experiences that trail his arrival. Jesus enters initially at the historical, observational, experiential level.

Afterward, any theory constructed about him, or the world that contains him, must be "up to" him. Never, for instance, did accounts of "accidental" or "epiphenomenal" events seem more hollow. Never did theories of "self-actualizing egos," of mind as a "cognitive processor," of "stimulus and response," or of religion as a "social projection" seem more facile. History flows through him, and any cosmic theory must provide a plausible account of a world, and the forces backing it, such as can generate this phenomenal Jesus. By such back-inference, the hypothesis of God offers the only adequate accounting for the emergence of Jesus in the world. God is Author and Actor in this passionate story.

The categories required for judgment here are transscientific. Science has no tools for the analysis of this idiomorphic Jesus, certainly not such sciences as physics or biology, and even psychology and the social sciences stutter before a figure of such historical moment. He perfectly loves in perfect freedom. He dies in witness to the power of suffering love. Such claims are neither scientific nor nonscientific; they are historically dramatic. Jesus is a commanding text to be interpreted. Even theology has but a faltering grasp on the *theologia crucis,* this logic of God on the cross. Much was going on at Calvary over the heads of those present, over the heads of interpreters then and now, over even the head crowned with thorns, the sacred head there wounded, with grief and shame weighed down. The story is of "Christ . . . having become a curse for us."[40] The trial he undergoes is not one of logic and illogic, not of biological or intellectual trial and error. His trial was conceived in tragedy, conducted in error by forces of evil grafted on to those of power.

Yet paradoxically that scaffold sways the future and he brings to focus a critical thrust in the world process. He is transfigured by his trial and cross into the suffering Messiah and risen Lord. There is a cross and a crown, and that a crown of thorns even when it is a crown of glory. Here is the survival of the fittest at an emergent level. He proves able to fit his disciples for living on, surviving in them, providing their survival. In earlier phases of organic evolution, organisms learned to pump out disorder, repairing injured organs or DNA breakdowns. On the cross Jesus launches a new, spiritual pumping out of disorder, a redeeming from moral evils. We have no reason to think that science can unlock the meanings in this triumph of suffering love, or that it can disprove their presence in these enigmatic, epic events.

Suffering through to something higher is always messianic. Transfigured sorrow is ever the divine glory. That was never more true than at Calvary, but it has always been true ever since the capacity for sorrow emerged in the primeval evolutionary process. The creatures "were always carrying in the body the dying of Jesus"[41] even before he came. J. B. S. Haldane found the marks of evolution to be "beauty," "tragedy," and "inexhaustible queerness."[42] But beauty, tragedy, and unfathomable strangeness are equally the marks of the story of this Jesus of Nazareth. It is a fantastic story, but then again, to recall the conclusion of a puzzled astronomer, Fred Hoyle, the universe itself is a fantastic story.[43]

To adapt the military metaphor of the paleontologist D. V. Ager (recalling how common conflict models are in theology), Jesus' career is a short period of eventful, strategic terror,[44] subsequent to (relatively at least) long periods of waiting, preparation, boredom in those who preceded him. In a better model still, he was "anointed," christened of God to struggle (as did Israel also "struggle") toward realization of the Divine Presence. There is in him a great and divine "yes" hidden beneath the apparent, harsh "no" of the world. But that is revelatory of the whole story, though it takes at a pitch the elsewhere gradual struggle upslope. God is standing in the shadows, the dim unknown.

Others will find this commitment to a historical individual archaic, but for the Christian this is *archaic* in a foundational sense, an archetype revealing the continuing plot. The selection of the singular Christ, anointed to bear sacred information, is thereafter reproduced and unfolded in the hosts who become his disciples. We continue "bearing on our bodies the *stigmata* of Jesus."[45] Thus, the Eucharist is done "for my recalling." The Last Supper takes the disciple back into those storied events so pivotal in Earth's history; it brings the story forward, empowering the faithful, on the telling edge of the present becoming future, to carry on the story in the Lord's presence.

In him we have the breakthrough of the divine Logos, the light coming into the world where it already was, yet explicitly in intense, pointed fullness, a light destined to shine in maturing brilliance over peoples in centuries to follow. A difference between dramatic history and scientific law lies in how the latter is repeatable and ever available for fresh access at each cross section of time, while the former finds directional currents taken of old. To these bearings one returns for orientation in current events, however much we also reach decision points and undergo conversion to new directions. The question "Who was he?" becomes the question "Who am I?" And vice versa.

God Is There in the Personal Life

Yahweh is the giver of meaning in each believer's life, where God's presence in history and in Jesus Christ is interiorized. Thus, the transscientific theist affirms all that the existentialist theist affirms about the presence of God as Spirit to spirit. He agrees with the process theist that love and freedom cannot be coerced but must be responsibly educated as the disciple is persuaded by these divine goals. Here there is demanded a far richer notion of causation than any known in natural science, a richer texture of meaning than any reached in the human sciences. We need the category of the presence of *grace.* The Prime Mover is the Prime Meaner, who graces life.

Under the prompting of this Spirit, one can decide, for instance, to reform a decaying marriage, or to forgive a brother, or to suffer for a friend, and one can believe in and receive divine grace in maintaining this decision. Science sometimes sideswipes such claims, especially in abnormal situations, revealing, perhaps—what theology already knows—that humans are prisoners of their ignorances, less rational, less autonomous, more flawed than they suppose themselves to be. At the same time, where there is genuine redemption in the personal life, as often there undoubtedly is, nothing known in science gainsays the claim that divine grace is superintending those psychological and social relationships that the human sciences can only partially study.

Or, again, in the claim that we are being crucified with him, like him, into newness of life, we confront a transscientific judgment about grace through tragedy. When we enter the realm of the universal Thou who draws particular thous into love and responsible freedom, a freedom able to respond to such grace, science fails in analytical power. But such religious encounter and power is dramatically there, and this demands explanation for the meaningfulness it supplies to the story.

"The Word became flesh and dwelt among us, full of grace and truth. . . . And from his fullness have we all received, grace upon grace."[46] That *Logos* is amply with its *logic,* rationality; but *Logos* means "word," speech, storytelling, and the Prologue to John, from which these words come, is the prologue to a gospel narrative. The narrative form of the passage of God through history is the *Logos* who dwells in the flesh of the world. That has taken place historically, eminently but not exclusively in Jesus, and it continues to take place in the storied personalities we enjoy. The logic is not that of an architect and his design, or a scientist and her theory, but a narration of the conflict of good and evil, of divine grace redeeming the story. Each of us is called to our roles in that struggle, and offered divine presence in acting our parts. This one God—the *Father* transcending nature, superintending history, the historically incarnate *Son,* and the immanent *Spirit*—this is, in Christian symbolism, the triune God, not three gods that add up to one ($1 + 1 + 1 = 3/1$), but one God to the third power ($1^3 = 1$), known as a unity in these multiplications of the divine presence.

God Is Freely, Effectively, Intelligibly There

Each adverb here is important, because each is keyed to a divine attribute, to freedom, power, and wisdom, as God is *objectively* present in the world. Each fits the open-ended nature that science now describes and the historical eventfulness we experience. With the crumbling of mechanistic, deterministic models of nature, there is room for the divine hand to be *freely* present—within superposed quantum states or random genetic mutations, or brain synapses, or in the unconscious mind, or luring human decisions, or launching conceptual innovations. If indeed this is divine freedom, we will not expect to capture its presence scientifically. The energies of nature, upon which is modulated all history, veil the divine power. Above all else, the world drama calls for enough imperative power, a call that has only been intensified by the history described in evolutionary science. But science nowhere supplies nature with enough power, with a principle able *effectively* to call the world into being, one able to command these causal processes to supply meaningfulness.

Nature repeatedly has the kind of underlying order that monotheism might expect, despite the frequent surprises. Even these surprises are partially handled by recalling the divine freedom and by supposing that God enjoys freedoms in the creatures, that God permits spontaneous novelties and insists that organisms pursue their own ends as they adventurously track through an open-ended nature. Perhaps God enjoys this "inexhaustible queerness." Yet the intelligibility found in nature is not self-contained. Nature nowhere supplies a rationale for its own orderliness, but only presents this as a given. This "given" leaves ample room, and sometimes seems to beg, for one who is *intelligibly* there.

God Is There in Righteous Love

God is most nobly revealed in freedom, power, and wisdom as these come to focus in righteous love. God is *subjectively* present in intersubjective human life. Righteous love, an emergent category, is not present across vast reaches of physical nature, not present in earthen biology; but we are not to conclude its unimportance from its rarity. Rather, it is the most startling fruit of the natural-historical process. The forces of causal attraction with which physics begins are supremely transformed when, in end result, one person is attracted to another in holy agape. Before this phenomenal (but not epiphenomenal) effect, science is speechless. But theology finds this the clearest of the tracks of God, seen pivotally in Jesus, where his royal freedom is his holy love. Through him, this regenerating life force is loosed across history, with intensity enough to keep his normative life present in his disciples.

We can conceive of no higher form of God's presence. Those who detect meaning at this level are prepared to predict that nothing will replace it. Sooner or later, others will be drawn to its lure. There may be limited truths in interpreting the world as an oscillating *yang* and *yin*, or a void Emptiness, or a motionless Plenum, or an illusory *māyā*. But righteous love is the key to the drama, and not only to Judeo-Christian portions of it. The other theories must be tested and healed by this interpretive pattern. Scientists hardly dare to predict what biochemistries living beings on other planets might have, or what neurophysiologies might sponsor their consciousness. But a Jewish or Christian theologian will prejudge that if they do not yet know righteous love, they do not know the final truth about the lofty potential of nature. This emergent conception that Judeo-Christian history has attained is a signal of the Kingdom of God.

God Is Always There

"The eternal God is your dwelling place, and underneath are the everlasting arms."[47] As Augustine said of God, nature is the medium in which we live, move, and have our being. But there is a passage from this natural womb to the omnipresence of God, a path permitted to faith. Nature, history, and persons are all "in God."

The study of physical nature has given us an introductory model for how this might be so, although only the spiritual life can incline us to believe this claim. Simple physical events, particles, are wave flows, warps in space-time, indentations in a pervasive plasma-ether. To recall a crude analogy from Chapter 2, they are like traveling dents in a partially inflated basketball. Complex physical events are compoundings of these, so that nature, history, and persons are, so to speak, enfoldings or implications in the plasma. These achieve their own integrities and enjoy emergent levels of spontaneity and downward causation, even while they well up from depths below.

The upper, later-coming reaches of the historical drama prompt us to think further of this great plasma in which all is conceived, and we can recognize it as a spiritual as well as a physical plasma. Everything crinkles the level below it, organisms over matter, mind over organisms, Spirit over persons, yet everything is a wrinkle in God, who is alike above and below. "Thou dost beset me behind and before."[48] We want a Presence in the plasma that makes for love and freedom. Rare though these outcomes may be, they are nevertheless highly revelatory. Physical aggregations are comparatively infrequent—for example, stars in interstellar space. Life is found but once, so far as we yet know, and is rare in any case, with human (or self-conscious) life rarer still. Righteous love is more exceptional still. Yet each emergent level increasingly shows what nature can do, progressively to reveal the Presence behind and before, below and above it.

The presence of suffering defeats this account only if over time it prevents the development and display of righteous love. The theist finds that it has not. To the contrary, we "learn through suffering."[49] Suffering is the logical and empirical obverse of caring, with caring the necessary precedent for righteous love. God is especially there in the power to suffer through and to gain love in freedom. Some caring for these highest values must run deepest in the nature of things, else we cannot explain the results we see manifestly delivered in ourselves, a historical and existential fact with which we have to reckon, and, more than this, ideals beyond the real that govern the shape of what *is* with what *ought to be.*

We are not here dealing with proofs for God. Proofs are unavailable in any hard sense, even in science. The movement from observation to theory is always a weak, backtracking one. If T, then O. If this theory is true, then what we observe follows in course. God is a "Cause" adequate to these effects. God brings complex nature out of simple nature, informs nature, informs culture, elects Israel to its history, sends Jesus with his impact, redeems the personal life by a constant power for suffering through. Such beliefs come by no linear proof; they detect a divine gestalt inlaid on the whole, a story that needs an Author. Transscientific theists will in candor admit that this finding is not as evident as they wish, but we are dealing with God and can expect only a glimpsing of the divine. The kind of *confirmation* to be looked for is not that of the *laboratory* but that of the altar, where believers are convinced of a Presence because they are conforming to the divine love in freedom.

Counterevidence to the Gracious Presence

Transscientific theism is impressive in its capacity to soar over nature and history, in its power to buoy up the personal life, but it also touches ground concretely to find many anomalies and misfits in nature and history. Its toughest challenges lie in biology and anthropology, and with the presence of suffering.

Perfect grace, mercy, patience, holiness, righteousness, and wisdom are certainly not the routine estimates one makes of the forces of *nature,* for instance, when touring a natural history museum depicting the agelong struggle of life. The relativity of time notwithstanding, it is hard to see why God spent so long with the trilobites. Perhaps what God is doing (since God has ample time!) is allowing, if also luring, the self-assembly of life. The self-assembling of the lower levels takes longer than the self-assembling of the higher levels, the development of complexity accelerating by a logarithmic rather than a linear scale. Still, why did God create the tapeworm, once give it eyes and legs, and then take them away and make it parasitic, the source of pain to its host, which it debilitates but (unless it is ill-adapted) does not kill?

Some projects in nature do look like "tinkering."[50] While we may not need a Perfect Craftsman-Architect, neither can we posit a Perfect Tinkerer God, even if this is a tinkering toward love and freedom. There does seem to be some broad-scale superintending, to be sure, which accounts for the long-range successes in construction, but of a remote and loose sort.

Local events are by no means optimal solutions; they are makeshift compromises, often tortuously twisting species into the ever-deepening ruts of overspecialization en route to extinction. The trials and errors by which all earthly creatures, ourselves included, develop and learn may also be true with God, but this ill fits *perfect* wisdom, power, and glory. Nor can these features of nature entirely be accounted for by a *perfect* God's insisting on spontaneity and self-actualizing in the creatures. Before Darwin, it was easier to believe of God that "ever since the creation of the world his invisible nature, namely his eternal power and deity, has been clearly perceived in the things that have been made."[51] There is still some truth in this, but the connections between God and biological nature are weaker than transscientific theism supposes. This is what process theism and scientific-existentialist theism have tried to admit.

Perfect grace, mercy, patience, holiness, righteousness, and wisdom are certainly not the routine estimates one makes of the forces of *culture* when touring an anthropological museum. Before its hopeless array of forgotten cultures, as wasted as the potsherds that are used to date them, it is hard to see why God, who selected the Israelites and sent the Christ, did not enlighten the nations sooner. Did he not care for the Neanderthals? This is not to say that Neanderthal lives were meaningless or without point, but there really is not much sense to all the non-Jewish and pre-Christian stories so far as they contain any relationships to the perfect monotheistic God. Whatever meaningfulness they managed to achieve, they too seem like "tinkering" at this point.

Of course, a biblical theist may say that God "did not leave himself without witness" to the gentile generations,[52] and that "what can be known about God is plain to them because God has shown it to them." But by the Bible's own admission this witness was largely "futile,"[53] futile, we now add, over several million years. This sits ill with the *perfectly merciful* God, no matter how much allowance is made for human sinfulness. God's creation is so broad-scale, the cultures are so abundant,

God's redemption is so local, late, and particular. Transscientific theism does not fully endorse the models of nature and culture that arise in the other, nontheistic, nonscientific cultures (and rightly so). The others cannot supply their own fully convincing models. But neither does Judeo-Christian theism supply one that links many of their affairs very closely to the presence of Yahweh. The Hebrews brought monotheism, but there were other storied developments elsewhere (the wheel, writing, the bow and arrow, the use of iron), which also had their worth. There was something godly in all those cultures. But there was no monotheism, Yahweh's presence was anonymous, and in this respect, if not also in others, they were "wandering in the wilderness."

Their religious life has no duration, but is mostly to be repudiated in the monotheism to which they are converted. Religiously speaking, theirs was a history of error and superstition. (Perhaps we should notice that other prominent world faiths find religious beliefs throughout history to be mostly mistaken: the Hindu *avidyā,* ignorance; the Buddhist *śūnya,* emptiness of all creeds; the Islamic *Jāhiliyyah,* days of ignorance). Of course, science too says those were ignorant, superstitious eras, and even more emphatically. Real truth has come only with our conversion from myth to science.

Meanwhile, it is hard to interpret all those years and cultures as experience with the *perfectly gracious* God, much as it is hard to interpret them as scientific years. Have their stories all that much coherence? They often seem like muddling through. Is there all that much evident grace available to common, not to say vulgar, human life? The upshot is that God must be loosely coupled to history. This does not mean that a divine influence is entirely to be eliminated, only that it is weak and indirect, as scientific-existentialist theism and process theism have tried to admit.

If we add to this the presence of suffering in both nature and culture, and try to feel its full weight, we further confirm a loose coupling of God to observed world affairs. Account for suffering though we may as necessary to the divine providence, we still have more of it than one would predict from the theory of a God who is perfectly gracious, merciful, patient, holy, just, loving, and wise. We were obviously pressed to make the cases of the pelican chick and the anemic African child conform to the monotheistic presence in wisdom and love. Too much seems brutal.

It seems rather that the divine presence has to struggle with suffering in some way that requires us to limit, to "soften" the divine perfections. Can the Holocaust in our own century be interpreted as the result of the providential activity of one who perfectly loves in perfect freedom? In any case, we must not portray nature as better than it is, or history as more meaningful than it is, to satisfy a religious doctrine. While certain broad lines of salient evidence are undeniably there, there is also more static in the background, more meaningless noise in the picture, than the classical picture of God can readily handle. In transscientific theism the dogma can grow too thick, the paradigm too hard-core to attend to the detail in the story, to the counterpoint and counterevidence.

Before this kind of evidence, transscientific theism becomes less confident. In the postscientific era, we can no longer make any sense thinking of a first creation and of a later fall, but perhaps there is always creating and always falling, upstrokes and collapses, goods and evils, with a steady suffering through. God loves in freedom, yet still the creatures wander. God provides, permissively. Alike in nature and human affairs, God creates by tolerating mistakes and slowly reconciling them over time, moving though the pains of growth, but not taking mistakes and pain away. God forms good out of evil long before moral agents arrive. To the notion of trial-and-error learning on the part of the creature we can add the notion of God patiently reconciling the mistakes of the Creator's sometimes clumsy children. Just as God lets humans do autonomous and foolish things, so God lets the creatures follow their queer paths. Perhaps God even enjoys this wildness. Perhaps we need a soft theism, to match a soft naturalism, giving up hard theism as well as hard naturalism.

Perhaps theism, in the arena of meaning, has to do what science had to do in the arena of causes—to give up an absolute requirement for meaning in each event (the perfectly present God) and to hold only statistical claims, admitting scattered areas of meaninglessness in the way that science admits scattered randomness. If in the games we humans create for ourselves we mix order, skill, surprise, chance, and find those games of most delight that interestingly balance these components, perhaps God the Father Almighty also enjoys mixtures of order, skill, surprise, and chance in the games of creation.

On the whole, meaningfulness can be gathered out of the processes of nature and history, although some episodes can be meaningless. We have a statistically present God, which does weaken the sense of the divine presence, but this can also be allocated partially to divine preference for freedom in the creatures, letting them do their own thing, fumbling though this may be. It is partially to be allocated to factors that we frankly do not now understand. Too much is going on over our heads.

Perhaps the most we can conclude from these theistic options is that theology is a multiple-paradigm science, where each theory has something to commend it, each fits some of the data of experience, but each has its unsatisfactory areas, partly in accounting for world events more or less anomalous to the paradigm, partly in getting clear on its internal logic. These theories to some extent feed off each other's weaknesses. Still, all of them have considerable plausibility and explanatory power. Whether or not we conclude that God is a process, we can at least conclude that theology is. And we can commend it, with no less embarrassment than with any of the sciences, as a noble attempt to make sense of nature and history.

We humans are creatures of a few dimensions trying to map a universe that has hundreds of dimensions; we are finite beings trying to map a universe that is infinite. We may need multiple and complementary models of God and of God's action in the world. We do not yet have a single, unified model of the electron wave cloud, or of evolutionary development, the psychology of the ego, society, or the sciences, much less of God. There is no particular reason to expect, now or ever, that a single, unitary account will handle all the divine mysteries. This need not make us agnostic, but it can caution us not to worry overmuch if our theologies are multidimensional, approximate, conflicting, and unsettled. Even our failures, instabilities, and insecurities can be a form of providence.

There may not be any straight lines from nature or history to God; there may not be any simple deductions from models of God to nature and history. For straight lines and simple deductions do not take risky adventures, make good stories, require much faith or logic to detect the plots. Any study of nature and history leaves one suspicious of monistic interpretations in a pluralistic world, however much one is still drawn to seek unity in the saga. With both good induction and good narrative explanation, the route to God will exemplify, not be exempt from, the sorts of travel we have previously known.

5. INSIGHT IN SCIENCE AND RELIGION: DOING THE TRUTH

Judgment in these matters often exceeds our capacities. Our creeds, like our sciences, are ever reforming. Yet we can draw some interim, minimal conclusions about what produces insights in science and religion, as these two great disciplines enable us to judge whether nature and history point to God. Our closing theme ties back to the question of Chapter 1 about methods in inquiry, enlarged as this has become by the attitudes we have taken up toward the sciences, nature, history, and theology in the succeeding chapters. Judgments about the content of a creed require a character of insight keyed to the character of the believer. We reach self-implicating judgments *par excellence,* a high pitch toward which we have steadily been rising.

Such judgments involve serious, objective claims about nature and God, but they are claims made by incarnate subjects who are, as it were, sandwiched in between nature and God. Humans find themselves on the battle lines of natural history, at the apex of the ecological pyramid and evolutionary process, so far as we know it, and charged with emergent qualities that portend the supernatural. Humans live at the intense junction of the natural and the spiritual, and their decisions rest on actions in their embodied lives. Answers lie hidden within ourselves. We do not discover these by ignoring where we are situated in the natural and cultural processes (a mistake religions have often made), but we discover them within ourselves as having leading roles in an evolutionary, historical, participatory Universe.

Correspondent Truthfulness

We can judge the worth of this drama only by experiencing our radical calling in it. Some crucial results of it all are currently underway within us. We are elected to *wrestle* with the question of the presence of God, to be "Israelites" in that etymological sense. We argue with our lives as well as with our heads. We take up the cross, follow the Suffering Messiah, learn who he is as we labor heavy-laden on his way. We are "christened," anointed to incarnate God's Spirit in the world. Truth based on experience is not something insisted upon in science and neglected in religion. To the contrary, this is more intensely demanded in religion as a critical science. We step up from experimental and empirical to experiential levels.

It need not follow that every failure to render decision here (for instance, between the denominations of theism just surveyed) is rooted in some morally faulty experience of the judge, since persons of good will disagree, and disputants must often rest content with unsettled views. Experience may be incomplete because of our juvenile knowledge, though we are not at fault because of this. The question of how closely coupled God was to the primeval evolution of life rests in part upon biological descriptions, upon facts (though theory-laden facts) that can be provided by scientists independent of the presence or absence of religious experience in their own lives.

But the question also rests upon judgments about appropriate levels for the divine permissiveness of spontaneity in the creatures, judgments of what is good and evil, of what is ungodly, queer, cruel; and these do rest upon religious sensitivities, even though sensitive persons can disagree about these. This is true also with judgments affected by psychology, anthropology, sociology, physics, and chemistry —for instance, in judgments about the quality of life in hunter-gatherer cultures, or the overriding of the individual by social forces, or the extravagance of the astronomically large universe.

As these issues integrate into longer-range judgments about the local stories in their larger contributions to the Earth story, we certainly need facts of the theory-laden sort that science can supply. We certainly need historical data, and we need them unbiased by religious experience in the individual or by dogma in the tradition. But we cannot begin to find meaningful patterns in the facts, to weigh the facts "for what they are worth," to "see them as" this way—suffering through to something higher—and not that way—an unfair, absurd universe—without empathy, without sympathy, that is, without a "pathos" in the interpreter who is sensitive to this, rather than that, mode of interpretation.

There are heavy participatory demands on all those qualified to judge at the story levels, requisites always necessary, though sometimes not sufficient, for competent decisions. Thus judgments about *nature,* as these can be made by scientists, are not of the same caliber as judgments about nature as a sacrament of *God,* made by theologians, owing to the way these latter judgments pass meaningfully through the life of the actor-judge. Much of the efficacy and content of belief in God arises from impacts within one's life course, personally and culturally. We see with what we have become, with the brains and sense organs we have evolved, but also with the persons we have become by tradition, experience, decision, and faith. Perhaps we cannot give a prescription for right judgments, but we can be sure that no one avoids wrong ones who stands off from obedience to truth.

Life is a pathway on which there can be no knowing without going. Even at biological levels, more informed genetic sets were formed in contexts of survival. Within psychology, learning is a product conditioned through trial responses to stimuli. The self matures by deciding for and defending a dynamic value system as it moves through the world. Across history, cultures evolve in challenge and response, where new truth has to be struggled for. This knowing-with-going is supremely true in religion. The casual, cool observer sees nothing conclusive about the meaningful plot, for he is inadequately sensitized to the realities that are expected to be observed. Even within science, truth comes to those who with zeal hunt for it, those who frame theories to catch the right data. Within religion, truth comes to those who in passion sacrifice for it, who compose lives to hear the Spirit-wind of God. Knowing requires an adventure in love and freedom. The truth lies on the way of the cross. The Logos story must once more become flesh.

Decisions about who and where we are depend on the sensitizing capacity of the roles we choose. Religious judgment is not uninterested in the causal connections found by the sciences, being often affected by them. But religious judgment is essentially the search for meaning. When such truth comes nearest home, judgments about good and evil in nature and history interlock with good and evil as decided for and against in the life of the judge. The presence of the Spirit of God is another of the *interaction* phenomena, exceeding those that take place when an organism achieves its identity by calling forth a life course from among the quantum states, or a self determines its character by modulating neural circuits, or a living cultural tradition superintends and educates organic, personal human nature. Lower-level phenomena have been gathered under higher-level superintending repeatedly before, and so again.

The Spirit of God enters life "from above," as a yet higher-interaction phenomenon, when the person is taken up, called to the divine destiny. God comes near as we elect a way through the superposed life states that confront us. God supplies those superposed life states, but becomes present in life, whether more or less, depending upon our interaction patterns. The life of sacrificial love lets more of God in, and confirms that Presence; the believer is thereby helped to see more of God in kindred phenomena around her in history, or to see more worth in the tragic struggle that so offends others. Alternatively, the life of selfishness and hatred, or the life of skepticism, indifference, and indecision, closes God out more and more, and blinds the nonbeliever to the presence of God near and far in the programs of history. God's Spirit interacts with human spirits, as human spirits interact with the world and each other. To know the presence of God one must embody that presence in one's own life. Our personal experiences are, nearest at hand, that *at* which we look, but they also become lenses *through* which we look at everything else.

To put the point boldly, one must live on the cutting edge of spirituality to make sense of what lies behind and around, because only at this focus can we form within the gestalt that decodes the drama. This happens appropriately (if also approximately) by appropriating the world course into one's own career, by living appropriately within the world. One will not, at the end, settle into conclusions simply by watching the world lines of mesons, the readout of DNA, the demise of a pelican chick, or that of an anemic African child. One will not decide by noticing what stimuli condition what response, or by sifting through how values function in others' lives, or how values guide a culture. The Freudian question of whether the divine father is an illusory belief is not concluded by scientifically analyzing the role of religion in a patient's life.

Rather, only by hungering and thirsting after righteousness in the Father's Kingdom can one here be satisfied. The ethics incorporated into one's story attunes one to the worth of the story, and here the brass laboratory instrument, the meter reading, the streaks on a photographic plate, or even the social science questionnaire tells us nothing of real interest. Rather, by one's own suffering, dying, and rising to newness of life are judgments of the worth of religion made. Explanation, evaluation, and existence are three in one. On this point existentialist theism has been most insistent.

The story, which for long epochs moved without us, and which the sciences can help relate, now moves through us. Though it earlier moved through us over our heads, and indeed still does so, we have recently come to fuller awareness how its headings lie in our human decisions, made on the frontiers on which we now stand. The world is still being made. Truth, which can sometimes be said in science and even in religion to constitute a reflection of nature, now comes to mean imaging God, reflecting over and becoming the noble consequence of the natural history, divine drama that this is. Truth lies in the creation of what *ought to be,* beyond what *is,* seen in visions that prescribe the conclusions for our descriptions of what, to this point, has been going on. A certain sort of being-in-the-world is required for knowing the world, since it is in ourselves that the story is taking place. We are coagulating the possibilities this way and not that way. We are writing the text.

There is, therefore, a limit to the correspondence theory of truth, a theory that some will say has been too much presumed throughout this book. Or, better, there is an active deepening of it, beyond the passive sense in which "correspondence" is usually taken. The mind does not merely mirror the world; the person moves through and evaluates the world. Truth must be functional and pragmatic, and thus our sciences, not less than our religions, are ways of getting things done. But neither do we want to reinterpret truth merely as a matter of what is useful for human purposes, of knowing nature as an instrumental resource for human self-actualizing.

The moment of truth is the moment of decision. How can humans play useful roles in the unfolding story? We want a *correspondent truthfulness,* where the actor is true to, corresponds to, becomes a faithful participant in, the drama he inherits from nature, history, and God. The *facts* call for *acts.* The *actors* form *characters.* The *is* demands an *ought* in the ontological and chronological sequence, although the *act* and the vision of *ought* are required, in the epistemic sequence, to know what is going on in the historical sequence. That is an operational view of truth in the richest possible sense. One needs a critic, one able to judge, but not a spectator critic; rather, a dramatic critic able to judge how she can make a nobler play. We do not have, or wish, the objectivity of the ideal observer, of the perfect, disembodied reason; nor, on the other hand, do we have or wish the subjectivity of the arbitrary existentialist, a self choosing whatever it wishes. Rather, we face the responsibility of the participant in the story in which she is elected, and elects, to take a meaningful part. What we learn from the Greeks is that the unexamined life is not worth living, but what we learn from the Hebrews is that the uncommitted life is not worth examining.

The Transformation of Science into Interpretive History

We can return to and modify a schematic methodology with which we began (Figure 7.4).[54] Progressively reforming and developing theories (T_1, T_2) are erected over observations. From these, further observations are deduced (O_1, O_2), observations that are also theory-laden facts (T-facts). This leads at a larger scale to progressively reforming and developing interpretive narrative models (I_1, I_2) erected over historical sequences (H_1, H_2). Thus, science is subsumed into story. The quest for causes passes over to the quest for meanings. The interpretive narrative hopes to *match* but also to *make* history, to reflect and to reform the storied sequence. We have meaning-laden history. The "deductions" (if T, then O; if I, then H) become

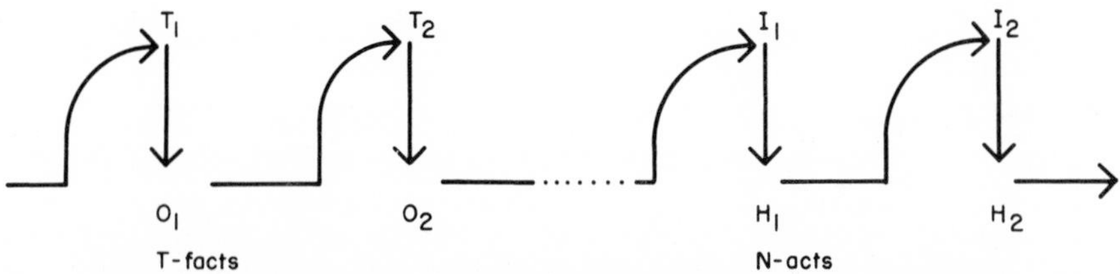

Figure 7.4 The transformation of science into interpretive history

life-orienting, life-implicating. They are experiential beyond empirical. A historian who *per impossible* merely recorded everything accurately, like a motion picture camera, would see everything and understand nothing. Only by framing narrative interpretations, in which one assumes a role oneself, does one begin to understand. In that sense, it takes a vision to have vision, really to see what one sees.

Confirmation is doing the truth. "Drawing conclusions" is acting out the consequence of the story. "Deduction" still includes thinking critically from interpretation to history, checking history against interpretation; but it further includes a being "led out" from interpretive creed to responsible action, so that the historical sequences (H_1, H_2) are narrative-acts (N-*acts*) as well as theory-laden facts (T-*facts*). One checks this theory by joining the story, trying to image God to decide whether there is a divine Author. We not only *have* a paradigm, we need ourselves to *be* the paradigm, the disciplinary matrix, the disciple who incarnates the truth. This rationality is not something abstracted from life, timeless and eternal, tested against experience and nature though rationality must be. Rationality is whatever it has developed to be at this point in our historical careers. The logic is not merely *inductive,* nor *deductive;* it is *productive,* as we weave a way through the superposed potential historical states.

Information and Reformation: Science, Values, and Truth

The story is ever reforming, but this now includes more than a revising of theories. Religious *information* comes with *reformation.* Whatever one makes of the details of the various concepts of sin and error, universally present in the world faiths, this much remains in outline: some off-centering of the self in favor of other-directedness cleanses the self for truth. Even in science, we found it desirable to recommend universal intent, past one's own stake in any professional research. In religion there is required a dying to self (not a dying of the self) and a regenerated life in the corporate realm of God. Precisely this rectifies the subjectivity that may have threatened to introduce error in the intense demand for existential participation.

Each subject is rededicated in his world ecology and his historical community. Like the divine Son, the self is martyred in behalf of the world. There is what Buddhists call the great renunciation. These are moments of truth. Revelations come when one no longer takes a commanding interest in one's own sector but in the whole. The visions we have depend upon the revisions of life we choose. In that sense, judgments about what *ought to be* feed back into our estimates of what *is.* Judgments become matters of conscience. But conscience, though self-implicating, does not permit one to do what pleases the self. To the contrary, conscience calls the self to duty, to transcending honesties and integrities, to charities, and therefore is self-involving so as to be self-denying. There is a norm beyond the self, a distance created between the self and what is right, just, true. This reformation of the self in the light of the larger story increases perceptive sensitivity.

Biological organisms have a kind of information (in genes and instincts) that defends only their own form of life, although this self-defense can be of value to others secondarily, perhaps as integrated into ecological webs, perhaps by providence or serendipity. We do not need to be taught to defend the self; that comes naturally. But through spiritual reformation persons can gain a kind of information that seeks and sees more holistic, less self-sectored truth. The self is emplaced in its total environment, is prepared to live and die following its role, defending its intrinsic values, and yet not apart from, but an instrumental part in, the whole. Our anthropocentrism is not "hard" but "soft" in the ecological and evolutionary story. The particulate self is a wave in the dynamic flow.

We reach what some psychologists recognize as penultimate and ultimate stages in the religious life. Against Durkheim, religion is not just a veiled concern about one's society or culture; it is a quest for ultimate understanding. One learns to love neighbors, more and more, other persons, other creatures, and in this one finds communion with God. One detects the Kingdom of Heaven at hand on Earth. There is a call to discipleship, to performative truth. Jesus accordingly says, "Whoever has the will to do the will of God shall know whether my teaching comes from him." "He who does what is true comes to the light." "If you continue in my word, you are truly my disciples, and you will know the truth, and the truth will make you free."[55] This logic must be lived as personal knowledge, making and keeping us loving and free. This word is found to be divine as it becomes incarnate.

When Jesus says that the "pure in heart . . . shall see God,"[56] he does not mean to disparage or replace clearheaded religious inquiry. God is to be loved "with all your mind."[57] But he cautions rather that only as the heart becomes pure can the head get any dependable clarity about ultimacy. Nor does this beatitude guarantee increasing answers in proportion to purity alone, and oblivious to logical rigor. Jesus does, however, direct us where and how to look for the divine disclosure, a direction that remains true despite what analytical capacities we may gain in science.

We may use this beatitude, first in a backward look, to preserve in part the integrity of prescientific religious inquiry, and, second, in a present and forward look, to worry about an ominous clouding within scientific inquiry. All the classical world faiths were founded by seers and saviors who were ignorant of most of what has been discussed in this book—celestial mechanics, relativity, quantum physics, molecular biology or evolutionary theory, behaviorism, psychoanalysis, humanistic psychology, social functioning, and often world history. They knew little of life's origins, little of external nature, and they were culturally provincial. It might seem that their truths must be archaic, and this is indeed often so. They did not have the right categories for reliably interpreting nature, and we must largely discard them here.

But in the areas of moral awareness and of ultimate issues in human experience, they were often on the cutting edge, and we may profitably test and often trust them here. Sometimes these seminal figures had passionate callings that empowered them to detect God's activity in their local world events, sometimes confusedly and under other names, sometimes with vision enough to see what meanings were normative for the whole. One needs no science to experience the categorical worth of justice or of suffering love, just as (what the two-language theory rightly insisted) one needs no biochemistry to know that one is waving to a friend. Such inquiries into life's meaning, couched in the mythologies of former days, may survive translation into a later, scientific world view.

Further, when science comes, it may not always be accompanied by purity of heart. Science can suppress what the saints had earlier known, and it can deliver error, or fail to deliver any ultimate truth, because of its naïveté about the loves and choices of the heart. Science may bring no better acting in the story; rather, it may yield less meaningful participation in the narrative of events, even bring failure, anomie, and tragedy. She who is doing science has yet to ask whether she is doing the truth, whether she is loving in responsible freedom.

Science is willy-nilly a value carrier. It may be value-free in itself, but science exists in persons, in societies, as surely as does religion. We use it to play our roles in the story. It is the product of actors traveling through their world. Science is always conducted in a value ambience. Pure science is always shadowed over by some social form of life. The owning of knowledge and power, especially as amplified in recent science, is a responsible process. In vitro science may be neutral; in vivo science is generated in the chasing and clashing of values. We must not let the laboratory analysis of a thing deceive us about what it is in the field from which it is inseparable.

Sociologists need to do with science what they sometimes think to do only with religion—tease out the latent behind the manifest functions. We must sometimes understand science in terms drawn from outside itself, just as sometimes science needs to understand religion in terms drawn from outside religion. Science too has a derivative status as social product. The good scientist has to ask not merely about effective inquiry, but about the effect of inquiry, and what inquiry itself is the effect of. Science lies *in situ,* quite as much as does religion. Neither is a noncultural discipline, despite the impressive universal and transcultural elements in both. Pure and applied science are part of a people's narrative story. Science is not only theory-laden but culture-laden.

This means that science serves cultural and individual perspectives in ways that can be self-serving, self-willed, materialistic, prejudiced, compulsive, and destructive. Where science seeks to dominate, manipulate, and control nature or persons, never asking about respect, submission, obedience, or righteousness, it operates in persons impure of heart, and blinds quite as much as it illuminates. Science can create the illusion that humans are alone, free to fulfill their desires. It can rationalize the belief that only humans count, with nonhuman nature valueless. It can destroy gods and replace them with wave clouds and randomness. It can seemingly justify and make inevitable or innate our selfishness.

We need not be surprised when a scientific culture leaves its citizens with their material needs better provided for, but lost in meaninglessness and alienation, divisiveness and angst. Indeed, these features have characterized scientific societies on a scale never previously known. Humans are increasingly competent and decreasingly confident in a sterile world. The fruits of unguided science are bittersweet, leaving us less sure than ever that the theoretical implications or the practical results of the next discoveries in physics, biology, or psychology, or social science will be beneficial, or even benign. Nothing in science ensures against philosophical confusions, against rationalizing, against mistaking evil for good. Science is a good servant but a bad master, and a futile tool for those who have no other master. The whole scientific enterprise of the last four centuries could yet prove demonic. We may be caught in a Faustian bargain, in a scientific sink.

Science bears on, affects value at the same time that it bears, carries what values we may have assigned to it. Science bears on value because it redescribes the world. We have to think only of the impacts of Copernicus, Newton, Darwin, Einstein, or Heisenberg to see how the passing of the science story through successive descriptions of the natural world brings tumult into theology. If we have been mistaken about how nature is operating, then, when science corrects this, values thought to be derived from God and found in nature must be reformed to fit the redescriptions. Our value judgments require an appropriate congruence with the way the world works in its cosmology and ecology. Similarly, in the human sciences, accounts of social functioning, or emotional drives, or social conflicts have their spillover into evaluative issues.

Further, science offers instrumental capacities and gives us the opportunity not just to understand the world, but to change it. A society can allocate funds and energies to eradicate malaria or to build thermonuclear weapons. Which should it do? We might undertake genetic engineering, but what optimal human genotype should we design? Science sets routes before us, between which it does not help us to choose. Over time, even the pure science we elect to do is with much eye to the uses society has for it. Here, not less than in religion, truth for truth's sake gets tugged over to truth for use's sake. Science presents vastly increased opportunities for self-actualizing, and yet, rightly understood and short of scientism, draws back in silence before which options we ought to take. We are left to consolidate and even choose roles through the chapters we write in a history in which science offers no paradigmatic plot.

Thus, for all the bearing of science on value, for all its being a value-laden enterprise, the conviction is well grounded that science of itself is value-free in the crucial sense that there is no such thing as the purely scientific guidance of life, whether of personal or social life, or of setting new directions in history. The point is that the values surrounding the pursuit of science, as well as those that govern the uses to which science is put, are not generated out of the science. These are rarely even launched from within the science proper, but we certainly do not select between them by science itself. Science gives us no resources with which to carry out fundamental decisions about good and evil, about meaningfulness and worthlessness, even though science casts upon us new questions and recasts old questions that it cannot answer.

The truth registered by the *is-ought* distinction is that there is no iron logic by which to move from purely scientific premises to evaluative conclusions. One must also own some valuational premises, which science alone seems never to provide. The boundary between *is* and *ought* is real but a twilight zone. Our values are formed and reformed in response to what we believe the world is like, and what we are prepared to believe about the world in turn, descriptive though it may seem, is linked into a feedback loop with our value sets.

Science affects values, is infected by values, and has its motor force in values. Where there is science, values lie in the offing. But science itself is barren for value generation. Francis Bacon complained against teleology (and, by extension, against theology) that, like a virgin consecrated to God, it bore no explanatory offspring.[58] We may cast a reverse complaint against science; lacking consecration, it bears no valuational offspring. The sterility of science at the cores of value production has kept religion fertile even in an age of science. Numbers and equations say nothing about joy and affliction. The problem of evil, which is often thought (probably rightly) to be the chief obstacle to belief in God, proves no less an obstacle to any scientific guidance of life. For evil does not merely linger after theism lapses. It grows worse, when science proves itself theoretically incompetent to handle questions of good and evil, and in practice able to multiply evils as readily as goods. Owing to the collapse of earlier theology, owing to the growth of past and present science, we have ceased to believe that the Fall of man lies behind us. But, owing to the growth of present and future science, we fear that it may lie ahead.

Regardless of one's scientific expertise—indeed, in extension of what good scientists know—no religious truths can come to us until we offer ourselves up to be controlled by the reality we seek to study. This means to open life up to the lure of the incoming Spirit, since we are called to a storied adventure from nature to spirit. It will no longer suffice to be prescientific, whatever insights of the prophets and saints we may continue to find canonical. Science is too much with us, and there is too much truth in it. But neither can one be merely scientific, since, in the ways just summarized, seeming "mere science" falls into the service of our unregenerate value sets, and is unable to generate nobler value sets, unable (in religious terms) to regenerate us. What is required is to be postscientific or (to adapt the term) transscientific. We must go beyond science. We see now what science cannot supply, and why it cannot supply it.

Science is the most powerful analytic tool yet developed, especially in its accounts of nature, perhaps less so of human nature and culture, and less so still in history. But it has proved steadily unable to tell us on its own resources what we most want to know about each of the four: how to value nature, how to guide human nature, or culture, and how to interpret and make history. We do not know what to believe, or how to behave. We do not know what text to write next. This needs carefully to be said, but it is importantly true.

One can find room for God in and beyond the sciences, in and beyond nature and history; but this room for God, though it is impressive, is not so unambiguous or commanding as to produce life-orienting faith, unless and until one finds room for God within one's own personal life. But this need not be cause for lament. This too is intelligible under the theistic model of a God who nurtures freedom, love, and faith. This too is part of God's design. God did not leave himself without witness in nature and history, but God leaves a chief witness in the person, to be found as, and only as, the person in daring expectancy reflects God.

Doing the Truth on the Cutting Edge of Nature and History

We must do the truth. We test, and are tested in, whether our self-actualizing can also actualize the divine presence. We must live at the eye of the storm. We must nurse a way through the possible scientific states of humanity (which at microlevels involves nursing a way through the quantum states), doing good and fighting evil, so as to let the divine Spirit, if such there be, come nearer in amazing grace. We see whether we can prophesy, that is, speak for God. We try to see whether we can image God. We act our parts in the story. The energy of our experiences here, biographically and culturally, will enable us to look behind and around at nature and history and to judge whether the cosmic and the earthen drama is a divine current. But we must be in the river to sense the flow. We thereby gain spiritually what physicists call a reference frame for participatory observations. We must look carefully at objective nature, in the light of the best sciences available, and we have sought to do this in the progressive chapters of this work. But that is necessary, not sufficient, for the formation of a creed.

In the final analysis the incumbent judge finds himself at the crossroads of nature, history, and God. Only caught in this grip can one know who and where one is. Perhaps such an emphasis on inwardness and participation will at times lead us to false estimates of what is going on in the world outside of us. But the deeper truth is that spirituality is just what the outside world has led to. We are the richest of the natural systems, its fruit called to continue the creation. Our performance here enables an accurate estimate of all our history and environment.

This posits a kind of privileged access to the religious viewpoint, but not one that ignores the human place in nature, or seeks an inwardness uncorrected by scientific redescriptions and philosophical criticisms. Rather, it notices, as a result of this, that in epistemic rank and in evolutionary place humans are a privileged species. They alone are called to this level of awareness. Further, within the possible states of human awareness, states of spiritual expectancy are still more privileged, with their vision into what in ourselves is taking place, deepening the world drama thus far.

In *coping* with our own world assignment religiously, we gain insight into whether our beliefs are *copying* the world order. Here the correspondence theory of truth (belief as a map of the world) must submit to a higher pragmatic view of truth (belief as instrument for living in and traveling through the world, which is what maps are for). To see whether our beliefs here are justified, or true, one has to see whether they can justify life, make life just, loving, free, spirited, and spiritual. One has to be born of blood, of fire, of the will of God. One has to be plunged under the floodwaters and raised up to see the descending power of the Spirit. In this sense, judgments about what *ought to be* not only enable us to detect what *is,* they are self-fulfilling and determine what comes to pass.

Any teaching has to be evaluated, certainly, on the basis of whether it is true; but it has to be evaluated "for what it is worth." In the end we turn to *truth* for *worth,* on the axiom that we should get some clues about truth from the worth of a teaching. We want to know what is so, whether we like it or not, whether it is satisfying or not. Yet a truth that is nonilluminating about values cannot be the final word in a world that willy-nilly we must evaluate. We are not simply after a truth's working, not mere pragmatism, but its worthiness, its overall serviceability as an evaluative account, its *truth-value,* with an emphasis on both those terms in inseparable conjunction. We want truth in some true-false sense—more or less, analogically and approximately. But we cannot be near the elemental word until the *truth* is carrying *value.* Unless a doctrine can tell us something about what the story is worth, and show us how to be worthy actors in the story, it has not yet achieved salient truth-value. Here science proves to lack ultimacy, while religion offers the pearl of great price.

NOTES

1. The term "supernatural" does not appear until the early Middle Ages, after the classical period, but before the rise of science. Earlier there were "marvelous events," "signs," or "mighty acts."
2. Anthony F. C. Wallace, "Rituals: Sacred and Profane," *Zygon* 1 (1966): 60–81, citation on p. 76.
3. Many process thinkers (as we later see) will soften the intensity of emergence portrayed here, finding analogues of mind and consciousness in all living things, and analogues of life and sentience in abiotic processes. An attenuated psychical pole perfuses physical objects. But such panpsychism is not an easy claim to understand or accept; it is hardly simpler than the startling emergence it seeks to soften.
4. Mark 4:28.
5. Ralph Waldo Emerson, "Compensation," *Essays* (New York: Thomas Y. Crowell, 1926, 1961), pp. 67–92, citation on p. 74.
6. 1 Cor. 13:12.
7. Prov. 27:17.
8. The position to follow is found among left-of-center Protestant theologians, eminently in Rudolf Bultmann, but also in John Macquarrie, Dietrich Bonhoeffer, John A. T. Robinson, Norman Perrin, mixedly in Paul Tillich, and approached in Karl Heim. Many precedents for it lie in Friedrich Schleiermacher and Immanuel Kant, going back to Descartes. Martin Buber is a Jewish representative (despite cryptic recognition of Thou relations with nature). Roman Catholic representatives are less evident—perhaps Leslie Dewart.
9. Rudolf Bultmann, "Is Exegesis without Presuppositions Possible?", in Schubert M. Ogden, ed., *Existence and Faith* (New York: Meridian Books, 1960), pp. 289–297, citation on pp. 291–92.
10. Rudolf Bultmann, "Faith in God the Creator," in Ogden, *Existence and Faith,* pp. 171–182, citation on pp. 174–175. Italics in the original.
11. Norman Perrin, *The Promise of Bultmann* (Philadelphia: Fortress Press, 1979), pp. 74–75.
12. Ibid., pp. 86–87.
13. For two-language accounts, see Paul L. Holmer, "Scientific Language and the Language of Religion," *The Grammar of Faith* (San Francisco: Harper and Row, 1978), pp. 54–80; Donald D. Evans, "Differences Between Scientific and Religious Assertions," in Ian G. Barbour, ed., *Science and Religion* (New York: Harper and Row, 1968), pp. 101–33.
14. Rudolf Bultmann, "The Meaning of the Christian Faith in Creation," in Ogden, *Existence and Faith,* pp. 206–225, citation on p. 207. Italics in the original.
15. William Temple, *Nature, Man, and God* (London: Macmillan and Co., 1935), p. 306.

16. The presiding genius of the position sketched here is Alfred North Whitehead, but he crystallizes frequent themes in evolutionary theism. Charles Hartshorne, John B. Cobb, Jr., and D. D. Williams are disciples; Ian G. Barbour is sometimes sympathetic, and Schubert M. Ogden is a process existentialist. There is independent development in F. R. Tennant, Nicholas Berdyaev, Bernard E. Meland, Samuel Alexander, Henri Bergson. One sort of Roman Catholic representative is Teilhard de Chardin; another is David Tracy. A form bordering on soft naturalism is found in Henry Nelson Wieman, and there are many precedents in G. W. F. Hegel.
17. Alfred North Whitehead, *Science and the Modern World* (1925) (New York: Free Press, 1967), p. 72.
18. Ibid., p. 107.
19. Edmund W. Sinnott, *Matter, Mind and Man* (New York: Atheneum Press, 1972), p. 153.
20. Chapter 3, p. 110.
21. Alfred North Whitehead, *Process and Reality* (1929) (New York: Harper and Row, 1960), p. 377; cf. p. 64.
22. Recall Chapter 6, p. 285.
23. Whitehead, *Process and Reality,* pp. 524, 526, 532.
24. Charles Hartshorne, "Process Philosophy as a Resource for Christian Thought," in Perry LeFevre, ed., *Philosophical Resources for Christian Thought* (Nashville: Abingdon Press, 1968), pp. 44–66, citation on p. 65.
25. Alfred North Whitehead, *Adventures of Ideas* (New York: Free Press, 1967), p. 277.
26. Whitehead, *Process and Reality,* p. 521.
27. Charles Hartshorne, "God and the Meaning of Life," in Leroy S. Rouner, ed., *On Nature* (Notre Dame, Ind.: University of Notre Dame Press, 1984), pp. 154–168, citation on p. 156.
28. Charles Hartshorne, *The Divine Relativity* (New Haven: Yale University Press, 1948, 1974), p. 59.
29. Whitehead, *Process and Reality,* p. 524.
30. Ibid., p. v.
31. Ibid., passim.
32. Temple, *Nature, Man, and God,* p. 259.
33. The definition and array is from Karl Barth, *Church Dogmatics,* vol. 2, part 1 (Edinburgh: T. and T. Clark, 1957), cf. p. 257, and forms the outline of the doctrine of God. Barth is eminently transscientific, but the discussion that follows is not meant to be Barthian, but rather a generalized theism as taught in most mainstream or conservative theological seminaries in Europe and America, Protestant or Catholic. Despite many differences among them, such theologians as the following have much in common: Karl Rahner, Reinhold Niebuhr, Jürgen Moltmann, Wolfhart Pannenberg, William Temple, T. F. Torrance, and Ian G. Barbour. With Christian subtractions, it fits much conservative Judaism. Two Protestant works oriented to science are Wolfhart Pannenberg, *Theology and the Philosophy of Science* (Philadelphia: Westminster Press, 1976), and Langdon Gilkey, *Maker of Heaven and Earth* (Garden City, N.Y.: Doubleday and Co., 1959). A survey by a Roman Catholic is Stanley I. Jaki, *The Road of Science and the Ways to God* (Chicago: University of Chicago Press, 1978).
34. Exod. 3:14.
35. Gen. 1:2, 20, 24.
36. George Wald, "Fitness in the Universe: Choices and Necessities," in J. Oró et al., eds., *Cosmochemical Evolution and the Origins of Life* (Dordrecht, Netherlands: D. Reidel Publishing Co., 1974), pp. 7–27, citations on pp. 8–9. See Chapter 2, p. 70, and Chapter 3, p. 114.

37. For discussion of narrative in Bible and theology, see Hans W. Frei, *Eclipse of Biblical Narrative* (New Haven, Conn.: Yale University Press, 1974); George W. Stroup, *The Promise of Narrative Theology* (Atlanta: John Knox Press, 1982); James A. Wiggins, ed., *Religion as Story* (New York: Harper and Row, 1975); Wesley A. Kort, *Narrative Elements and Religious Meanings* (Philadelphia: Fortress Press, 1975); Gabriel Fackre, *The Christian Story,* rev. ed. (Grand Rapids, Mich.: William B. Eerdmans Publishing Co., 1984); Brian Wicker, *The Story-Shaped World* (Notre Dame, Ind.: University of Notre Dame Press, 1975); Stanley Hauerwas, *A Community of Character* (Notre Dame: University of Notre Dame Press, 1981); Stephen Crites, "The Narrative Quality of Experience," *Journal of the American Academy of Religion* 39 (1971): 291–311; Julian N. Hartt, "Theological Investments in Story" (with discussion), *Journal of the American Academy of Religion* 52 (1984): 117–156; James L. Mays, ed., *Narrative Theology* (a thematic issue), *Interpretation* 37, no. 4 (October 1983): 339–401; Paul Ricoeur, *Time and Narrative,* vol. 1 (Chicago: University of Chicago Press, 1984).
38. Hence, what we look for in Genesis 1–3 is the defeat of polytheism. Israel is struggling to envision the one true God, launching monotheism. We should not look to Genesis 1–3 for science; that was not to be launched for two thousand years, not until Europe —though when science comes, the monotheism that came earlier must be reconciled to its offspring.
39. Heb. 2:10.
40. Gal. 3:13.
41. Cf. 2 Cor. 4:10.
42. Chapter 6, p. 245.
43. Chapter 6, p. 245.
44. Chapter 6, p. 275.
45. Cf. Gal. 6:17.
46. John 1:14, 16.
47. Deut. 33:27.
48. Ps. 139:5.
49. Cf. Heb. 5:8.
50. François Jacob, "Evolution and Tinkering," *Science* 196 (1977): 1161–67.
51. Rom. 1:20.
52. Acts 14:16.
53. Rom. 1:19, 21.
54. See Figure 1.1. p. 2.
55. John 7:17 (New English Bible), 3:21, 8:31–32.
56. Matt. 5:8.
57. Luke 10:27.
58. Francis Bacon, *Of the Dignity and Advancement of Learning (De Augmentis),* in *The Works of Francis Bacon,* vol. 4, ed. James Spedding (New York: Garrett Press, 1968), p. 365.

INDEX